THE GULF WAR AFTERMATH

ENVIRONMENT & ASSESSMENT

VOLUME 4

The titles published in this series are listed at the end of this volume.

The Gulf War Aftermath

An Environmental Tragedy

edited by

Muhammad Sadiq

Research Institute
King Fahd University of Petroleum and Minerals
Dhahran
Saudi Arabia

and

John C. McCain

Research Institute
King Fahd University of Petroleum and Minerals
Dhahran
Saudi Arabia

and

Texas Christian University
Department of Geology
Ft. Worth, Texas 71629, USA

Springer Science+Business Media, B.V.

Library of Congress Cataloging-in-Publication Data

Sadiq, Muhammad, 1949-
The Gulf War aftermath : an environmental tragedy / by Muhammad Sadiq and John C. McCain.
p. cm. -- (Environment & assessment ; v. 4)
Includes index.
ISBN 978-94-010-4737-1 ISBN 978-94-011-1685-5 (eBook)
DOI 10.1007/978-94-011-1685-5
1. Oil spills--Environmental aspects--Persian Gulf Region. 2. Oil wells--Fires and fire prevention--Environmental aspects--Persian Gulf Region. 3. Persian Gulf War, 1991--Destruction and pillage. I. McCain, John C. (John Charles), 1939- . II. Title. III. Series.
TD196.P4S24 1993
363.73'82'0916535--dc20 93-15315

ISBN 978-94-010-4737-1

Printed on acid-free paper

To the Innocent Victims of the Gulf War

Table of Contents

Acknowledgements

The Meteorology and Environmental Protection Agency (MEPA) of Saudi Arabia was the focal point within th Kingdom for activities related to the environmental crisis caused by the 1991 Gulf War. MEPA provided the substantial funding to the Research Institute of King Fahd University of Petroleum and Minerals (KFUPM/RI) to study the environmental impacts of the Gulf Oil Spill and Kuwait Oil Fires. During the Gulf Crisis, the MEPA Planning Committee under Dr. Nazar Tawfiq, appointed a Scientific Committee to coordinate scientific and technical matters relating to the Gulf War impacts. Dr. Abdullah E. Dabbagh, the Director of KFUPM/RI, was the Chairman of the Scientific Committee and Dr. Mohamed B. Amin acted as Co-chairman. Their staff at the KFUPM/RI formed the basis of the Committee with representatives from MEPA, the National Commission for Wildlife Conservation and Development (NCWCD), Saudi Arabian Oil Company (Saudi Aramco), and Saudi Arabian Bechtel Company (SABCO) providing pertinent scientific and technical information to the Committee. Many of the discussions held in the Committee broadened our knowledge related to the environmental crisis at hand. We wish to thank the members of these organizations for their contributions during the Gulf Crisis.

Seventy-three scientists, technicians, and managers at the KFUPM/RI studiously worked day and night during the Gulf Crisis. In addition, a number of KFUPM faculty members and graduate students worked on a part-time basis. We are deeply indebted to these highly dedicated persons, for it is by their diligent work that we had the necessary data available to produce much of this book.

Advisors from around the world came to Saudi Arabia to become part of the International Environmental Team (IET), including scientists from the European Community Council (EC), the United States National Oceanographic and Atmospheric Administration (NOAA), the United States Fisheries and Wildlife Service (USFWS), and the United States Environmental Protection Agency (USEPA), Dr. Ibrahim A. Alam of KFUPM/RI was instrumental in coordinating the activities of this team during the crisis. Many of these advisors made significant contributions to our understanding of the environmental crisis in the Arabian Gulf. We sincerely thank Dr. Alam and the members of the team.

Much of the data used for the environmental damage assessment was generated as a result of studies carried out and still being conducted for the Environmental Engineering Division of Saudi Aramco. Saudi Aramco's past and continuing contributions to the assessment of damage caused by the 1991 Gulf Oil Spill deserves special mention. The position of the KFUPM/RI as the focal point for scientific research on the 1991 Gulf Oil Spill was largely due to the continued long-term sustaining research support by Saudi Aramco. Saudi Aramco, in fact, is a major contributor to the damage assessment and to this book.

We garetfully acknowledge the contribution of the field staff of Earthtrust (25 Kanohe Bay Drive, Kailua, Hawaii 96734, USA) who took the cover photographs. Earthtrust sent the first emergency environmental assessment team into Kuwait and assisted the Kuwait government in the effort to control the burning oil wells and to protect the vital shoreline from inundation by the oil spill. Earthtrust is a relatively small, extremely effective environmental organization which depends upon contributions for funding. This organization has paid heavily in funding and in the health of its staff for work on the environmental crisis caused by the Gulf War. Their work during the Gulf War and continued efforts to control whaling, to stop dolphin killing, to ban the use of destructive driftnets, and to eliminate trade in endangered species are sincerely appreciated.

We cannot hope to list all those who have contributed in some way to this book; however, there are several people who were particularly helpful and we wish to acknowledge their contributions. Dr. Yusef Fadlallah, Mr. David J. Lennon, and Dr. Andy Y. Kwarteng of KFUPM/RI were of great assistance during the preparation of this book, not only through their research but also by providing editorial comments and reviews. Dr. Blair P. Bremberg (Texas Institute for Applied Environmental Research, Tarleton State University, USA) and Mr. Akhtar Ali Azhar (University of Dallas, USA) conducted literature searches for the authors which proved invaluable. Mr. Abdulmuhsen A. Al Sunaid (Saudi Aramco, Environmental Engineering, Saudi Arabia) was generous in his support by providing air quality and meteorological data. Dr. Khalil M. Al Thagafi (Ministry of Health, Saudi Arabia) provided considerable help by sharing his knowledge of health related literature.

In addition, we owe sincere thanks to many of our friends and colleagues who directly or indirectly contributed towards the successful completion of this work. We extend our gratitude to the management of the Research Institute in particular, and to the King Fahd University of Petroleum and Minerals (Dhahran, Saudi Arabia), in general, for their encouragement during this work.

Foreword

In 1962 Rachel Carson warned of the consequences of man's pollution in her book *Silent Spring*, a book that some feel marks the real beginning of our environmental awareness. *Silent Spring* told of the consequences of our increasing pesticide use to birds. Almost 30 years after her warning, the western Arabian Gulf experienced its "silent spring" when approximately 100,000 to 250,000 waterbirds died, along with millions of other organisms, due to the massive oil spill that resulted due to Gulf war. The magnitude of our environmental problems has continued to grow during the last thirty years to a point where even the "doomsday" environmentalists could hardly have envisioned back in 1962.

It seems the death of yet uncounted thousands of humans was not sufficient for Saddam Husain. His desire for power and infamy led him to unleash environmental war on mankind. At the end of the Gulf war he set ablaze the oil fields of Kuwait and released more oil into the sea than had been spilled at any time throughout history. These actions were despicable and an affront to civilized man. A quality environment should be a right of all mankind, and to wage war by deliberately polluting the earth cannot be tolerated.

In discussing the impacts of the Gulf war with the authors of this book, we found that one of their major concerns was that environmental consequences seemed almost trivial compared to the loss of so many lives in Kuwait, Saudi Arabia, and Iraq. A quality environment can have meaning only if it nurtures human development. In the Gulf region, we have lost that nurturing environment for the moment but we are struggling to regain it. We must strive to maintain a quality environment. By documenting the environmental disaster caused by the Gulf war, this book may help mankind rethink its approach to environmental problems.

We lived through the agnony of the Gulf war. We saw our air and sea polluted due to the greed and twisted logic of a cruel man. We worked hard to combat the pollution caused by a needless war. We hope that this book will help to set the stage for definitive and resolute action to protect and enhance our environment.

Dr. Abdallah E. Dabbagh, Director
Research Institute
Kind Fahd University of Petroleum
& Minerals

Dr. Mohamed B. Amin, Manager
Water Resources and Environment Division
Research Institute
Kind Fahd University of Petroleum & Minerals

Preface

The ecological tragedy in the Gulf region started with the invasion of Kuwait and will probably continue into the distant future. When the world realized that Sadam Hussain, the President of Iraq, was serious about his threat to use oil as a weapon, the international scientific community warned the world of dire ecological consequences – and many predicted nothing less than a doomsday for the Gulf region and possibly the world. Before Iraqi forces were evicted from Kuwait they released perhaps as much as 11 million barrels of oil into the Arabian Gulf and sabotaged more than 800 oil wells which emitted approximately 6 million gallons of crude oil into the atmosphere. In reponse to the world's largest ecological disaster (the oil spill in the northern Arabian Gulf and the Kuwait oil fires), many local and foreign organizations rushed to the region to try to minimize adverse effects. As a result, a large body of literature has emerged. By the end of 1992, there were more than 30 books written on various aspects of the Gulf war, mostly pictorial presentation of Iraqi atrocities without many scientific facts. The available scientific information on the environmental consequences of the Gulf crisis are scattered, fragmented, and often confusing. There was a compelling need to formulate a comprehensive and cohesive scientific assessment of environmental damages sustained by regional and global environments. The objective of this book is to accomplish this task.

This book addresses impacts of the Gulf crisis on the marine and terrrestrial environments. The effects of the oil spill on marine communities such as mangroves, seagrasses, corals, and benthic fauna and flora, seawater chemistry, and bioaccumulation of organic and inorganic pollutants by fish are discussed. The Gulf War was short but very intensive. The environmental consequences of the Kuwait oil fires, particularly on air quality in the upper atmosphere and at ground levels are described. Efforts were made to assess the damage sustained by the land resources in the Gulf region. Special emphasis was placed on the evaluation of the Gulf crisis on human health in the region. The available data on each of the above were collected, synthesized, and summarized.

The world community was shocked by the scorched earth policy of Iraqi leaders during the Gulf conflict. Anyone who has even the smallest interest in environmental conservation would like to know how badly the environment in the Gulf region was actually damaged by Saddam's actions. Millions of people were trapped in the Gulf war. People would like to know the effect of the war on the health of those who lived through this crisis. According to one of the UN Security Council Resolutions, Iraq is liable to pay for war damages to individuals, organizations, and governments. Ecological damage will be given due shares in claims. It is, therefore, envisioned that many individuals, organizations, and governments will be able to make use of the information gathered in this book. The main beneficiary of this work will no doubt be the environment and those humans who suffered due to the Gulf crisis.

CHAPTER 1

Introduction

1.1. Historical preview of the Gulf war

The Gulf crisis started with the Iraqi invasion of Kuwait on August 2, 1990. The world was not ready to let Iraq annex Kuwait. Soon after the occupation, the United Nations called for Iraq's unconditional, immediate, and complete withdrawal from Kuwait. The Iraqis were equally adamant about legitimizing the occupation of Kuwait. Responsible authorities in Iraq described the occupation as "the child [Kuwait] has come back to mother's lap [Iraq]". Instead of retreating from Kuwait, the Iraqi army swiftly bolstered their forces and assiduously erected defensive structures in Kuwait. Saudi Arabia felt threatened and the government asked friendly countries to help defend its territory. The United States was the first country to heed to this call and sent the 82nd Airborne to the Kingdom in the second week of August to defend Saudi Arabia from possible Iraqi aggression. Later about 40 countries joined the United Sates directly or indirectly and formed an allied command.

The world's insistence on Iraqi eviction from Kuwait grew steadily and United Nations' resolutions demanded that Iraq must leave Kuwait. However, Iraq stubbornly declared that dire consequences would occur if attacked. The allies as well as Iraqi commands sped up their preparation for the eventual armed conflict. Finally, the world's patience grew thin and the United Nations gave an ultimatum to Iraq to leave Kuwait by January 15, 1991. Up to the deadline, the allies massed over 700,000 troops along with an incredible array of modern sophisticated weapons and ordinance. Iraq also beefed up their defenses by deploying 300,000–500,000 troops in and around Kuwait. At about 2 AM on January 16, 1991, the Allies Air Force struck a decisive blow on Iraq and kept pounding relentlessly for next 43 days. The Allied ground forces joined the war operation on February 25, 1991. The ground war was quick and decisive. It lasted for only 100 hours. Kuwait was at last liberated on February 28, 1991. Upon the conclusion of the war, the allies were rewarded with a historically magnificent victory and the Iraqis were evicted in a humiliating defeat. By the end of the Gulf war, Iraq had sustained a staggering loss, both in military hardware and personnel. About 100,000 Iraqis were made prisoners of war

(POWs), and the majority of the remaining troops were either dead or wounded (the casualties are estimated to range between 200,000 and 400,000) as compared with 215 Allied forces dead and 350 wounded.

1.2. Environmental devastation of the Gulf war

The environment is always the victim of war, and none of its components suffers more than humans. Human suffering started on the day of invasion and continued well beyond the Gulf war. Throughout the occupation, innocent people of Kuwait were slaughtered, women and young girls were raped, youths were tortured, and many of the victims were left penniless. Several hundred thousand refugees crossed into Jordan. The plight of the these people was unbelievable. Many of those who were left behind in Kuwait and Iraq were rounded up, harassed, and used as human shields in case of an Allied attack. During the Gulf war, hundreds of thousands of people were killed, wounded, or taken prisoners. Indeed this was a great tragedy for mankind.

The other victim of the war was a part of our fragile planet. Starting with the invasion, through the troop deployment, and during the war, the land was abused mercilessly. Armored vehicles, tanks, and other military hardware were roaming in the once peaceful desert. Construction of defensive structures and other military related activities took a heavy toll on the integrity of the desert soil. Ground surfaces were further insulted by the solid wastes, such as scattered live and spent ammunition, mines, craters from the explosion and associated wastes, damaged military hardware, and sanitary wastes. Desert vegetation faired no better than the ground surface. It was trampled, up-rooted, and destroyed.

The atmosphere, not only in the war zone but for several hundred kilometers away, was filled with the smell of explosives. The air was polluted with a few known and many unknown chemicals. The remaining residents in around the war zone were forced to breath polluted air. Emissions from the jets, military armor, and other vehicles were scattered over vast areas. A large portion of the solid wastes generated was buried thus raising the potential for future groundwater contamination.

During the troop build-up, Saddam Hussain promised the world that if he had to be evicted by force, Kuwait would be burned. He was determined that this would happen at any cost. During their retreat, Iraqis troops set fire to over 600 oil wells in several Kuwait oil fields. This created a scene dreadfully resembling hell fires. Many million barrels of crude oil were shooting up and burning in the air. The Kuwait oil fires showered oil mist and soot, not only over Kuwait but also over parts of Saudi Arabia, Bahrain, Qatar, Iran, and many other countries. Emissions from the fires choked the population, polluted the atmosphere, and changed regional meteorology. This nightmare finally ended when the last oil well fire was extinguished on November 6, 1991.

The Gulf war showed no mercy to the Gulf ecosystem. Perhaps as much as 11 million barrels of oil were released to the Gulf. More than 800 km of Kuwaiti and Saudi Arabian beaches were heavily oiled. Marine wildlife was devastated and thousands of birds lost their lives. Other marine biological resources were also stressed. The oil slick threatened many coastal and offshore industrial and commercial facilities. Working for several months, the Saudi government, with the help from many countries, successfully contained the largest oil spill in history.

Fortunately, the acute effects of the Gulf war do not appear as severe as predicted earlier by many scientists. However, the scientific community is still skeptical regarding the chronic effects of the Gulf crisis. Ironically, humans started it, suffered form it, and now humans are demanding mitigative measures. Despite all the environmental devastation caused by war, humans continue to engage in wars.

1.3. Assessment of environmental damage

Assessment of environmental damage is an intricate problem which necessitate multidiciplinary involvement. By the end of 1992, there were about 30 books written on the different aspects of the Gulf war. These books highlighted the environmental disaster that followed the Gulf crisis as pictorial presentations of Iraqi's atrocities. Few covered but a single aspect of the war. The literature is replete with scientific and general articles describing a specific environmental damage. There was a compelling need to collect such information, synthesize the data, and formulate a comprehensive and cohesive assessment of the environmental damage sustained by regional and global environments. The objectives of this book are to accomplish this task.

This book is divided into seven chapters. The devastation of the marine environment is summarized in Chapter 2. The environmental impacts of Kuwait oil fires, both on the upper atmosphere and pollution at ground level, are discussed in Chapters 3 and 4. Attempts have been made to address environmental damage from the military operations in Chapter 5. Chapter 6 is dedicated to the assessment of the insults to land resources caused by the Gulf war. Dire consequences to human health were predicted during the early assessment of the impacts of Iraq's scorched earth policy. Fortunately, many of these predictions have been proven wrong. The health effects of environmental disasters that followed the Gulf crisis are summarized in Chapter 7. Efforts have been made to include as much scientific data as possible. The authors have endeavored to justify the objective of this book.

CHAPTER 2

Impact of the war on the marine environment

2.1. Summary

The Gulf War Oil Spill released up to 11 million barrels of oil into the Arabian Gulf during from the period January 20, 1991 through May 1991. Since the spill, scientists from many countries have come to Saudi Arabia to assist with the evaluation of its impact on the Gulf ecosystem. What follows below is little more than a non-critical review of the findings of these scientists. Critical review can only follow years of painstaking research and quantitative analysis so that we may begin to understand the nature of the Gulf. Basic research on many ecological aspects of the Gulf is urgently required if we are to ever gain an understanding of the impact of the spill. Until then, we can only report changes without an in-depth knowledge of the impact of these changes.

Those changes to the Gulf ecosystem caused by the Gulf War Oil Spill are listed in Table 2.1. Many people first learned of the spill by watching CNN (Cable News Network) coverage of the Gulf war. Birds covered with oil and struggling for their life were featured. This was fitting since birds were apparently hit the hardest of any group of organisms. Saudi Arabia is on the West Asian Flyway, an important flyway for birds migrating between western Asia and eastern Africa.

2.2. Introduction

Hawley (1992) described the impact of the Gulf war on the Arabian Gulf (Persian Gulf or ROPME Sea Area) as follows: "During most of 1991, the Gulf was transformed into a stinking sea of death, fouling its own shores, bed, and even the air above." This graphic description of impacts, produced well after impacts were being studied and quantified, echoes similar pronouncements which appeared in the press during the Gulf war just after the oil spills and the ignition of the Kuwait oil well fires.

The Gulf is not a dying sea because of the oil spill as predicted by those not familiar with its unusual nature. In time, it will recover and be exposed to more released oil. The Gulf's importance as a oil producing region assures this.

Table 2.1. Summary of documented impacts on the Arabian Gulf ecosystem as a result of the Gulf War Oil Spill and Kuwait Oil Well Fires.

Affected Entity	Effect or Evaluation	Reference
Seawater	Temperature – 2.5°C lower	McCain et al. (in press)
Shoreline	Oiling – 777 km	MEPA (1991a)
Sediment	Oiling – 1,367,000 m^3	MEPA (1991d)
Supratidal Zone	Key supratidal species absent – Saudi beaches	Jones and Richmond (992)
Seawalls and piers	Oiling – 13 km	MEPA (1991a)
Rocky shores	Oiling – 55 km	MEPA (1991a)
Riprap	Oiling – 12 km	MEPA (1991a)
Fine sand beaches	Oiling – 25 km	MEPA (1991a)
Coarse sand beaches	Oiling – 292 km	MEPA (1991a)
Sandy shores	Intertidal infauna reduced from 739 to 415 organisms per m^3 in highly contaminated area – Kuwait	Faraj et al. (1992)
Exposed tidal flats	Oiling – 20 km	MEPA (1991a)
Sheltered tidal flats	1. Oiling – 166 km	1. MEPA (1991a)
	2. No infauna in high tide zone, mid-tide to shallow subtidal zones not impoverished, no distinct community structure adjacent to oiled zone	2. Prena (1992)
	3. Intertidal infauna density 418 and 511 organisms per m^3 after spill compared to 467 organisms per m^3 before the war	3. Faraj et al. (1992)
Salt marshes	1. Oiling – 149 km	1. MEPA (1991a)
	2. Up to 100 dead crabs per linear meter of beach; extensive damage to marsh plants, particularly at lower intertidal elevations	2. Reported herein
	3. Progressive decrease in living plant cover in mid and high tide zone, lower tide zone plants show no recovery	3. McCain et al. (in press)
Mangroves	1. Oiling – 14 km	1. MEPA (1991a)
	2. Great abundance of dead organisms in lower intertidal, continual decline in percentage of living trees	2. McCain et al. (1992)
Seagrass beds	1. Low infaunal densities of 263 to 2868 organisms per m^2	1. Jones and Richmond (1992)
	2. Infaunal densities in oiled areas 510–8650 organisms per m^2 compared to unexposed area densities of 500 to 25,200 organisms per m^2	2. Reported herein
Coral reefs	Generally healthy condition with no evidence of oil pollution	KFUPM/RI (1991c) Jones and Richmond (1992) McCain et al. (1992) Downing (1991) Fadlallah (in press)

Table 2.1. Continued

Affected Entity	Effect or Evaluation	Reference
Fisheries	1. Kuwait fishing fleet and processing plants devastated	1. EPC (1991)
	2. Saudi shrimp production less than 1% of pre-war level	2. SAFISH (1992)
	3. Saudi prawn fishery loss about US$55 million, October 1991 catch rate 14 kg/hr compared to 32 kg/hr in 1990	3. Mathews et al. (1992)
	4. Gill net and trap oiling	4. Reported herein
	5. Saudi Arabia closed fisheries during war period, shrimp smaller with 60–80 per kg as compared to 20–30 per kg prior to Gulf war	5. Al Nokhathah (1992)
	6. Bahrain closed fisheries during Gulf crisis, artisanal and industrial landings similar to pre-war period	6. RDA (1992)
	7. Shrimp landings showed 30–40% decline in 1991, red-spot disease kills scores of fishes off Bahrain	7. *Arabian News* (1991)
Fish eggs and larvae	Significantly reduced abundances of fish eggs and larvae at sites with visible slicks or sheens, abundance of eggs and larvae of penaeid shrimp also reduced; fish eggs and larvae densities not reduced in areas removed from slicks and sheens	McCain and Hassan (in press)
Birds	1. Shorebird population reduced from 260,000 to less than 100,000 in pre- to post-war counts; only south facing bays now supporting representative selection of waterbirds; 25–30,000 pelagic species probably died due to spill,perhaps as many as 100,000–250,000 waterbirds died	1. Harbard and Wolstencroft, 1992
	2. About 77% of waders oiled in Dawhat ad Dafi, 50% in Ras Tanajib; over 200 dead birds along 4 km of shoreline	2. Evans (1992)
	3. Total of 89 dead birds along 200 m of open Saudi coast	3. Walsh (1992)
	4. Approximately 1385 birds treated at Jubail Rescue Center, 507 released	4. Abu Zinada and Belcher (1991)
Marine turtles	1. Island beaches used for nesting were oiled	1. KFUPM/RI (1991f)
	2. Some green turtles died	2. Pellew (1991)
	3. Many green turtles on Karan Island had lesions	3. Greenpeace (1992)
Marine mammals	At least 93 marine mammals died from late February through mid-April 1991, including 14 dugongs, 57 bottlenose dolphins, 13 humpback dolphins, 1 finless porpoise, and 8 unidentified cetaceans	Preen (1991)

2.3. The Arabian Gulf environment

There are several reference books discussing the general nature of the Arabian Gulf; these include Basson et al. (1977) and Sheppard et al. (1992). The information contained in these books is not repeated herein, rather this chapter deals with those aspects of the marine environment that were affected by the spill or those factors that may have influenced the nature and magnitude of the impact of the spill.

In a recent article appearing in the *Arab News* (Kawash, 1992), Mr. Saad Nimeiri of the United Arab Emirates Higher Education Committee estimated that 143,000 tonnes (over 1 million barrels) of crude oil are released into the Arabian Gulf each year. Earlier, Golob and Brus (1984) estimated that approximately 144,000 tonnes of oil were released into the Gulf in 1979. Since 1979 innumerable spills have occurred including two large spills, the 1983 Nowruz Oil Spill and the recent Gulf War Oil Spill. These spills released many million barrels of oil to the Gulf and undoubtedly Mr. Nimeiri's estimate is quite conservative. According to Linden et al. (1990), oil pollution in the Gulf represents 3.1% of the total world oil pollution, "...47 times the average estimated amount for a marine environment of a similar surface area." In line with the degree of oil released to the Gulf, Coles and Gunay (1989) reported 10 to 100 times more tar on Saudi Arabian Gulf beaches than reported elsewhere in the world.

In many coastal areas, the disposal of untreated or only slightly treated sewage and industrial wastewaters in the Gulf is a common practice. ROPME (1986) estimated that the current sewage treatment facilities in the Gulf coastal region was sufficient to treat only 58% of this waste. According to Linden et al. (1990), "Completion of all known proposals, would result in a total treatment capacity in the coastal urban centres, sufficient to treat domestic wastewater from a population of approximately 7 million". This capacity would be sufficient for the treatment of the majority of the population residing along the coastline of the Gulf.

In addition to having a reputation as the most heavily oil polluted body of water in the world and receiving a considerable quantity of untreated sewage and industrial wastewater, the western Arabian Gulf environment is extreme in both salinity and temperature variation. Water temperature in the Gulf generally ranges between 15 and 35°C; however, temperatures below 10°C to as high as 38°C have been reported in the literature (Kinsman, 1964; Shinn, 1976; Coles, 1988; Coles and Fadlallah, 1991).

Evaporation (-350 km^3 per year) exceeds freshwater input (precipitation 24 km^3 per year and runoff 5 km^3 per year) to the Gulf, creating a net loss of 321 km^3 per year (Linden et al., 1990). This water loss is made up by the inflow of seawater through the Strait of Hormuz. Salinity in the northwestern portion of the Arabian Gulf is influenced by runoff from the Tigris-Euphrates river. Surface salinities during the winter vary from 37 ‰ (parts per thousand) at the Strait of Hormuz and 40.5 ‰ near the Kuwait/Saudi coast to over 70 ‰ in the southern Gulf of Salwah (Al-Rabeh et al., 1992).

General circulation in the Gulf is counter-clockwise. Indian Ocean/Arabian Sea surface water enters the Gulf via the Strait of Hormuz while higher salinity water exits along the bottom. Water flowing into the Gulf moves north along the Iranian coast, then south along the Kuwait, Saudi Arabian, and other eastern Arabian coasts. Prevailing winds blow from the northwest at an average intensity of approximately 1–4 m per second (MEPA, 1987). These winds tend to generate a consistent surface drift toward the southeast in the central Gulf with speed ranging from 0.2 to 0.7 knots (Linden et al., 1990).

The Arabian Gulf is extremely shallow with an average water depth of only about 35 meters (Linden et al., 1990). In the western Arabian Gulf the average depth of the area inundated by the oil spill is considerably less with estimates of less than 15 meters (NERC, 1991). Bottom sediment in the subtidal zone of the shallow western Arabian Gulf consists primarily of sand or silty sand. Seagrass beds are plentiful nearshore, in depths of less than 10 meters. Coral reefs fringe the offshore islands and numerous, small patch reefs are present nearshore. Large areas of the subtidal zone consist of consolidated sandstone covered by a thin layer of sand with scattered patches of hard substrate.

Intertidal areas exposed to the open sea consist primarily of sand beaches with an occasional wide tidal flat. Extensive tidal flats have developed in the sheltered water of the numerous embayments along the coast of Saudi Arabia. The intertidal zone within these embayments is conducive to the development of salt marsh and mangrove communities. Seagrass beds and algal covered rubble are commonly found within these embayments.

Evidence suggests that western Arabian Gulf organisms are extremely tolerant (Sheppard et al., 1992). Coral reefs of this area are exposed to a greater annual temperature range than that found elsewhere in the world (Coles, 1988). The principal inshore reef building coral, *Porites compressa*, and various faviid corals survived lengthy exposure (four days) below 11.5°C, a period longer than reported elsewhere (Coles and Fadlallah, 1991).

The waters of the Saudi Arabian Gulf are generally low in nutrients except for a few isolated areas such as Tarut Bay. Table 2.2 shows nutrient concentrations for water samples collected in the Northern Area of Saudi Arabia near the Kuwait border and similar concentrations in other tropical and subtropical environments (KFUPM/RI, 1987). The data suggest that Northern Area planktonic communities could be limited in both nitrogen and phosphorus. Phosphorus concentrations are far lower than is characteristic for most marine environments. The lack of sufficient nutrients in the water column may limit the biodegradation of oil.

2.4. Sources and quality of oil spilled

It is estimated that from 1.9 to 11 million barrels of oil were intentionally released to the waters of the Arabian Gulf from January through May 1991 by

Table 2.2. Nutrient concentrations in Northern Area and other tropical and subtropical environments (from KFUPM/RI, 1987)

Location	Depth (m)	Concentration (μg-at/l)				
		PO_4	Total P	NH_4	NO_2 + NO_3	Total N
Northern Area, Saudi Arabia	0	0.02	0.45	0.15	0.11	15.3
Truk Lagoon	1–15	0.16	0.63	0.11	0.10	6.8
Mamala Bay, Hawaii	15	0.18	0.40	0.47	0.17	5.9
Barbers Point, Hawaii	1–10	0.13	–	0.15	0.15	–
Lahaina, Hawaii	1–5	0.20	–	0.24	0.13	–
Palau	1–5	0.23	0.48	0.08	0.11	8.5
American Samoa	0.5	0.14	–	0.93	0.63	–
Canton Island	0	0.50	–	1.20	2.50	–

retreating Iraqi troops (Al Rabeh et al., 1992). The discrepancies in estimates were time dependent, early estimates tended to be lower and more conservative.

Most of the oil was released at or near Mina Al-Ahmadi in southern Kuwait on the northwestern coast of the Arabian Gulf (Figure 2.1). Oil was released from three export terminals and a number of tankers moored off Kuwait. The major oil release, which began on January 20, 1991, was from the Mina Al-Ahmadi Terminal through the single point mooring at the sea island terminal (Table 2.3). This major leak was stopped by the Allies on January 28, 1991; however, minor releases continued much longer. Both the North and South Piers at Mina Al-Ahmadi were damaged and released oil at a rate of approximately 3000 barrels per day until after March 31, 1991. Two other Kuwait facilities, the Mina Al-Bakr South Terminal and the Mina Abdalla Refinery, released minor amounts of oil (Al Rabeh et al., 1992).

Table 2.3. Terrestrial sources of spilled oil (from Tawfiq, 1991)

Source	Amount (barrels)
Mina Su'ud Refinery	100,000
Mina Al-Ahmadi (Sea Island Terminal)	6,000,000
Mina Al-Bakr	700,000
Total	6,800,000

Table 2.4 shows the estimated amount of oil released by tankers in Kuwait waters after January 19, 1991. Oil releases from these tankers continued until after March 25, 1991 when leaks were still detectable. Some of the tankers and terminals were burning during the period of oil release, thus making it difficult to estimate the total volume of oil spilled. An estimated 1500 to 3000 barrels per day of oil was released until May 1991 from damaged pipelines and tankers.

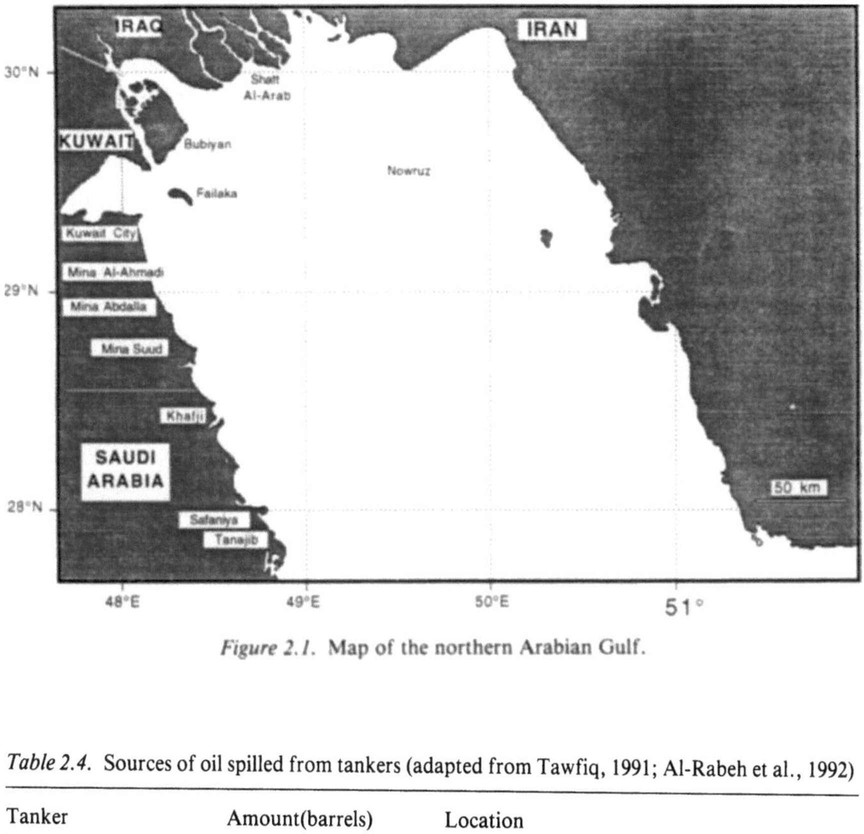

Figure 2.1. Map of the northern Arabian Gulf.

Table 2.4. Sources of oil spilled from tankers (adapted from Tawfiq, 1991; Al-Rabeh et al., 1992)

Tanker	Amount(barrels)	Location
Al Qadasiyah	1,086,477	Ahmadi North Pier (29°07' N, 48°09' E)
Hitten	1,086,477	Ahmadi North Pier (29°07' N, 48°09' E)
Amuriyah	217,295	East of Al-Bakr (29°41' N, 48°48' E)
Tarik Ibn Ziyad	826,973	Ahmadi South Pier (29°05' N, 48°09' E)
Al Mutanabbi	911,863	Shuaiba Fuel Pier (29°03' N, 48°10' E)
Al Fao	31,213	Mina Abdalla (29°03' N, 48°10' E)
Ain Zalah	49,543	Khawr Abdalla (29°59' N, 48°11' E)
Unknown Tanker	56,000	Khawr Abdalla (29°59' N, 48°11' E)
Total	4,265,841	

Approximately 80 ships were sunk during the Gulf war, many around the Kuwait Island of Bubiyan (EPC, 1991). Many of the sunken vessels carried not only oil but munitions, and some were booby-trapped, therefore delaying their salvage. These vessels will remain a chronic source of contamination of the Arabian Gulf for many years along with those still remaining on the seafloor as a result of the Iran-Iraq War.

2.5. Movement and containment of the spill

2.5.1. Oil spill movement and fate

Movement of spilled oil is primarily wind driven. Prevailing winds during the early months of 1991 were from the northwest. The majority of the floating oil moved down the Kuwait and Saudi Arabian coasts of the Gulf making landfall in Kuwait and Saudi Arabia south to the north shore of Abu Ali Island (Figures 2.1 and 2.2). Oil reached the beaches of most offshore Saudi Arabian Gulf islands and slicks and sheens were observed well to the south of Abu Ali Island but most broke up before making landfall. In the northern Gulf, the coastline of Kuwait from Mina al Ahmadi to the Saudi Arabian border was heavily oiled (EPC, 1991). El-Baz (1992) reported heavy oil on the northern coastline of Qatar, 300 miles south of Kuwait. Tawfiq (1991) estimates that 1,000,000 barrels of oil came ashore in Iran and Qatar. Tar balls reportedly washed

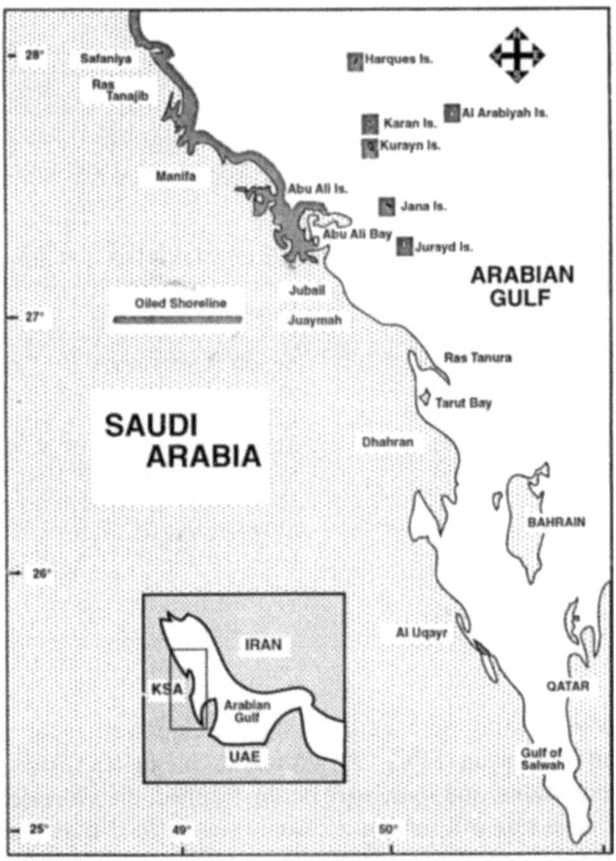

Figure 2.2. The oiled shoreline of Saudi Arabia south of Safaniya.

up on the beaches of Bahrain, Qatar, and Iran (Horgan, 1991). Estimates of the length of coastline heavily oiled vary due to scaling differences; however, it is reasonable to conclude that at least 800 km of shoreline was oiled as a result of the Gulf war.

An oil slick covering approximately 2000 km² was reported (EPC, 1991). The dates of sightings of the leading edge of the oil slick at various locations within Saudi Arabia are shown in Table 2.5. During the period of major inundation of the Saudi Arabian coastline, from the latter part of January through February, tides were exceptionally high as illustrated for Safaniya Pier in Figure 2.3. The datum (LAT – lowest astronomical tide) for Safaniya Pier is 127 cm below mean sea level (MSL) with the average annual tide LAT + 128 cm and the average spring tide LAT + 157 cm. During major oil

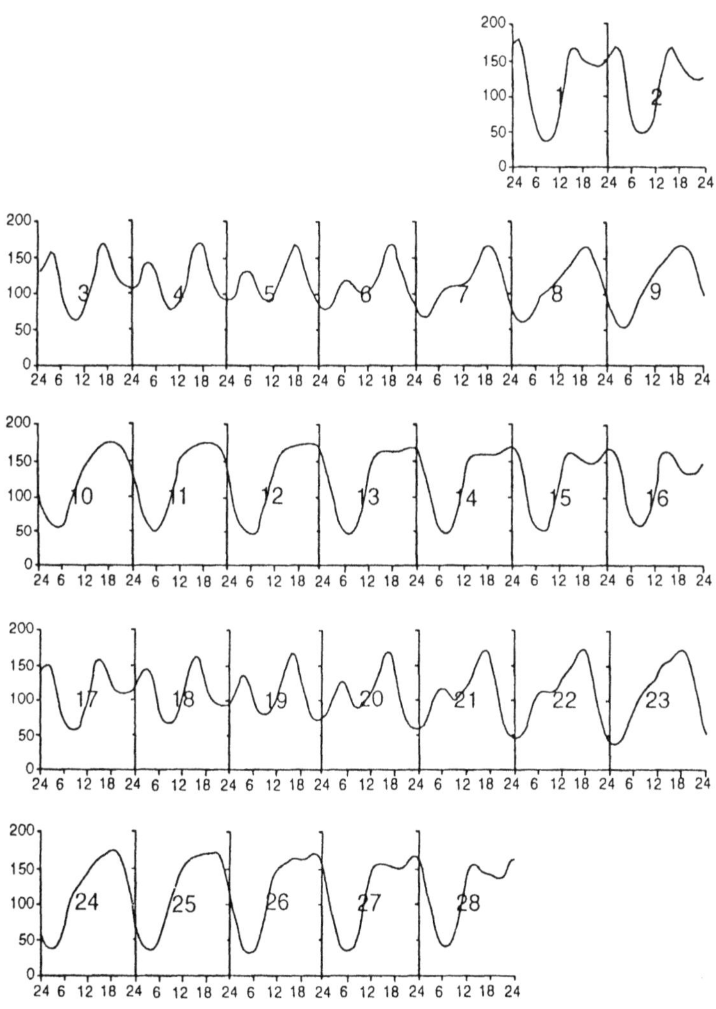

Figure 2.3. Tide table for February, 1991 (from Saudi Aramco).

Table 2.5. Dates of oil sightings along the Saudi Arabia coast (adapted from Al-Rabeh et al., 1992).

Location	Date
Khafji	January 25, 1991
Safaniya	January 29, 1991
Ras al Ghar	February 8, 1991
Abu Ali Island	February 14, 1991

inundation, tidal heights frequently exceeded the average spring tide and tides above LAT + 180 cm were common.

In many areas, the extremely high tides carried the oil quite high into the intertidal range, often into the supra-tidal zone (Figure 2.4). Tidal oscillation exposed lower intertidal elevations to oil but substantial amounts of oil were resuspended with the following high tide, carried into the upper intertidal zones, and deposited there. This may have helped in the collection of oil from the sea surface in shoreline recovery pits and it tended to decrease the exposure of organisms on the extensive tidal flats which are exposed during low tide periods.

Modeling of the movement and fate of the Gulf War Oil Spill was accomplished at KFUPM/RI using the models GULFSLIK II for trajectory

Figure 2.4. N.E. Tanajib Bay intertidal zone showing oiling of the supratidal zone. Photograph by John C. McCain.

analysis and OILPOL for transport and fate (Al Rabeh et al., 1992). Representatives from the National Oceanic and Atmospheric Administration came to Saudi Arabia just after the oil spill and, along with KFUPM/RI, provided trajectory analyses. Trajectory analysis was used to support tactical response to the oil spill crisis by the Saudi Arabian National Oil Spill Response Coordinator (Meteorology and Environmental Protection Administration). The OILPOL model results indicated that after 80 days, approximately 45% of the oil impacted the coastline from latitude 28°44′ to 27°10′ N., with 8% of the initial oil on the sea surface, 15% suspended in the water column or deposited on the seafloor, 33% lost to the atmosphere due to evaporation, and less than 1% dissolved in the water (Al Rabeh et al., 1992).

The OILPOL model suggests that a substantial amount of oil was suspended in the water column or deposited on the seafloor. At the official release volume of 11 million barrels, over one million barrels of oil may have been in contact with subtidal benthic habitats along the Saudi Arabian Gulf coast. In addition, oil deposited on the shoreline was observed to mix with sand forming pellets which rolled in the surf and were probably transported offshore, thus contributing more oil to the subtidal zone.

2.5.2. Oil spill containment efforts

Thirty jetties and dikes were constructed along the Saudi shoreline to intercept oil moving along the coast (Figure 2.5, MEPA, 1991d). A number of companies and agencies participated in collecting the intercepted oil including the Meteorology and Environmental Protection Administration, the Royal

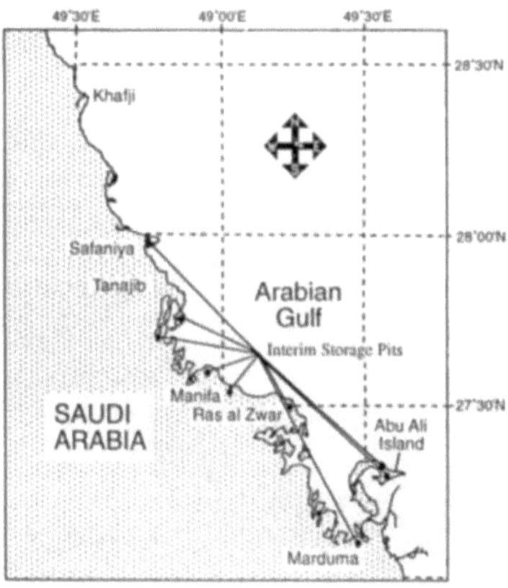

Figure 2.5. Location of interim storage pits for oil collected from the sea surface.

Commission for Jubail and Yanbu, Saudi Arabian Oil Company (Saudi Aramco), Tank Cleaning Amsterdam, Crowley Maritime Corporation, Martech USA, O'Brien Oil Pollution Services, and the International Maritime Organization. These organizations removed over 1.7 million barrels of oil from the sea surface in mid-June 1991 when only a small amount of free floating oil remained (MEPA, 1991a, 1991c). Booms were used to contain the oil where it was recovered from the sea surface using skimmers, diaphragm pumps, and dredges (Figure 2.6). Tidal traps consisting of shoreline pits with weirs were also used to skim surface oil. Tidal recovery in the Marduma area peaked at about 28,000 barrels per day (OGJ, 1991).

Interim storage sites for oil removed from the sea surface were developed at ten sites including Abu Ali Island, Marduma, Ras Tanajib, Ras al Zwar, and Manifa. The sites on Abu Ali Island were not used (Figure 2.7). By

Figure 2.6. Oil containment boom in harbor near Jubail. Photograph by John C. McCain.

Figure 2.7. Oil being release from tank truck into an interim storage pit at Marduma.

mid-November 1992, approximately 250,000 barrels of oil had been removed from the storage sites for processing and disposal under a contract with Saudi Aramco.

2.6. Physical and chemical characteristics of the spill

Since the beginning of the oil spill, efforts were made to contain the spill to minimize socioeconomic impact and environmental damage. Characterization of the spilled oil was required to make the cleanup program more effective. A knowledge of physical and chemical characteristics of the spill was important in the selection of cleanup and disposal options and to predict the toxicity of the oil and its degradation products to marine biota. Oil spill samples were collected from different locations and at different times. Physical and chemical properties of these samples were determined and are summarized in Table 2.6. Characteristics of the spilled oil as shown in Table 2.6 suggest the following:

1. With the passage of time, the spill became a mousse. The mousse increased in density, pour point, and viscosity. The water content of the collected mousse samples varied between 46 and 70% by volume.
2. Concentrations of all metal gradually increased with time in the mousse samples. Concentrations of nickel, vanadium, and the vanadium/nickel ratio were found to be closer to Kuwait Export crude oil than other oils examined.

Table 2.6. Physical properties and metal concentrations in mousse samples collected from the Arabian Gulf during the Gulf war Oil Spill (KFUPM/RI (1991f, 1991g)

Physical Characteristics

Oil Spill Sample	API at 60°F	Density at 60°F	Kinematic Viscosity at 140°F (CS)	Pour point Point (°C)	Water + Sediment (%)
OS-7	14.5	0.9684	123	10	40
OS-8	9.6	1.002	683	11	70
OS-9	-	-	291	-	65
OS-11	-	-	1514	18	64
OS-14	-	-	1500	17	66
OS-18	7.0	1.021	1507	18	68

Metal Concentrations (μg/l)

	Nickel	Vanadium	Iron	Lead	Copper	Zinc
OS-7	9.3	33.7	10.0	-	-	-
OS-8	10.3	37.9	9.3	-	-	-
OS-9	9.6	32.9	26.17	2.4	2.4	1.5
OS-11	9.7	27.6	53.1	3.1	3.5	1.7
OS-14	11.5	37.2	40.4	3.0	11.7	1.9
OS-15	12.5	40.1	44.5	5.5	4.7	2.7
OS-16	12.6	39.2	39.2	4.3	2.5	2.2
OS-17	11.1	35.6	51.8	4.1	7.8	2.8
OS-18	13.3	42.9	65.2	3.5	9.5	2.6

- Missing information or not determined because of small quantity of mousse sample.

	Sampling Dates	Station Locations
OS-7	07/02/91	Tanajib Marine Facilities
OS-8	07/02/91	Safaniya Pier
OS-9	11/02/91	Safaniya Shoreline
OS-11	14/02/91	Northwest of Abu Ali Island
OS-14	17/02/91	Abu Ali Island
OS-15	17/02/91	Abu Ali Island
OS-16	17/02/91	Abu Ali Island
OS-17	17/02/91	Abu Ali Island
OS-18	19/02/91	Ras Al Zwar

3. Concentrations of toxic metals in the mousse, especially copper, may play an important role in determining its toxicity to marine biota.
4. The metal concentrations in Kuwait crude and that found in seawater suggest significant inputs due to atmospheric fallout, particularly that of iron, copper, lead and zinc. Metal concentrations increased with time which further suggest atmospheric input.

2.7. Effect of oil on the marine ecosystem

2.7.1. Shoreline assessment survey

The Saudi Arabian Meteorology and Environmental Protection Administration (MEPA) acted as the Saudi Arabian National Oil Spill Response Coordinator during the spill. As part of this response, MEPA formed a team to survey the impacted coastline in order to quantify the amount of various habitats which were oiled, to document the extent of oiling, to establish a cleanup priority for sites, and to recommend cleanup procedures. This team consisted of representatives from the Research Institute of King Fahd University of Petroleum and Minerals, the United States Coast Guard, the National Commission for Wildlife Conservation and Development, the European Community Council, Saudi Arabian Bechtel Corporation, and Crowley Maritime Corporation.

The assessment survey was conducted along 777 km of Saudi Arabian shoreline from the Kuwait border to the northern shore of Abu Ali Island (MEPA, 1991a). A total of 148 sites were surveyed during late May and early June 1991 using off-road vehicles. The entire affected coastline was surveyed except for approximately 20% of the coastline where access was extremely difficult. The length of each oiled shoreline type is presented in Figure 2.8.

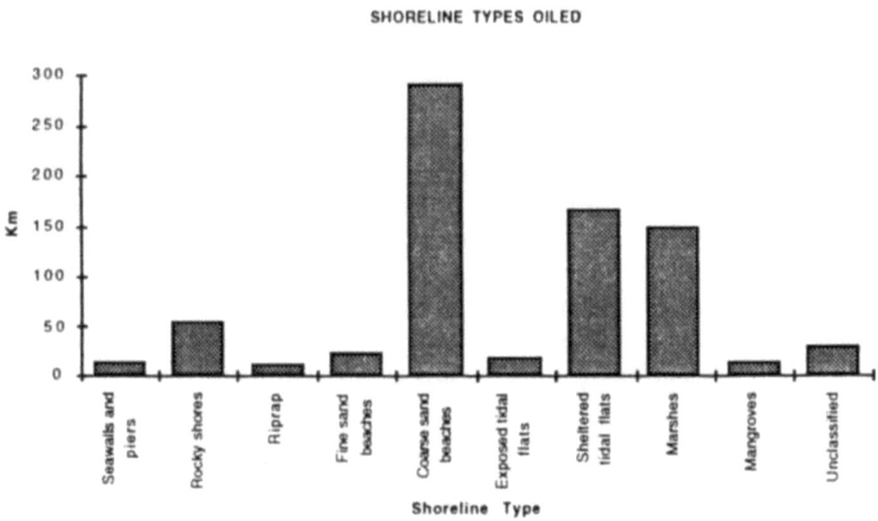

Figure 2.8. Estimated lengths of shoreline types oiled (from McCain et al., 1992).

The percentage of oiling and depth of penetration of oil at each site were estimated. Oil penetration of the sediment varied with sediment grain-size. Most of the sites surveyed by the MEPA Shoreline Assessment Team had oil penetration depths of less than 10 cm (MEPA, 1991a). The results of the shoreline assessment indicate that the affected 34 km^2 of intertidal area contained between 0.9 and 1.9 million barrels of oil with over 1.4 million m^3 of contaminated sediment (MEPA, 1991c).

2.7.2. Effects on intertidal communities

The intertidal zone of the Saudi Arabian Gulf coast north of Abu Ali Island was almost completely inundated with oil. Along the open, wave exposed coast, coarse sand beaches were the dominant habitat, accounting for 292 km of the total 777 km of oiled shoreline (MEPA, 1991a). Coarse sand beaches are relatively porous and quickly absorb oil deep into the sediment. Deeply buried oil can create an anaerobic zone which can substantially delay natural biodegradation (Gundlach et al., 1983). The depth of oil penetration during the Shoreline Assessment Survey in May and June 1991 was on average 10 cm. On some beaches oil has been found as deep as 0.5 m; however, the very deep oil is thought to be from previous oil spills.

Inside embayments along the coast, productive habitats such as mangrove stands, salt marshes, and tidal flats were oiled. These habitats were estimated to contain 1,293,900 m^3 of oiled sediment which is about 95% of the total of 1,367,000 m^3 oiled by the spill. Most of these embayments are relatively low energy environments which results in the deposition of fine grain sediments. Habitats with fine-grain sediment tend to recover from pollution far more slowly than those with larger grain sediment (Neff, 1987).

Jones and Richmond (1992) surveyed 37 intertidal sites in the Dawhat al Musallamiyah and Dawhat ad Dafi area using "rapid assessment techniques" about one year after the oil spill. Their sites included a variety of rocky shores, sandy shores, mangroves, and mud shores. Their conclusion was that "In general the intertidal habitats were found to be in a reasonably healthy condition." However, further into the text, they state that "As a result of the oiling of the upper shore, key supratidal species are now absent or virtually extinct at most sites within the study area". Similarly, McCain et al. (1984a) attributed decreased infaunal densities in the higher intertidal elevations at some sand beach stations sampled along the northern Saudi Arabian Gulf coast, to the accumulation of oil residues at these elevations.

2.7.2.1. Mangroves

The roots of mangrove trees (*Avicennia marina*) trap fine-grain sediment to form a nutrient rich substrate for the development of a complex, productive community of organisms. Juvenile fishes and shrimp inhabit the mangroves for protection and food during their nursery period. Evans (1992) considers mangrove-associated mudflats to be probably the richest wader (shorebirds) feeding habitat in the Gulf. The trees provide shelter and cover from predators and the fine-grain substrate provide an abundance of organisms for food.

Mangroves, particularly their root system, are especially sensitive to contact with oil. Recovery of mangrove stands may take decades if trees are killed (API, 1985). There has been tremendous pressure on the few remaining mangrove stands of the Arabian Gulf, particularly from landfilling in Tarut Bay where the largest mangrove stands are present. It was because of their importance and sensitivity that MEPA chose to designate mangroves as the impacted community with the highest priority for protection, cleanup, and restoration.

Seven sites with mangrove trees or seedlings were identified during the shoreline survey conducted just after the oil spill in Dawhat ad Dafi and Dawhat al Musallamiyah (MEPA, 1991a). Mangroves stands were limited to only a few seedlings or trees at all of these sites with the exception of the relatively luxuriant stand on Gurmah Island in Dawhat ad Dafi (Figures 2.9 and 2.10).

The mangrove stand on Gurmah Island was heavily oiled when visited on March 13, 1991. Oil in many areas was about 2–3 cm thick. Tidal channels had turned into oil pools. Mangrove branches, leaves, and pneumàtophores were coated with oil due to the high tide elevation. Other plants such as holophytes in the stand were also heavily oiled. Dead animals covered with oil littered the lower intertidal elevations. These included fiddler crabs, gastropods, bivalves, shrimp, and juvenile fishes.

During early April 1991, yellowing and defoliation of leaves were noted. Intensive cleanup operations using fire hoses and sprinklers were in progress from April through July. Many dead trees were noted during July 1991. The stands have shown continual decline since then as illustrated in Figure 2.11 (McCain et al., 1992).

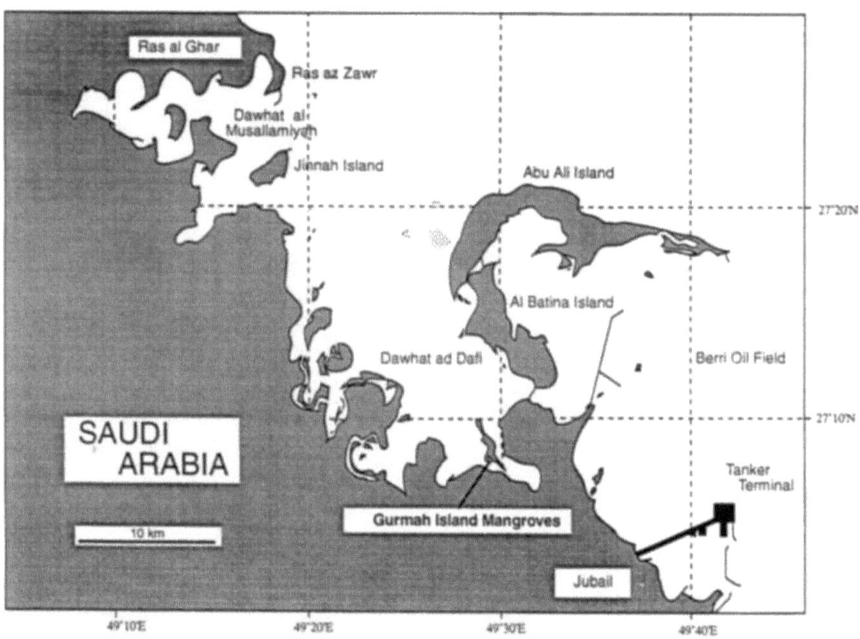

Figure 2.9. Gurmah Island mangroves (from McCain et al., 1992).

Figure 2.10. Sprinkler cleaning of the Gurmah Island mangroves. Photograph by David Lennon.

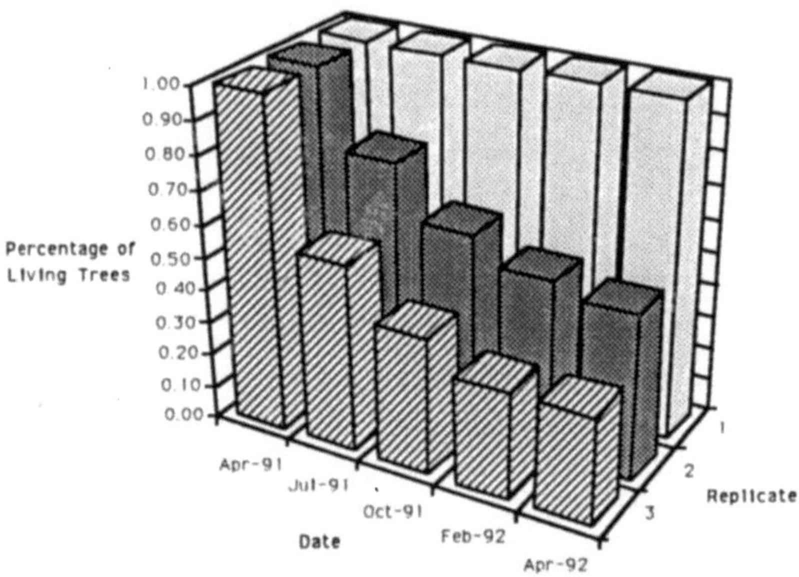

Figure 2.11. Mean percentage of living trees in mangrove quadrats on Gurmah Island. Inland (1); shoreline (2); gradient (3) (from McCain et al., 1992).

2.7.2.2. Salt marshes

Salt marshes are generally considered highly productive communities and of particular importance to waterbirds for feeding and resting during migrations (API, 1985). Evans (1992) considers salt marshes as favored roosting areas for wading birds. Sheppard and Price (1991) recently questioned the importance of the salt marshes along the western Arabian Gulf although there is no published data to support that assertion. Algal mats several centimeters thick cover much of the substrate between macrophytes in the middle and upper intertidal zones. These mats may contribute a substantial amount of productivity to the marshes and export much of that productivity to the Gulf ecosystem in general.

Forty eight separate salt marshes were identified during the MEPA (1991a) Shoreline Assessment Survey of the oiled Saudi Arabian shoreline (Figure 2.12). These marshes covered 149 km of shoreline. The estimated amount of oiled sediment in these marshes was 677,700 m^3, the largest amount of oiled sediment estimated for any shoreline habitat impacted by the Gulf War Oil Spill (MEPA, 1991b).

Figure 2.12. Heavily oiled Saudi Arabian salt marsh. Photograph by John C. McCain.

Counts of dead crabs (primarily *Scopimera scabricauda*) littering the shoreline at the high tide elevation were made during the remediation tests (see Section 2.10.1). As many as 1000 dead crabs per linear meter of beach were counted in these marshes. Other dead invertebrates included large numbers of small snail *Cerithidea* sp. The plant *Halocnemon strobilaceum* was abundant in the upper intertidal zones of the salt marshes while at lower intertidal elevations, *Salicornia herbacea* was the dominant macrophyte. At most sites inundated by the oil spill, all *Salicornia herbacea* were killed, whereas many of the *Halocnemon strobilaceum* plants were only partially oiled and a substantial percentage of the plants survived.

2.7.2.3. Sand beaches and tidal flats

Both sheltered and exposed tidal flats exist along the Saudi Arabian Gulf coastline. Many sheltered flats are associated with salt marshes and their sediment is generally very soft silt. Exposed tidal flats tend to have larger-grained sediments than their sheltered counterparts, but some fine-grained flats occur along the open Gulf coast. Sheltered tidal flats account for 166 km of the oiled Saudi Arabian coastline while exposed tidal flats account for only 20 km (MEPA, 1991a). Sheltered tidal flats were underestimated due to their association with salt marshes which were considered the primary shore type due to their relative importance to the system. Estimates of the amount of oil-contaminated sediment in tidal flats was approximately 556,900 m^3 (MEPA, 1991b).

Open coast sand beaches are important components of the Gulf ecosystem. They are the predominant shoreline habitat when only linear extent is considered (Figure 2.13). Coarse sand beaches account for 292 km of oiled shoreline and fine sand beaches 25 km (MEPA, 1991a). Due to their relatively steep slope and narrow width, the amount of oil contaminated sediment in sand

Figure 2.13. Heavily oiled sand beach. Photograph by John C. McCain.

beaches was only estimated at 90,100 m^3 (MEPA, 1991b).

Faraj et al. (1992) conducted a post-war survey of the intertidal infauna inhabiting several sand beaches and mud flats of Kuwait. They reported densities of 418 and 511 organisms per cm^3 in the mud flat at Sulaibikhat where the previously reported minimum was 467 organisms per cm^3. In the highly contaminated sand beach at Fahaheel, a maximum density of 415 organisms per cm^3 was reported while the previously reported minimum density was 739 organisms per cm^3.

The macrofauna of the tidal flats in Dawhat al Musallamiyah and Dawhat ad Dafi were sampled in November and December 1991 by scientists working at the recently created Marine Habitat and Wildlife Sanctuary near Jubail (Prena, 1992). Cores were collected at various intertidal elevations and in the shallow subtidal zone. The results of this study indicated that no infauna occurred in the high tide zone while "... The assemblage of the mid tide zone does not seem to be impoverished compared to the shallow sublittoral zone. No distinct community structure occurs adjacent to the oiled zone."

McCain et al. (1984a) estimated infaunal densities for organisms greater than 0.5 mm in size in the sand beaches of the Northern Area of Saudi Arabia to be as great as 413,000 individuals per m^2. There can be little question that the number of infaunal organisms living in the beaches of the Northern Area is indicative of a productive system. The loss of a tremendous amount of this production due to the oil spill may ultimately prove to be quite detrimental to the ecosystem of the western Gulf.

Bodin (1988) examined the impact of the *Amoco Cadiz* Oil Spill on the meiofauna (organisms living between the sand grains) in sandy beaches on the northern Finistère coast, Brittany, France. The meiofauna went through two phases, degradation and impoverishment for two or three years, then recovery or reconstitution which took about five years. This study amply illustrates the long time period necessary for the recovery of intertidal communities after an oil spill.

2.7.3. Effects on subtidal communities

Net primary production in the Gulf is thought to come almost entirely from subtidal communities (Sheppard and Price, 1991; MEPA, 1992). According to MEPA (1992), gross phytoplankton productivity (57.9×10^{11} grams of carbon per year) in the Arabian Gulf exceeds that of subtidal communities (42.4×10^{11} grams of carbon per year); however, net phytoplankton productive is thought to be zero while some subtidal communities contribute excess organic carbon.

Extensive algal flats and seagrass beds cover a substantial portion of the bottom in the relatively shallow waters of the Saudi Arabian Gulf, and they undoubtedly account for a major portion of net production (MEPA, 1992 – 1.3×10^{11} and 0.3×1011^{11} grams of carbon per year, respectively). Coral reefs are considered more productive than seagrass beds (MEPA, 1992 – 0.36×10^{11} grams of carbon per year) even though they cover a relatively small portion of the subtidal zone. Huge expanses of sandy substrate, essentially devoid of any hard substrate or macrophytes, cover the majority of the shallow subtidal zone. These areas contain a relatively high density of infaunal organisms (McCain, 1984b); however, MEPA estimates that the shallow sand areas (less than 5 m) remove excess carbon from the system (-1.4×10^{11} grams of carbon per year).

Jones and Richmond (1992) surveyed 55 subtidal sites in the Dawhat al Musallamiyah and Dawhat ad Dafi area "mostly" during January 1992. Based on this survey, they concluded that "In general these subtidal communities [bare sand/silt bottom, mature seagrass beds, rocky flats, coral reefs, coralline algae, and macroalgal beds] appeared to be in a healthy condition with no evidence of oil pollution. A single, small tar ball was found in benthic samples taken in the channel along Qurma Island."

2.7.3.1. Coral reefs

Nearshore platform reefs along the Saudi Gulf coast are abundant in the coastal zones which were heavily inundated with oil. Some reef platforms and aggregations are found less than 100 m from beaches where oil made landfall. No verified record of direct coral oiling exists. This is fortunate since Guzmàn et al. (1991) pointed out that subtidal coral reefs "...may suffer extensive damage from chronic exposure after major oil spills."

KFUPM/RI (KFUPM/RI, 1991c; McCain et al., 1992) is monitoring changes in areal cover of the dominant corals and macroalgae and changes in the abundance of fishes on 13 reefs to assist in determining if the Gulf War Oil Spill caused damage to the reefs (Figure 2.14). Several of these reefs have been monitoring periodically since 1982. No evidence of oiling or direct impacts from the oil spill on the Saudi Arabian reef corals, fishes, or macroinvertebrates

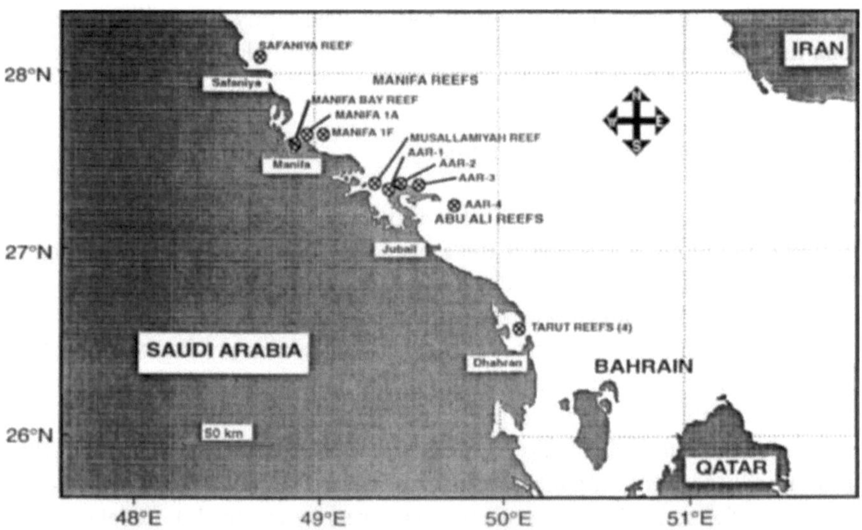

Figure 2.14. Coral reef survey stations (from McCain et al., 1992).

has been found during this study.

Jones and Richmond (1992) conducted a survey of six patch reefs and two fringing reefs in the Dawhat al Musallamiyah and Dawhat ad Dafi area during December 1991 and January 1992. Their survey showed that "No clear evidence of coral death or bleaching was observed, nor was there any evidence of oil pollution on the reefs."

The US National Oceanic and Atmospheric Administration (NOAA) conducted a series of oceanographic cruises in the Arabian Gulf after the oil spill. These cruises aboard the research vessel Mount Mitchell afforded scientists the opportunity to visit ten coral reef systems in Kuwait, Bahrain, Saudi Arabia, and Qatar to assess their "health" (Cushman, 1992; Fadlallah et al., in press). During a cruise in May 1991, substantial damage including tissue necrosis and bleaching was noted on two reefs near the port of Shuaiba in Kuwait. This damage has not been directly linked to the oil spill; in fact, scientists visiting reefs in Kuwait six months prior to the Mt. Mitchell visit noted no damage (Downing, 1991). Fadlallah et al. (in press) conclude that "Coral reefs in the central and northern ROPME Sea Area, appear not to have suffered catastrophic community losses as a result of the Gulf War Oil Spill."

2.7.3.2. Seagrass beds

Seagrass beds provide much of the primary production in the shallow waters of the Arabian Gulf. These beds are utilized as feeding and nursery grounds for many species of fishes and invertebrates and the seagrass itself is important as a food source for marine turtles and dugongs. Although only a small amount of seagrass was exposed to direct oiling, the acute and chronic toxic effects of exposure to dissolved and particulate petroleum hydrocarbons may well reduce overall production levels in the western Gulf nearshore zone.

Seagrass beds in Dawhat ad Dafi were oiled. A seagrass bed off the western tip of Abu Ali Island is within the intertidal zone and is exposed during some low tides. Gurmah Island and western Abu Ali seagrass beds are quite shallow, many less than 1 meter in depth. Studies are now underway to determine the impact of the oil on these beds.

Jones and Richmond (1992) quantitatively sampled three seagrass beds in the Dawhat al Musallamiyah and Dawhat ad Dafi area. They found densities from 263 to 2868 organisms per m^2. These density values appear very low when compared with data for Tarut Bay seagrass beds where the range varied from 2603 to 55,757 organisms per m^2(KFUPM/RI, 1990) and that reported for Northern Area seagrass benthos of 5,760 to as high as 113,480 organisms per m^2 (McCain, 1984b; KFUPM/RI, 1992).

Preliminary sorting of seagrass benthos samples collected around Abu Ali Island before and after the oil spill indicate that the density of organisms in the unexposed areas of Abu Ali Bay range from 500–25,200 organisms per m^2 (KFUPM/RI, 1992). Benthos densities for seagrass beds in areas exposed to the oil spill but not directly oiled range from 510 to 8650 organisms per m^2.

2.7.3.3. *Sand areas*

Subtidal sand flats are the predominant benthic habitat in the shallow waters of the western Arabian Gulf. Quantitative samples were collected by KFUPM/RI in sand areas potentially exposed to the oil spill; however, these have not been analyzed.

2.7.4. *Indirect effects*

2.7.4.1. *Seawater temperature*

McCain et al. (in press) examined seawater temperature data recorded in the Manifa area of Saudi Arabia and concluded that the average temperature during the period June 4 through December 10 was 2.5°C lower in 1991 than the mean temperature for that period during 1986–90 (Figure 2.15). During 1991, solar radiation was only 79% of that recorded during 1990 (Figure 2.16). The relationship between solar radiation during 1991 and the combustion of oil in the Kuwait Oil Fields is evident.

Ambient air temperature was 0.8–4.0°C lower during the period February through September 1991 than during the same period in 1990. The change in air temperature due to the smoke plume from the Kuwait Oil Well Fires is discussed in Chapter 6.11.5.

Lower than normal seawater temperatures may have directly affected the reproduction and growth of organisms in the Gulf. Mathews et al. (1992) state that no sexually mature hamour (a favorite food fish in the Gulf) were found in

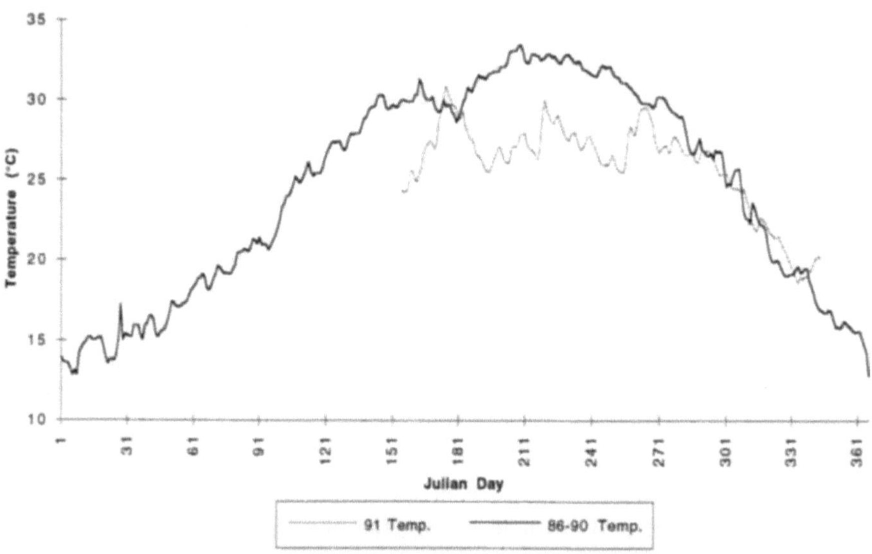

Figure 2.15. Plots of daily mean seawater temperature for 1986–90 and 1991 recorded at Manifa, Saudi Arabia (from McCain et al., in press).

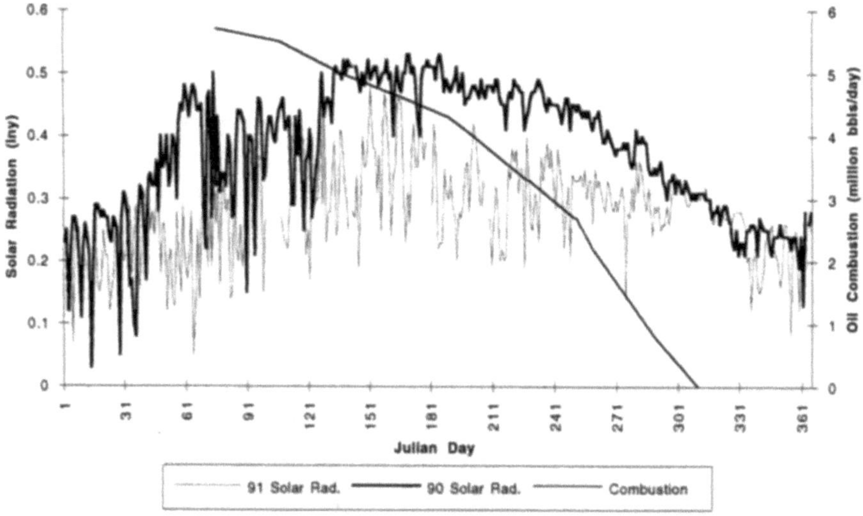

Figure 2.16. Plots of solar radiation for 1990 and 1991 (Rahimah) with an overlay of oil burned during the Kuwaiti Oil Well Fires (from McCain et al., in press).

Saudi catches since the Gulf war. Other organisms may have been similarly affected.

Scientists, based on numerical calculations, predicted a drop of 4 to 10°C in the ambient temperature in the Gulf region during the early stages of the Kuwait oil fires. Several investigators measured ambient air temperatures in the Gulf region. The decrease in ambient air temperatures in Kuwait from March to September, 1991 was of the order of 3–4°C (EPC, 1991) as compared to 0.8–2.4°C in Bahrain and Saudi Arabia. Beyond the distance of about 1000 km, the decrease in the ambient temperature was insignificant (Chapter 6).

Hobbs and Radke (1992) reported the absorption of over 90% of solar radiation in the thickest part of the smoke plume and about 75–80% in the super-composite region of the smoke plume. In Saudi Arabia, solar radiation was reduced to as much as 36% during the period from January through June 1991. Shaw (1992) also reported a similar decrease in solar radiation for the Island of Bahrain (Chapter 6).

2.8. Effect of the Gulf war on fisheries

2.8.1. Industrial and artisanal fishing

The western Arabian Gulf has supported a traditional, artisanal fishery for centuries. The fishery was largely confined to small dhows (traditional fishing vessels) until the 1950's when offshore fishing grounds were opened using larger dhows and ultimately industrial trawlers (Mathews et al., 1992). Today Saudi

Arabia, Kuwait, Bahrain, and Qatar operate large fishing fleets in the western Gulf. It is estimated that the total landings during the 1986 reached about 14,000 tonnes of prawns and 335,000 tonnes of shellfish and finfish (Sheppard and Price, 1991).

The nearshore areas of Kuwait and Saudi Arabia exposed to slicks and sheens from the Gulf War Oil Spill contain some of the major shrimp fishing and spawning areas of the Gulf. The areas off Safaniya, Manifa, and Abu Ali Island are generally recognized as major spawning and nursery grounds for shrimp (Price and Jones, 1975; Price, 1979b). Although perhaps not oiled directly, the shallow subtidal habitats in these areas undoubtedly were exposed to elevated petroleum hydrocarbons levels either through suspension of oil in the water column, dissolved hydrocarbons in the water, or through the movement of oil from the intertidal zone out into deeper water in the form of oil/sand pellets and from atmospheric fallout.

The commercial fishing fleet in Kuwait was destroyed. The Environmental Protection Council of Kuwait (1991) report on the Gulf environmental crisis indicates that a large number of dhows and small boats were stolen or destroyed during the occupation of Kuwait. Only five of 35 shrimp trawlers registered in Kuwait remained after the Gulf war. Approximately 90% of the mines deployed in the Arabian Gulf by Iraqi forces have been removed, but until the remainder of the mines are removed, the steel-hulled Kuwait shrimp trawlers will not operate (EPC, 1991). In November 1991, the Agricultural Affairs and Fish Resources Authority (AAFRA) reported approximately 10 dhows and 50–60 small boats were fishing in Kuwait waters, primarily with fish traps.

Processing plants, restaurants, and retail outlet stores owned by the largest fishing company in Kuwait, United Fisheries of Kuwait (UFK), were seriously damaged or destroyed during the occupation of Kuwait. The loss of the Kuwait Fisheries Administration and Kuwait Institute for Scientific Research buildings, libraries, records, and most importantly staff, will undoubtedly effect the fishing industry in Kuwait for many years.

Recent reports indicate that the fisheries of the Arabian Gulf were severely affected. The Saudi Fisheries Company announced that shrimp production is now less than 1% compared to the level before the Gulf war (SAFISH, 1992). The Saudi Arabian shrimp fishing season is from July 15 through October 15. The Saudi government closed the fishing season early during 1991, on September 21. Most of the trawlers from Saudi Fisheries Company were sent to Sri Lanka for repair until they returned in June 1991. Shrimping began again in September 1991 to complete the season. Due to small catches, Saudi Fisheries Company voluntarily stopped fisning for shrimp in September 1992 and the trawlers moved to the Red Sea.

Al Nokhathah (1992), the Saudi Fisheries Company newspaper states that the 1992 shrimp season in Saudi Arabia was approximately 33% of the normal harvest. They report a decline not only in total catch but in the size of the shrimp. Most shrimp caught in 1992 were in the range of 60–80 per kg whereas normally there are about 20–30 shrimp per kg. Saudi Fisheries Company has suspended exports of shrimp and resorted to purchasing and importing shrimp at higher prices.

Mathews et al. (1992) estimate losses of about US$ 55 million to the Saudi Arabian prawn fishery, and they indicate that the reproductive cycle of hamoor (*Epinephelus suilis*), a favorite food fish in the Gulf, may have been interrupted. Mathews et al. (1992) state that the October prawn catch rates for 1991 were less than half those for 1990 (14 kg per hr compared to 32 kg per hr). Seawater temperatures in the Gulf may be one of the major factors which contributed to this decline in the prawn fishery since the spawning activity of many marine organisms is intimately linked to temperature.

KFUPM/RI conducted an interview survey of fishermen who utilize the Safaniya, Manifa, and Abu Ali areas of Saudi Arabia. Most fishermen did not feel that the oil spill caused much disruption to either their finfish fishing practices or the size of their catch, except for closure of the fishery for approximately one year. Several fishermen surveyed believed that consumers were unwilling to buy fishes caught in the Gulf after the spill and that the taste of the fishes was tainted. Gill net and trap oiling were common complaints.

The State of Bahrain closed fishing during the Gulf crisis. Industrial and artisanal sector data for the period 1979 through 1991 are presented in Table 2.7. Total annual landings of commercial fishes and shellfish by the industrial sector did not drop below those levels reported during the early 1980's and the artisanal fishing sector landings during 1991 even exceeded those of 1990 (RDA, 1992). However, Mr. Al-Qaseer, Fisheries Director, stated in an article in the *Arab News* (1991) that shrimp landings had declined 30–40% in 1991 and that the shrimping season was delayed for three weeks to give shrimp more time to grow. Al-Qaseer indicated that "A mysterious disease [red spots] that may be linked to Gulf war pollution is killing scores of fish off the island of Bahrain".

Table 2.7. Annual commerical fish and shellfish landings (tonnes) in Bahrain (adapted from RDA, 1992).

Year	Artisanal	Industrial
1979	3008	1305
1980	3262	2237
1981	3950	2149
1982	4732	1509
1983	3302	1509
1984	4904	1358
1985	6185	1578
1986	6130	1926
1987	5729	2115
1988	4972	1764
1989	7009	2199
1990	6142	1962
1991	6533	1170

2.8.2. Fish and prawn eggs and larvae

McCain and Hassan (in press) reported the results of the analysis of 1073 ichthyoplankton tows (500 μ mesh net) made at 49 stations along the Saudi Arabian Gulf coast from 1984 to 1991. The results of analyses of these samples indicate that exposure to the slicks and sheens from the Gulf War Oil Spill reduced the abundance of fish eggs and larvae significantly ($p<0.05$) as compared to samples collected nearby in unexposed areas (Figure 2.17). The abundance of the eggs and larvae of penaeid shrimp was also reduced.

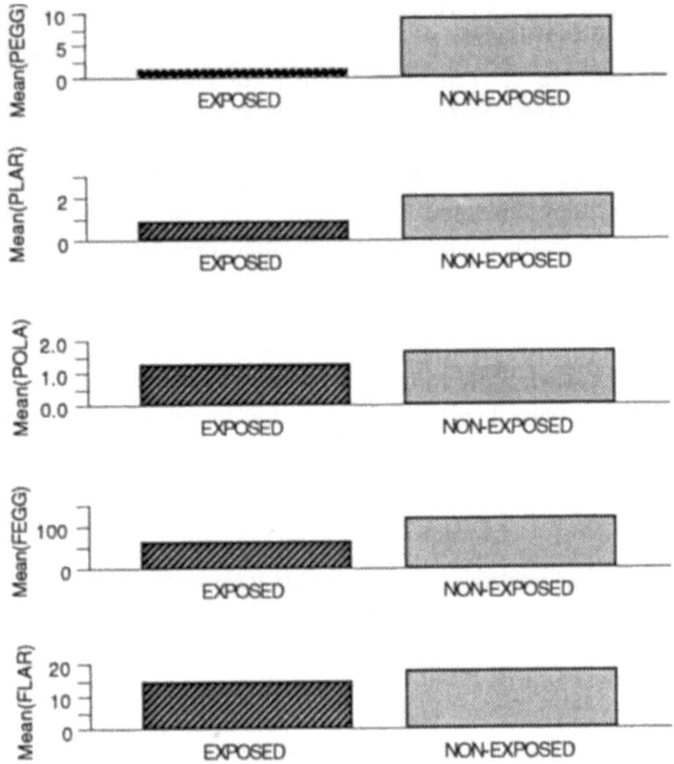

Figure 2.17. Mean abundance (individuals per 100 m³) of penaeid eggs (PEGG), penaied larvae (PLAR), penaeid post-larvae (POLA), fish eggs (FEGG), and fish larvae (FLAR) at stations exposed and not exposed to oil slicks and sheens (from McCain et al., 1992).

Several authors (Enomoto, 1971; Price and Jones, 1975; Price, 1979b) state that prawn eggs (*Penaeus* spp.) are most abundant in plankton from December to May, a period which encompasses the time when the oil spill was observed as slicks and sheens along the affected coastline. Offshore of this coast are several very important shrimp spawning areas at Safaniya, Manifa, and Abu Ali (Price and Jones, 1975; Price, 1979a). If the reduction in penaeid eggs and larvae observed at stations near slicks and sheens were extrapolated to the entire shallow subtidal area off the inundated coast, the impact on the northern Gulf shrimp fishery could be devastating.

2.8.3. Fish and shrimp quality

One of the early public concerns was the probable deterioration of seafood quality due to the oil spill and atmospheric fallout from the Kuwait oil fires. KFUPM/RI was the first organization to collect fish and shrimp samples from the Gulf in order to determine concentrations of toxic organic and inorganic chemicals in their edible tissue. Sampling started in March 1991, and by the end of December, over 500 samples of fish and shrimp had been analyzed for petroleum hydrocarbons and toxic metals. Except for a few fish, concentrations of polycyclic aromatic hydrocarbons were below the detection limit. Concentrations of n-alkanes varied from below the detection limit (only one sample) to 663 ng/g wet tissue. Mean concentrations of n-alkanes in hamour (*Epinephelus suillus*) and shaeri (*Lethrinus sp.*) were comparable, but lower than those found in safi (*Siganus rivulatus*) and seabream. The data on n-alkanes concentrations in fish were summarized and are plotted on Figure 2.18. It is obvious that safi contained the highest concentrations of n-alkane,

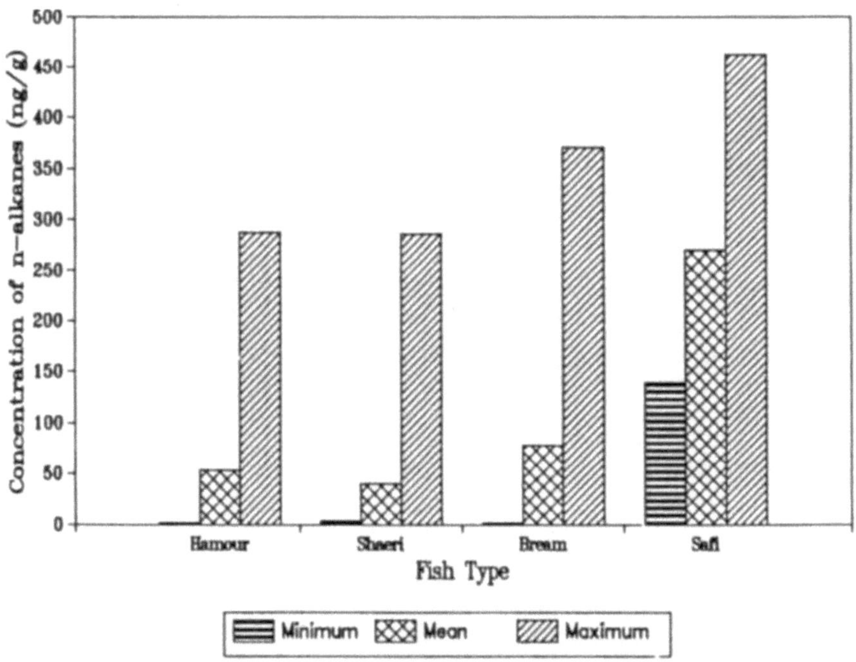

Figure 2.18. Concentration of n-alkanes found in the tissue of fishes collected from the Saudi Coast of the Arabian Gulf.

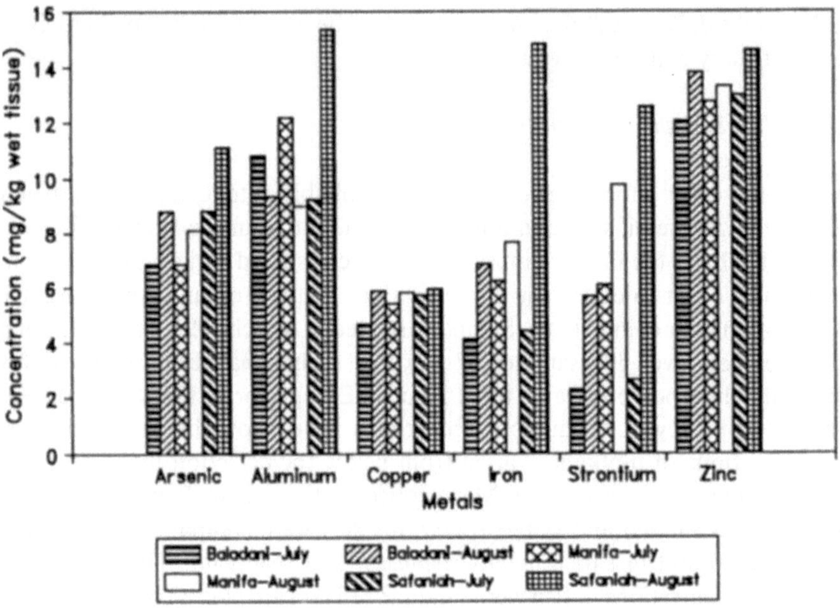

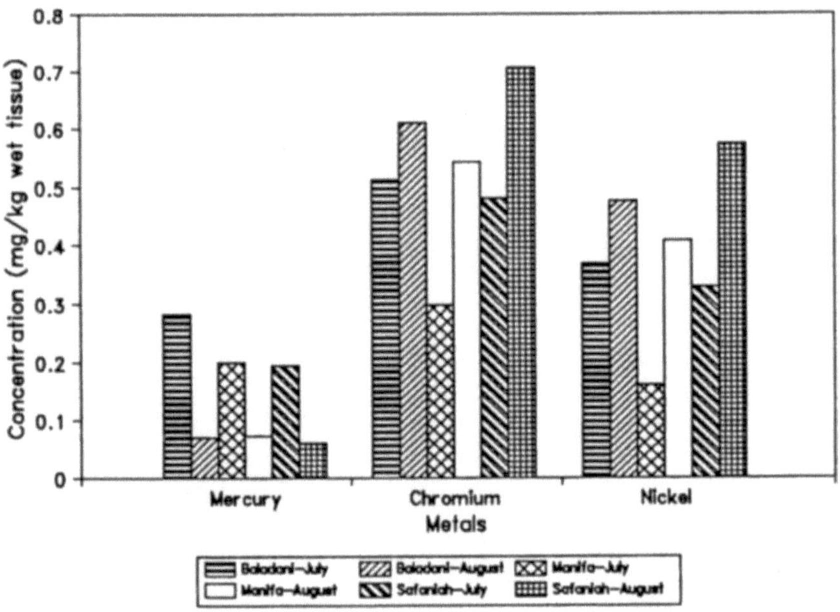

Figure 2.19. Selected metal concentrations in Arabian Gulf shrimp tissue.

followed in descending order, by bream, hamour, and shaeri. The concentrations of n-alkanes found in the collected fish samples were similar to previously known concentrations found in these fish (KFUPM/RI, 1983b, 1988b). These data suggest that the quality of fishes in the northern Gulf was not adversely affected by the oil spill, at least as far as tissue hydrocarbon concentration is concerned. Oil contamination of other organs was not examined.

The concentrations of toxic metals such as arsenic, cadmium, chromium, mercury, nickel, vanadium and lead were determined in the edible fish and shrimp tissue collected from the northern Arabian Gulf. Metal concentration data collected during this study are summarized in Figure 2.19. Metal concentrations in all the fish were found to be within the permissible limits with few exceptions (KFUPM/RI, 1991i). No geographic trend was observed in the metal distribution in fish. Metal concentrations found in this study were statistically similar to established data (KFUPM/RI 1983a, 1983b, 1988). Metal concentrations in fish and shrimp could not be linked to the oil spill and any other episode during the Gulf crisis. It seems that quality of fish and shrimp were not affected by the oil spill.

2.9. Effects of oil spill on marine vertebrates

2.9.1. Birds

The Arabian Gulf is an important part of the West Asian Flyway used by wintering western asian birds during their migration to eastern Africa (Summers et al., 1987). Zwarts et al. (1991) believe that the intertidal zone of the Gulf coast is one of the five major wintering areas for waders.

The Saudi Arabian National Commission for Wildlife Conservation and Development (NCWCD) and the International Council for Bird Preservation (ICBP) surveyed the waterfowl and shorebirds along the oiled coast of Saudi Arabia (Heneman, 1991; Evans, 1992; Harbard and Wolstencroft, 1992). Evans (1992) estimated that 77% of waders in the oiled area of Dawhat ad Dafi and about 50% in the Ras Tanajib area were oiled to some extent.

Based upon shorebird surveys conducted during November and December 1991, Harbard and Wolstencroft (1992) concluded that the total number of shorebirds along the Saudi Gulf coast was probably less than 100,000, with less than 25,000 waders utilizing the oiled coastline. Similar estimates made before the Gulf war by Zwarts et al. (1991) indicated that 260,000 shorebirds spend winter along the Saudi Arabian Gulf coast with about 100,000 waders in the area oiled during the Gulf war. A comparison of shorebird populations before and after the Gulf war is shown in Table 2.8.

Harbard and Wolstencroft (1992) observed that only those embayments with entrances to the south (south facing) supported a representative selection of waterbirds. Smaller shorebirds were notably less abundant in the oiled areas than they were before the Gulf war.

Table 2.8. Comparison of shorebird populations utilizing the Saudi Arabian Gulf coastline before and after the Gulf war (adapted from Harbard and Wolstencroft, 1992).

Shorebird	Before War	Spring 1991	Winter 1991
Dunlin	116,000	2856	17,121
Little stint	51,000	4165	3533
Lesser sand plover	28,000	5885	3766
Redshank	9000	810	5776
Ringed plover	8000	326	456
Greater sand plover	8000	75	1279
Grey plover	7000	1937	2606
Broad-billed sandpiper	6000	533	490
Curlew	6000	762	2658
Bar-tailed godwit	6000	6777	5350
Black-tailed godwit	5000	4	247
Curlew sandpiper	4000	3015	407
Terek sandpiper	4000	3591	1727
Turnstone	3000	1885	1321
Oyster catcher	1000	347	645
Kentish plover	800	1865	6547
Greenshank	500	124	201
Sanderling	200	1230	110
Total	263,500	36,187	54,240

Evans (1992) reported that over 200 dead birds were found in a short, 4 km stretch of shoreline at Ras Tanajib in May 1991 (Table 2.9). Walsh (1992) in his NOVA television documentary *Saddam's War on Wildlife* showed graphic scenes of numerous dead birds floating in a thick layer of oil at the shoreline on an open stretch of Saudi coast. He found 89 dead birds along a 200 m length of beach. Harbard and Wolstencroft (1992) estimate that approximately 25–30,000 pelagic species of birds died based upon the number of oiled birds brought into the Rescue Center and those found dead on the shoreline. They anticipate the total number of dead waterbirds to be in excess of 100,000 and perhaps as great as 250,000. In addition to death, the breeding success of oiled birds could be reduced which will further reduce their populations in the future.

The Wildlife Rescue Center located at Jubail collected oiled birds along the oiled Saudi Arabian Gulf coastline during the spill crisis. This center was a collaborative effort between the Royal Commission for Jubail and Yanbu and the National Commission for Wildlife Conservation and Development (NCWCD). The Rescue Center was manned by volunteers with specialists supplied by the NCWCD to provide training (Al Obayyed and Khan, 1992). Approximately 1385 birds were brought to the center for care and cleaning. Of these birds, 507 were ultimately released south of the oiled shoreline (Abu Zinada and Belcher, 1992).

Maki (1991) gives the release rate for sea birds treated during the *Exxon Valdez* spill to be 50 percent. He cites release rates for other spills between 30

Table 2.9. Dead oiled birds recorded along a four km section of coastline near Ras Tanajib in May 1991 (adapted from Evans, 1992)

Bird Type	Number of Deaths
Socotra cormorant	114+
Great cormorant	15+
Black-necked grebe	22
Great crested grebe	20
Lesser crested tern	6
Spotted crake	5
Warbler (*Acrocephalus* sp.)	4
Corncrake	3
Sandwich tern	3
Common sandpiper	2
Garganey	1
Little crake	1
Oyster catcher	1
Kentish plover	1
Little stint	1
Curlew sandpiper	1
Wood sandpiper	1
Turnstone	1
Armenian gull (?)	1
Gull (*Larus* sp.)	1
Caspian tern	1
European scops owl	1
Lesser short-toed lark	1

and 60 percent. The release rate for the Jubail Rescue Center was quite good considering the number of birds treated and the relatively small expenditure of funds since over 400 bird rescue personnel were involved in the *Exxon Valdez* spill program which cost more than $25 million and treated 1600 birds.

2.9.2. Sea snakes

Burchard and McCain (1984) reported that large numbers of sea snakes and other air-breathing organisms were killed during a mass mortality incident in the western Arabian Gulf during March and April 1983. This period corresponded in time to the Nowruz Oil Spill caused by the Iran-Iraq War. Sea snakes littered the beaches by the hundreds. Although the dead organisms found on the beaches cannot be linked directly to the Nowruz incident, carcasses began appearing in abundance on the beaches coincidental with the appearance of tar balls of Nowruz origin. Petroleum hydrocarbon burdens in the tissues of the sea snakes examined were consistent with acute poisoning rather than a gradual accumulation process (Burchard and McCain, 1984).

Only a few dead sea snakes were reported during the survey of the beaches after the Gulf War Oil Spill. Carcasses of other dead air-breathing organisms were not observed in numbers approaching that of the Nowruz spill.

2.9.3. Marine turtles

Four species of sea turtles are known to be present in the Arabian Gulf, including the hawksbill (*Eretmochelys imbricata*), the green (*Chelonia mydas*), the leatherback (*Dermochelys coriacea*), and the loggerhead (*Caretta caretta*). The green and loggerhead turtles are cited as threatened and the hawksbill and leatherback as endangered in the United States Endangered Species Act. The International Union for Conservation of Nature and Natural Resources (IUCN) considers all marine turtles as endangered except for the loggerhead which is considered vulnerable.

Both hawksbill and green turtles utilize the offshore islands of the Gulf as nesting sites with only infrequent nesting on the mainland beaches (Miller, 1989). In Saudi Arabian waters, Karan Island appears to be the primary nesting site with nesting known to occur on other islands including Jana, Kurayn, Harqus, Arabiyah, and Jurayd. Nesting generally occurs between June and August or September with mating usually occurring about a month earlier.

Burchard and McCain (1984) reported large numbers of dead green and hawskbill turtles killed during a mass mortality incident in the western Arabian Gulf during March and April 1983. The appearance of carcasses was coincidental with the appearance of tar balls arriving on beaches from the Nowruz Oil Spill.

The shoreline of Karan Island was surveyed by the International Environmental Team on March 14, 1991 (KFUPM/RI, 1991f). The team found the entire shoreline to be heavily oiled, more so than the shorelines of the other islands. The team cleaned oiled debris from the beach and later, the International Maritime Organization removed most of the remaining debris and oiled sand from the shoreline to either pits located towards the middle of the island or to barges.

The National Commission for Wildlife Conservation and Development (NCWCD) brought in experts to monitor the successes of sea turtle reproduction and nesting. These experts were assisted by NCWCD personnel and volunteers from Greenpeace and other organizations (Greenpeace, 1992). Pellew (1991) reported that some turtles died; however, quantitative data are not yet available. Many of the green turtles on Karan Island had lesions (Greenpeace, 1992). The reproductive success of the turtles returning to the Gulf Islands to breed and nest during 1991 is still unknown.

2.9.4. Marine mammals

Dugongs have been reported as far north as Ras al Zwar on the Saudi coastline; however, they are principally restricted to the area along the southern Saudi Arabian coast to Abu Dhabi (Preen, 1989). Since the oil slick from the Gulf War Oil Spill was essentially limited to the area north of Abu Ali Island, they were not exposed to any great extent to the oil spill. Dolphins, particularly the humpback and finless porpoises which frequent the nearshore areas and bays of the Gulf, are especially vulnerable (Greenpeace, 1992).

Preen (1991) estimated that at least 93 marine mammals died in the western Arabian Gulf from late February through mid-April 1991. These included 14 dugongs (*Dugong dugon)*, 57 bottlenose dolphins (*Tursiops truncatus*), 13 humpback dolphins (*Sousa chinensis*), 1 finless porpoise (*Neophoecaena phocaenoides*), and 8 unidentified cetaceans. This estimate was based primarily on beached carcasses found south of the oiled area.

The marine mammal losses reported above are relatively low in comparison to those reported during the 1983 Nowruz Oil Spill caused by the Iran-Iraq War. Approximately 50 dugongs and over several times as many dolphins were found dead on the beaches of Saudi Arabia after that spill.

Preen (1989) discusses four types of impacts to dugongs potentially caused by oil contact including asphyxiation due to inhaled oil; acute poisoning due to contact with fresh oil; lowering of tolerance to other stress due to the incorporation of sub-lethal amounts of petroleum fractions into body tissues; and nutritional stress through damage to food sources. This list of potential impacts could be applied to any air-breathing animal. Dolphins do not avoid oil slicks and sheens as witnessed during the Gulf War Oil Spill (Pellew, 1991; Preen, 1991; Greenpeace, 1992).

2.10. Effect of the Gulf war on the sediment and seawater quality

Two sources could cause seawater and sediment contamination: atmospheric fallout and the oil spill. The data given on the physical and chemical characteristics of oil spill samples suggest appreciable atmospheric inputs to seawater metal concentrations. These observations are further supported by data on dry deposition. The mean monthly rates of atmospheric dry deposition of sulfate and nitrate at Dhahran during the month of May 1991 are approximately 19 mg nitrate per m^2, 65 mg non-sea sulfate per m^2 and 24 mg sea sulfate per m^2 (KFUPM/RI, 1991h). Published reports indicate that concentrations of metals and organic chemicals were enhanced in ambient air particulates in the Gulf region and that the enhancement could be associated to Gulf crisis (Chapter 4). The total input from atmospheric fallout over the Arabian Gulf may be higher than that of the oil spill since it was confined to only about 800 km shoreline.

Crude oil contains metals, especially nickel and vanadium (Chapter 4). Therefore, the oil spill was expected to enhance metal contamination and organic pollution. With reference to the Gulf crisis, it is extremely difficult to differentiate between pollutants released by the spilled oil and those introduced by atmospheric fallout. Therefore, combined effects of the Gulf crisis on the seawater quality of the northern Arabian Gulf will be discussed. The Gulf war was still continuing when the oil spill occurred. The northern Saudi coastal areas of the Arabian Gulf were under military control. Therefore, the scientists and authorities assigned to spill containment program had to wait for the spill to reach a safe area well south of Kuwait. Data on the impact of spill on water quality are limited because the scientific infrastructure in Kuwait was completely destroyed and the majority of the Saudi scientists were out of the country at the time of the spill.

Few water samples were collected from the northern Arabian Gulf during the oil spill and analyzed for selected metals. A summary of the available data is given in Table 2.10. Very high concentrations of metals were found in the water samples collected from Mina Abdullah in Kuwait. This water sample was collected near to a partially sunk oil tanker and the leakage of oil from the tanker was probably responsible for the elevated metal concentrations. The data are limited, but they strongly suggest that the Gulf crisis and especially the oil spill has caused a deterioration of water quality.

Table 2.10. Metal concentrations (μg/l or ppb) in seawater samples.

Metal	OS-12 1991	OS-13 1991	OS-43 1991	Safaniya* 1990
Cadmium	0.7	0.9	1.8	0.3
Copper	1.5	1.4	3.9	4.1
Iron	15.6	16.4	78.9	12.8
Lead	1.5	1.3	15.6	2.1
Molybdenum	0.32	0.27	0.36	10.4
Nickel	0.9	0.9	2.5	0.6
Vanadium	2.01	1.9	3.6	1.0
Zinc	8.2	7.8	13.7	7.6

*Mean of 10 seawater samples collected from about 1 km north of Safaniya Pier in December, 1990.

OS-12 – From north side of Abu Ali causeway, 17/2/91

OS-13 – From north side of Abu Ali causeway, 17/2/91

OS-43 – From Mina Abdullah, Kuwait, next to a partially sunk oil tanker.

The northern side of Abu Ali Island was oiled but the southern side remained clean. Water samples collected concurrently from both sides were analyzed for metal concentrations (Sadiq, in press). These sampling stations are shown in Figure 2.20. Seawater samples were collected during March–April, 1991 (immediately after the spill impacted the northern side of Abu Ali Island), during July–August, 1991 (when large quantities of oil were still visible in the impacted areas), and during October–November, 1991 (about six months after the spill). The mean concentrations of metals in these seawater samples collected from the oil impacted and clean areas were computed and are shown in Figure 2.21. In general, concentrations of the selected metals were higher in the seawater samples collected from the northern side of Abu Ali Island (oil contaminated) than those collected from the southern side (clean). The magnitude of variations in metal concentrations in the samples from the oil contaminated areas and the clean areas decreased with time. Sadiq (in press) concluded that the Gulf War Oil Spill increased metal concentrations in seawater. Habashi (1991) collected seawater samples from 20 coastal stations in Kuwait during September–October, 1991 and analyzed for metals and hydrocarbons. These results were compared with pre-war observations. Based on the limited data, he concluded that levels of trace metals were much higher than those reported by the Kuwaiti Environmental Protection Department before the Gulf war. He attributed this increase to the large quantity of barbed wire, cross bars, and nets used in the coastal areas by the Iraqi army, and sunken vessels and ammunition that fell into the sea during the War.

Concentrations of petroleum hydrocarbons were determined in seawater

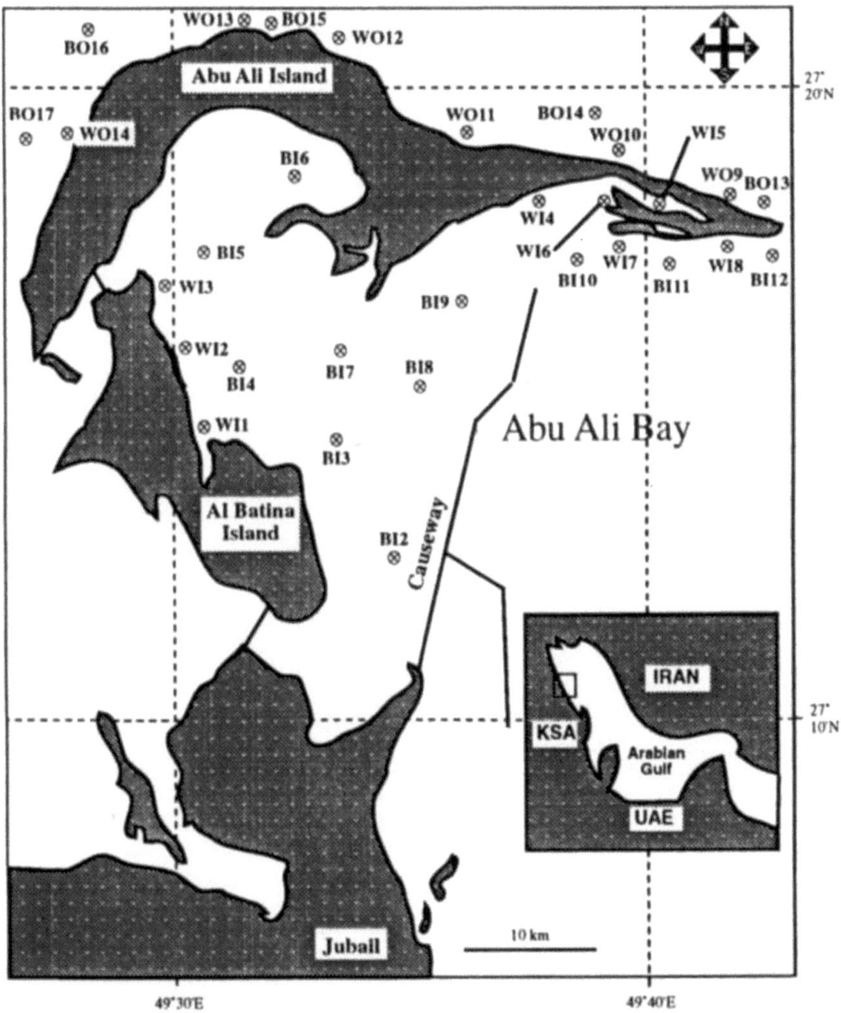

Figure 2.20. Seawater and sediment collection stations in the vicinity of Abu Ali Island.

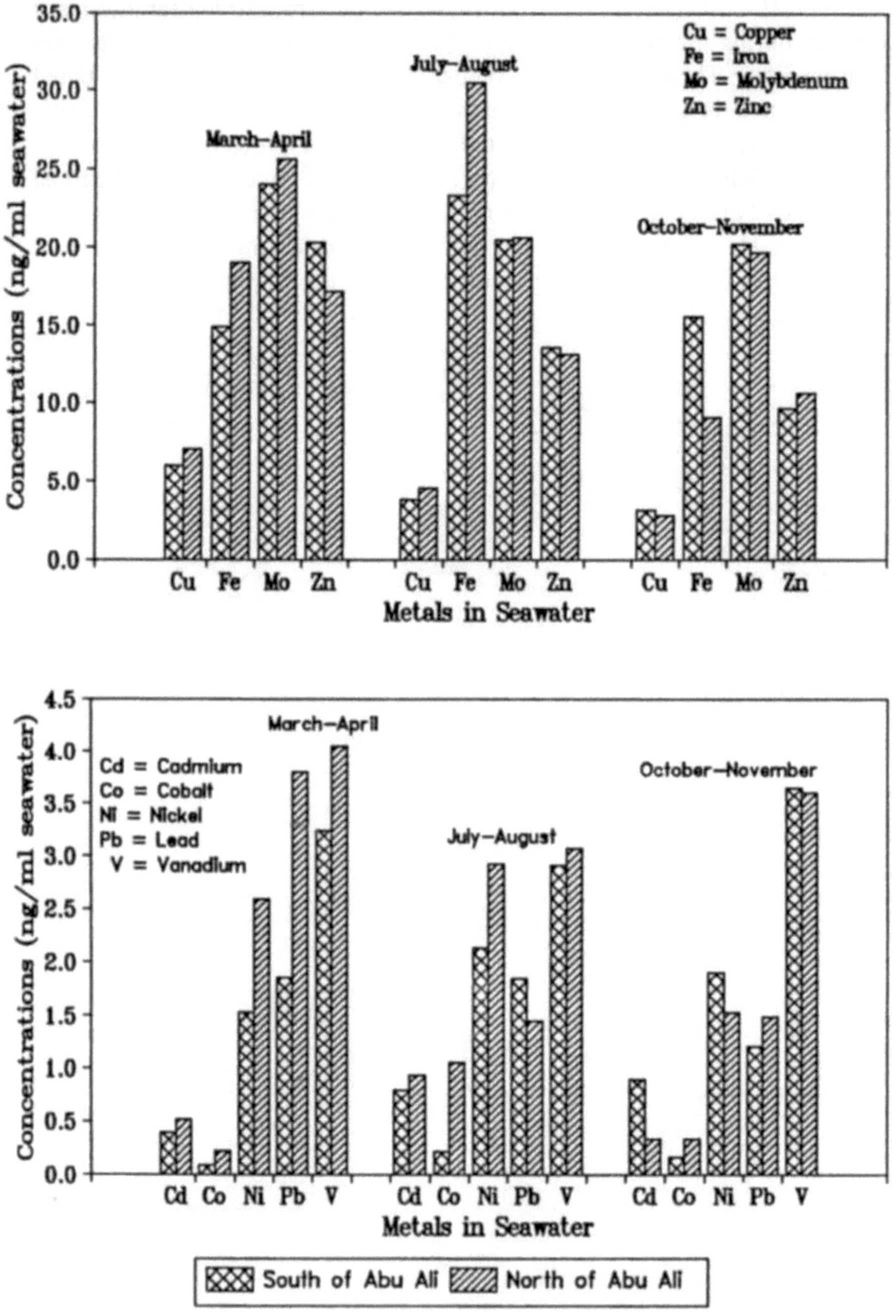

Figure 2.21. Concentration of metals in seawater from the vicinity of Abu Ali Island. North exposed to oil spill, south not exposed.

samples collected from the Kuwait coastal areas in September and October, 1991. Concentrations of petroleum hydrocarbon were elevated at a location where either a tanker had sunk or near an oil leaking source. Hydrocarbons were higher in October than September, 1991. Comparison with previous reports shows that concentrations of petroleum hydrocarbons were significantly higher in 1991 than those found in 1989 and 1988. The levels of of petroleum hydrocarbons were especially high in the areas near sunken vessels, oil terminals, or marinas with a number of sunken boats. The Kuwait Environment Protection Department resumed seawater sampling in September 1991, about six months after the spill. The Department found enhanced concentrations of nutrients as compared with 1989 (EPC, 1991).

Significantly higher concentrations of copper, lead, and iron, somewhat higher concentrations of nickel, and slightly lower concentrations of vanadium were found in seawater samples collected in September, 1991 as compared to those collected in 1989. Seawater samples were collected from the intakes of the Jubail and Al Khobar desalination plants from February and September, 1991 (Nomani et al., 1992). The collected water samples were analyzed for trihalomethanes, mineral oil, benzene, toluene, polynuclear aromatic hydrocarbons, phenols, chloroform, and total organic carbon. They concluded that, except for occasional spikes in concentrations of specific organic chemicals both in the intake and output water, none of the parameters exceeded drinking water quality limits set by the Saudi Standard Organization. The quality of the desalinated water at both plants throughout the monitoring period remained unaffected in spite of the spill. This suggests that organic chemicals from the spill were not dissolved and transported by current to as far as the Industrial City of Jubail which is located immediately south of Abu Ali Island.

Coastal sediments in Kuwait and along the northern Saudi coastline were heavily oiled. Oil was visible in spill contaminated areas. Because of the obvious presence of oil in the sediments, available data on petroleum concentrations in the sediment in spill contaminated areas were not reviewed. Sediment were collected from the coastal areas of Qatar and were analyzed for oil hydrocarbons. Concentrations of oil hydrocarbons in the samples ranged between 2.8 and 248 μg/g (Dahab et al., 1992). The concentrations found in Qatar sediments were higher as compared with the historical data reported for other areas of the Gulf.

Al Omran (1992) reported PAH concentrations in surface sediments from Kuwait in the range of 12.21–1318.5 μg/g wet weight (oil equivalent). High concentrations were found close to Ras Al Jleyaah and Jleyaat al Ahrar (naval base activities) and Mina al Shuaiba (oil refinery and terminal, industry, and shipping).

Readman et al. (1992) examined petroleum hydrocarbon concentrations in sediments and pelecypod molluscs in Kuwait, Saudi Arabia, Bahrain, the United Arab Emirates, and Oman. As expected, sediment from the northern shoreline of Saudi Arabia was highly contaminated; however, they found that sediment and oyster samples showed a decrease in hydrocarbon concentration as compared to pre-war samples. They reported that sediment concentrations carcinogenic polycyclic aromatic hydrocarbons (PAH) which would have been produced by the burning Kuwait oil wells, "...were relatively low, and comparable to levels reported in the Baltic Sea, coastal locations of the northeastern United States and United Kingdom."

2.11. Recovery and cleanup

2.11.1. Remediation of salt marshes

Testing of the environmental effects and effectiveness of a variety of oil spill cleanup methods was initiated at salt marsh test sites in Dawhat al

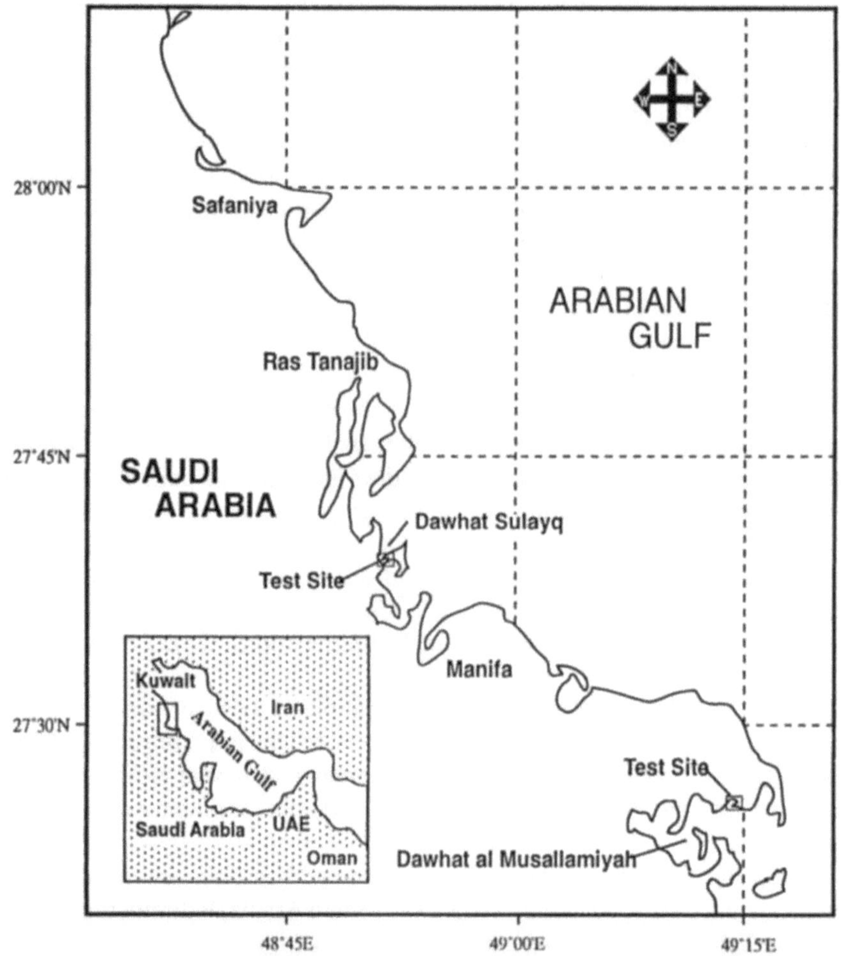

Figure 2.22. Chart showing the locations of the Dawhat Sulayq and Dawhat al Musallamiyah Test Sites (from McCain et al., 1992).

Musallamiyah and in Dawhat Sulayq (Figure 2.22) (KFUPM/RI, 1991a,b; McCain et. al., 1992; McCain et al., in press). The International Maritime Organization conducted the various cleanup trials at the Dawhat al Musallamiyah test site and Crowley Maritime Corp. conducted the marsh flooding trial at Dawhat Sulayq.

2.11.1.1. Dawhat al Musallamiyah test site

Table 2.11 shows the various oil spill cleanup techniques utilized at the Dawhat al Musallamiyah Test Site and Figure 2.23 illustrates the test plot layout for the marsh. The living above-ground biomass of the plant *Halocnemon strobilaceum* has consistently declined over the entire test site since the spill (Figure 2.24). Figure 2.25 illustrates the mean percentage of living tissue on *Halocnemon strobilaceum* plants for the various plots. Uncleaned plots (Plots

Table 2.11. Test plots at the Dawhat al Musallamiyah Test Site (from McCain et al., 1992).

Plot	Treatment	Test Protocol
A	Flooding with helicopter downdraft agitation of water	Plot surrounded by sandbags and flooded with seawater. Plot remained flooded for eight days. A back-pack air blower was used for water agitation and to push any free oil into catchment channels.
B	Flooding using sprinklers with soluble nutrient additions	No cleanup was attempted on this plot since it was set aside for later application of nutrients through sprinklers.
C	Flooding using sprinklers	An above ground sprinkler was set in the upper, middle, and lower intertidal elevations. The sprinklers operated for 12 hours on each of two days.
D	Flooding using perforated hose or pipe	A perforated pipe (3 inch diameter) was placed across upper area of plot. Perforations provided a light spray of water. Water was pumped through the pipe for 12 hours a day for one week.
E	Flooding	Flooding as described for Site A with light raking in the western portion of plot to break algal mat cover.
F	Control	Plot set aside, without treatment.
G	Control	Plot set aside, without treatment.
H	Algal mat removal	Algal mat removed in 1-meter wide strips with 25 m devoted to strips parallel to shoreline and another perpendicular to shoreline. A one meter space was shore and another, 25 m left between strips. Removal of the algal mat was accomplished by raking between plants then grubbing most of the remaining algal mat from the plant clumps by hand.
I	Algal mat removal	Algal mat completely removed. Removal of algal mat was accomplished as described for Plot H.
J	Nutrient application (oleophytic)	Plot set aside for later application of oleophytic nutrient.
K	Nutrient application (pelletized granules)	Plot set aside for later application of granular nutrient.

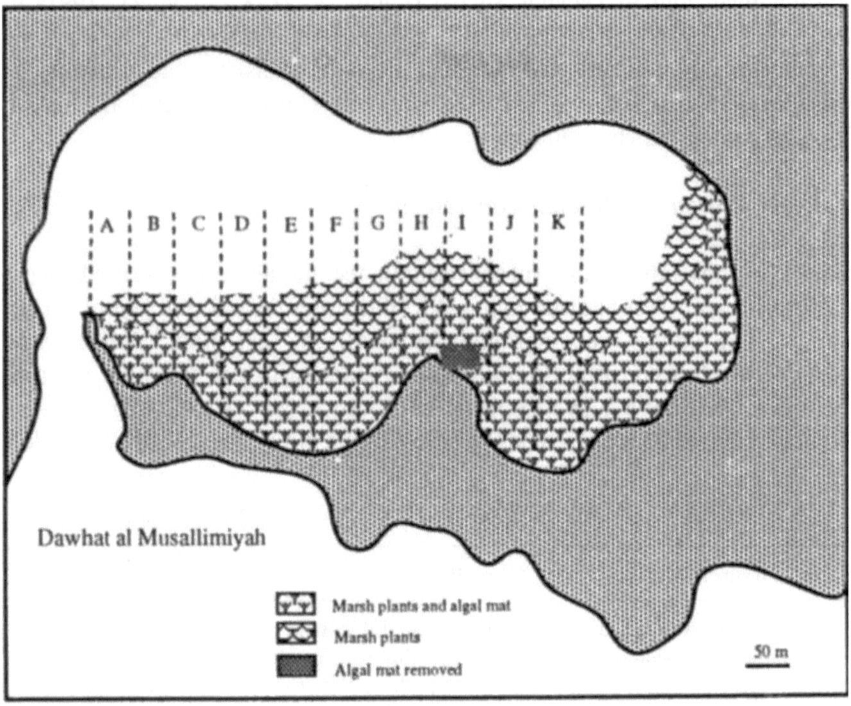

Figure 2.23. Test plots at the Dawhat al Musallamiyah Test Site (from McCain et al., 1992).

B, F, G, J, and K) showed little or no recovery. Living plant tissue in the photoquadrats of Plots J and K declined substantially whereas those of Plots F and G remained almost stable. The percentage of living tissue remained stable at Plots H and I where the algal mat was removed. The algal mat acted like a sponge to absorb much of the stranded oil. The removal of oiled algal mat in salt marshes may be an effective means of stimulating recovery or at least slowing the death of adjacent macrophytes. More time is required to determine the consequences to the ecosystem of the removal of the algal mat.

2.11.1.2. Dawhat Sulayq test site

The mouth of the Dawhat Sulayq (Figure 2.26) salt marsh was blocked by a sand dike on June 1, 1991, then filled with seawater to approximately the elevation of prior oiling which covered an area of about 9.83 hectares (24.2 acres). Over 1000 barrels of floating oil was then skimmed from the water surface in the southeast portion of the marsh using a vacuum truck. The dike was removed on June 16, 1991, and the entrance returned to its original bathymetry.

The mean living tissue estimated for the 11 sampling stations is plotted in Figure 2.27 for each sampling date. The percentage of living tissue has consistently decreased since the oil spill impacted this salt marsh.

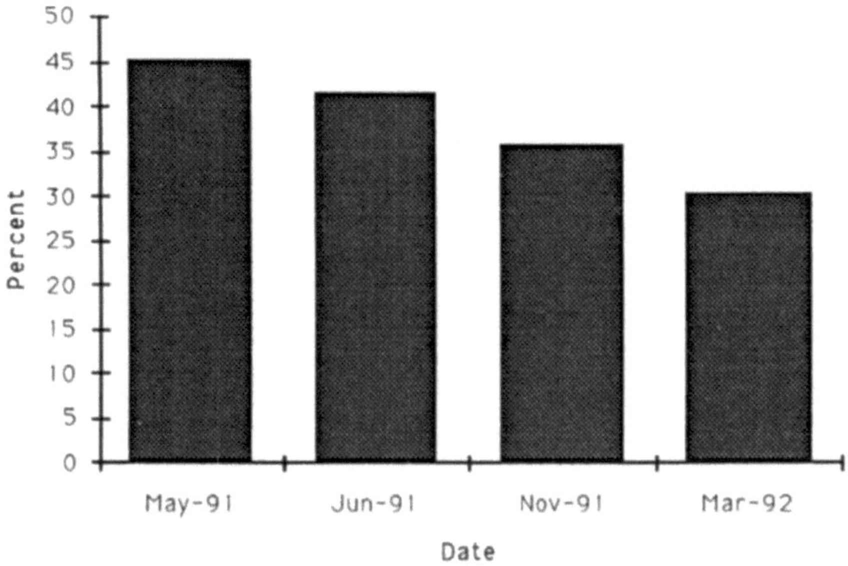

Figure 2.24. Mean percentage of viable above-ground biomass on *Halocnemon strobilaceum* plants in Dawhat al Musallamiyah Test Site photoquadrats (from McCain et al., 1992).

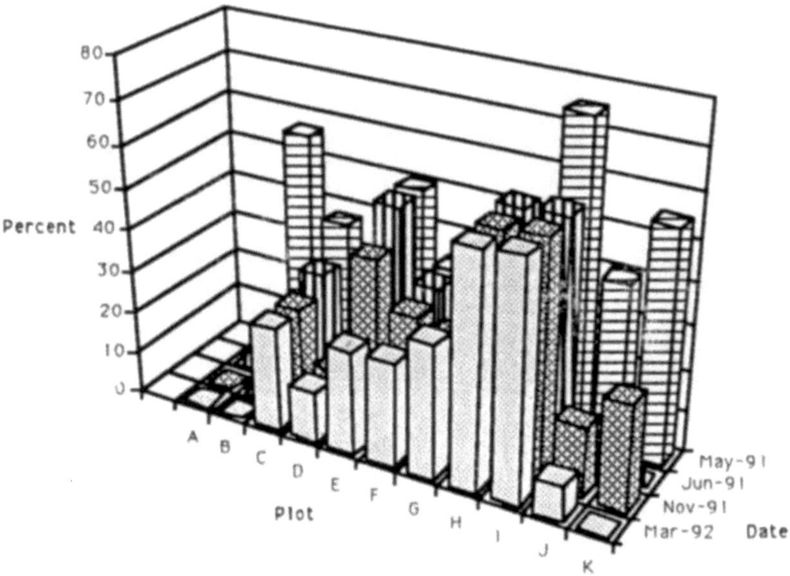

Figure 2.25. Mean percentage of viable above-ground biomass on *Halocnemon strobilaceum* plants in various Dawhat al Musallamiyah test plot photoquadrats (from McCain et al., 1992).

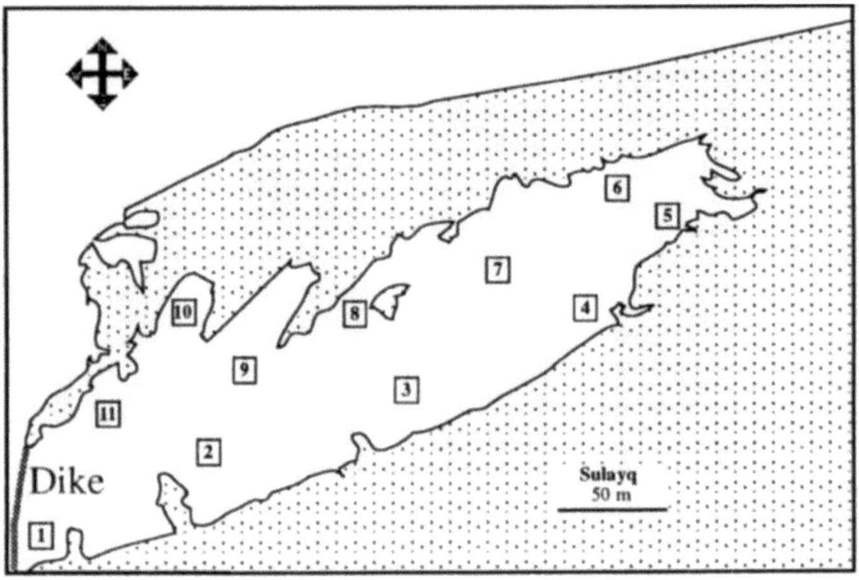

Figure 2.26. Plot of Dawhat Sulayq Test Site area showing sampling stations (from McCain et al., 1992).

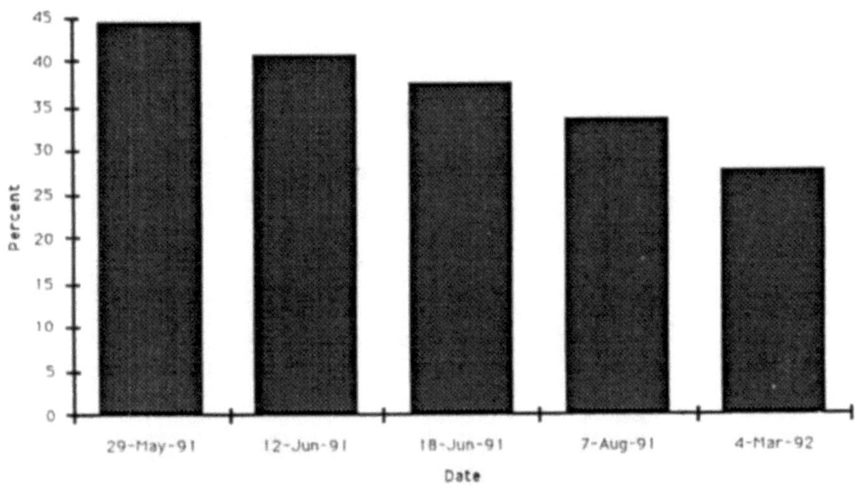

Figure 2.27. Mean percentage of viable above-ground biomass on *Halocnemon strobilaceum* plants at the Dawhat Sulayq Test Site (from McCain et al., 1992).

Flooding of a salt marsh is an effective method of recovering stranded oil. Flooding should be considered only for marshes which are heavily oiled and which do not show any potential for unassisted recovery.

2.11.2. Oil spill cleanup and recovery

No major cleanup effort has been implemented. Only a few sites have received the necessary approval for cleanup from MEPA, these being funded primarily by the International Maritime Organization. MEPA indicated that shoreline cleanup has been delayed due to the lack of funding by the Saudi Council of Ministers (OSIR, 1992). The cost of cleanup of the oil spill impacted coastline of Saudi Arabia has been estimated at $450 million (MEPA, 1991c; OGJ, 1991; OGJ, 1992).

It appears likely that natural cleaning processes will be the primary method of cleanup. Waves and associated abrasion will act to remove oil from the shoreline, particularly along the open coastline. In sheltered areas, other cleansing actions take place which may be just as effective. Jahns et al. (1991) followed natural cleaning processes which occurred after the *Exxon Valdez* spill in Alaska. Sediment interactions such as the formation of flocculated aggregates of a solids-stabilized emulsion were important in natural cleansing. Microscopic mineral flakes were noted to coat oil droplets, similar to larger sand/oil pellets noted during the Gulf War Oil Spill. In this form, oil can be moved easily by currents or waves since it does not tend to stick to the underlying sediment.

Ultimately, oil which is incorporated into the sediment of the Gulf must be broken down by microbial action. The Gulf has a thriving set of oil-consuming bacteria (Horgan, 1991). Biological degradation can be rapid in the aerobic surface sediments, but take a number of years with burial in the subsurface anaerobic layers (Gundlach et al., 1983).

The doomsday predictions by those unfamiliar with the Gulf ecosystem and its chronic exposure to oil made at the time of the Gulf War Oil Spill has been replaced by more tempered insight and scientific judgment. Dr. Sylvia Earle, previously the Chief Scientist at NOAA, remarked during a recent R/V Mt. Mitchell cruise that the Gulf will recover from the oil spill, but it will be different after recovery (Cushman, 1992). Mielke (1990 from Maki, 1991) after reviewing of the environmental impacts from several oil spills concluded that "To date, pollution from offshore petroleum activities has not appeared to be a significant threat to the survival of various species...Despite short-term media attention to the catastrophic nature of major oil spill events, the chemicals contained in petroleum have long been part of the marine environment and physical impacts are likely to be temporary in the dynamic natural flux of the coastal environment."

Neff (1987) states that "It may take a few decades for certain marine ecosystems to recover, even though they are not particularly susceptible to the spilled oil, if they are severely damaged by an oil spill or other pollution event." Boehm et al. (1985 from Neff, 1987) graphically illustrated the long recovery time required for some marine communities as shown in Figure 2.28. MEPA (1991b) estimated that about 95% of the oiled sediment on the shoreline of Saudi Arabia consists of fine grain sediment (1,293,900 or 1,367,000 m^3). Neff

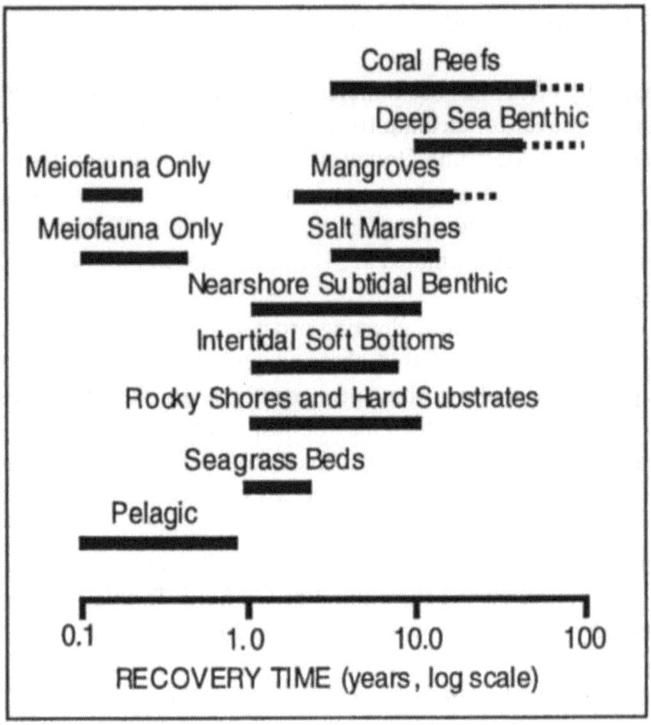

Figure 2.28. Ecosystem recovery time after a pollution event (after Neff, 1987).

(1987) points out that the recovery of habitats with fine-grain sediment tends to be slow.

Working in Panama, Jackson et al. (1989) carried out what is perhaps the best documented assessment of the ecological effects of an oil spill on biological communities. Detailed data were available for the affected coral reef and mangrove communities before the spill. Seagrass data were limited to that collected after the spill but comparisons were made between oiled and unoiled sites. Mangroves, seagrasses, algae, coral reefs, and seagrass infauna were covered with oil and died. In 1989, 1.5 years after the spill, only minor recovery was found in areas exposed directly to the sea.

Signs of recovery along the oiled Gulf coastline are few and limited primarily to the formation of new algal mats. Algal mats were observed in the Abu Ali Island and Dawhat al Musallamiyah areas in September 1992. Based upon the time estimates given by Boehm et al. (1985 from Neff, 1987) illustrated in Figure 2.28, it will probably be another decade before the Gulf recovers from the Gulf War Oil Spill.

2.12. Future response and research needs

According to an editorial which appeared in the May 27, 1991, edition of *Business Week*, only about 500 workers were devoted to the cleanup of the Gulf War Oil Spill which was financed at about $60 million. In Alaska, $2 billion and 11,000 workers were utilized to cleanup the *Exxon Valdez* spill, a spill less than 1/40th the volume of the Gulf War Oil Spill. The reasons behind the lack of a major cleanup response should be reviewed and an action plan to combat future spills in the Arabian Gulf should be developed and fully implemented. The Gulf Area Oil Companies Mutual Aid Organization (GAOCMAO) has been involved in the development of such a plan. This plan and adequate action must be implemented in the future.

The development of detailed habitat maps is crucial for adequate protection and containment response. KFUPM/RI, supported by the Saudi Aramco, is currently producing a series of maps depicting the nature and extent of the various intertidal and shallow subtidal habitats (at depths less than 10 meters) found along the Saudi Arabian coast of the Arabian Gulf. Mapping of the intertidal zone has been completed; however, shallow subtidal zone mapping will take another two years. Habitat maps should be developed for the entire Arabian Gulf coastline and offshore islands. It is recommended that these maps and the data generated to produce them be entered in a geographical information system (GIS) such as the Arc/Info system and that their production and standardization be coordinated between the various Gulf countries.

The Natural Environment Research Council of the United Kingdom made an important recommendation which, if followed, would have prevented a very large portion of the damage caused by the Gulf War Oil Spill (NERC, 1991). NERC recommended to "Minimise the smothering effect of the oil slick by preventing it from coming ashore wherever possible." The use of dispersant after the oil had been at sea for a few days was advised.

Greenpeace (1992) sent a scientific team to the Gulf area to investigate the environmental damage caused by the Gulf war. One of their recommendations regarding cleanup was to mechanically clean selected areas along the impacted coast to create "cleaner islands". This action would provide "seed" areas for the re-establishment of intertidal communities and might aid in the more rapid recovery of the ecosystem. This recommendation should still be considered.

Saudi Arabia resisted proposals to introduce oil-consuming bacteria to degrade the spilled oil, primarily on the grounds that the efficacy and safety of the technology were unproved. At the request of MEPA, KFUPM/RI reviewed the available literature on numerous biodegradation products and conducted a controlled experiment to determine the efficacy of Alpha BioSea, a bioremediation agent produced by Alpha Environmental Corp. (KFUPM/RI, 1991e; Fayad et al., 1992). In general, the reviews indicated that insufficient data were presented to allow a determination of either efficacy or safety. However, the experiment showed that under laboratory conditions, BioSea degraded some fractions of the oil. Bioremediation is a promising technology and the Gulf States should participate in research to obtain the necessary data to demonstrate the efficacy and safety of a variety of products.

A program to determine the toxicity of oil dispersants to critical habitats is important to the Gulf region and should be instituted as soon as possible. The use of dispersants has been rejected for many spills due to the potential toxicity of the dispersants and dispersed oil to important habitats. It may prove far better to either disperse or even sink spilled oil in relatively deep water than to allow it to come ashore and inundate productive intertidal and shallow subtidal communities.

Much of the marine research conducted at Saudi Arabian institutions has been directed towards the acquisition of general baseline data on the Gulf. Attention should turn from gathering general baseline data towards obtaining a more complete understanding of the processes which effect biological communities. A major shortcoming of our research in the Gulf, that of a total lack of data on the productivity and contribution to the Gulf ecosystem of the intertidal zone, became very evident during the oil spill. The Marine Habitat and Wildlife Sanctuary being developed by the National Commission for Wildlife Conservation and Development and the European Community Council is now funding some of the more specific types of research necessary to gain an understanding of the system. The Regional Organization for the Protection of the Marine Environment and local institutions of the Gulf should consider directing their efforts in a coordinated manner towards this goal.

References

Abu Zinada, A.A. and Belcher, N. 1992. Report on the effects of the 1991 oil spill on the wildlife of the Arabian Gulf and the efforts of the National Commission for Wildlife Conservation and Development. Paper presented at the Fourth General Conference, Third World Academy of Sciences, Kuwait, 23–26 November.

Akielaszek, J.J. and Haines, T.A. 1981. Mercury in the muscle tissue of fish from three northern Mainelakes. Bull. Environ. Contam. Toxicol. 27: 201–208.

API, 1985. Oil Spill Response: Options for Minimizing Adverse Ecological Impacts. Prepared by the American Petroleum Institute Task Force on the Ecological Impacts of Oil Spill Cleanup.

Arab News 1991. Mysterious disease kills fish in central Gulf. Arab News, Monday, August 19, 1991.

Barber, R.T. and Whaling, P.J. 1983. Mercury in marlin and sailfish. Mar. Pollut. Bull. 14:395–396.

Basson, P.P., Burchard, J.E., Hardy, J.T., and Price, A.R.G. 1977. Biotopes of the Western Arabian Gulf: Marine Life and Environments of Saudi Arabia. Arabian American Oil Company Department of Loss Prevention and Environmental Affairs, Dhahran, Saudi Arabia, 284 pp.

El-Baz, F. 1992. The war for oil: Effects on land, air, and sea. *Geotimes*,:37(5):12–15.

Blumer, M. 1969. Oil pollution of the ocean, pp. 5–13. In: *Oil on the Sea*, Plenum Press, New York.

Bodin, R. 1988. Results of ecological monitoring of three beaches polluted by the 'Amoco Cadiz' oil spill: development of meiofauna from 1978 to 1984. Mar. Ecol. Prog. Ser. 42:105–123.

Boehm, P.D. et al. 1985. Contaminant Resident Times and Ecosystem Recovery Rates for Shelf

and Estuarine Ecosystems. Final Rept. to National Oceanic and Atmospheric Administration (from Neff, 1987).
Braune, B.M. 1987. Mercury accumulation in relation to size and age of Atlantic herring (*Clupea harengus harengus)* from the southwestern Bay of Fundy, Canada. Arch. Environ. Contam. Toxicol.16:311–320.
Burchard, J.E. and McCain, J.C. 1984. Marine life kill in the Arabian Gulf, March/April 1983. Proc. Symp. Coral Reef Environment Red Sea, Jeddah, January 1984:392–393.
Cekirge, H.M., Al Rabeh, A.H. and Gunay, N. 1989. Determining the wind driven surface currents for prediction of movement of oil slicks in the Arabian Gulf. Comp. Math. Appl.17:1449–1453.
Clark, R.C., Jr. and Blumer, M. 1967. Distribution of n-paraffins in marine organisms and sediments. Limnol. Oceanogr.12:79–87.
Coles, S.L. 1988. Limitations in reef coral development in the Arabian Gulf: Temperature or algal competition? Proc. 6th Intern. Coral Reef Symp. 3:211–216.
Coles, S.L. and Fadlallah, Y.H. 1991. Reef coral survival and mortality at low temperature in the Arabian Gulf: New species specific lower temperature limits. Coral Reefs, 9:231–237.
Coles, S.L. and Gunay, N. 1989. Tar pollution on Saudi Arabian Gulf beaches. Mar. Pollution Bull. 20:214–218.
Cushman, J.H. 1992. Most marine life in Persian Gulf thrives despite huge oil slick. New York Times, May 26, 1992, p. B6.
Cutshall, N.H., Naidu, J.R. and Peracy, W.G. 1978. Mercury concentrations in Pacific hake, as function of age and latitude. Sci. 200:1489–1491.
Dahab, O.S., Al Madfa, H., and Saleh, I. 1992. Petroelum hydrocarbons in Doha coastal sediments. First Bahrain International Conference on Environment. pp. 149–156.
Downing, N. 1991. A Survey of the Coral Islands of Kubbar, Qaru and Umm al Maradem, Kuwait, 13–19 July 1991. Report to the International Union for the Conservation of Nature and the World Wildlife Fund.
Eisler, R. 1987. Polycyclic Aromatic Hydrocarbon Hazards to Fish, Wildlife, and Invertebrates: A Synoptic Review. United States Department of Commerce. NITS Rept. no. PB87–189825.
Enomoto, Y. 1971. Oceanographic survey and ecological study of shrimp in the waters adjacent to the eastern coast of the State of Kuwait. Bull. Tobai Reg. Fish. Res. Lab. 66:1–73.
EPC, 1991. State of the Environment Report: A Case Study of Crimes Against the Environment. Environment Protection Council, Kuwait.
Evans, M.I. 1992. The ICBP/NCWCD Wader Survey of the Gulf Coast of Saudi Arabia during April-May 1991. Final Report to the National Commission for Wildlife Conservation and Development and the International Council for Bird Preservation.
Fadlallah, Y.H., Eakin, M., Al-Sofyani, A., Allen, K.W., Jaubert, J., Rahim, S.A., Earle, S.A., and Reaka-Kudla, M. (in press). Assessment of coral reef ecosystems in the ROPME Sea Area following the 1991 Gulf war oil spill. Abstract of Paper Presented at the ROPME Scientific Workshop on Results of the R/V Mt. Mitchell Open Sea Cruise, January 24–28, 1993, Al-Ain, United Arab Emirates.
Faraj, M., Al-Kady, S., and Al-Kandari, A. 1992. Biological indicator of the impact of the Gulf environmental catastrophy on Kuwait beaches. Poster Paper presented at the Fourth General Conference, Third World Academy of Sciences, Kuwait, 23–26 November.
Fay, J. 1970. Physical processes in the spread of oil on water surface. Proc. Conf. on Prevention and Control of Oil Spills, American Petrol. Inst., Washington, D.C., pp. 463–467.
Fayad, N.M., Edora, R.L., El-Murbarak, A.H., and Polancos, Jr., A.B. 1992. Effectiveness of a bioremediation product in degrading the oil spilled in the 1991 Arabian Gulf war. Bull. Environ. Contam. Toxicol. 49:787–796.
Galt, J.A., Payton, D.L., Torgrimson, G.M., and Watabayashi, G. 1984. Trajectory analysis for the Nowruz oil spill. In: El-Sabh, M.I. (ed), Oceanographic Modelling of the Kuwait Action Plan, UNESCO Reports in Marine Sciences, no. 28.
Golob, R. and Brus, E. 1984. Statistical Analysis of Oil Pollution in the Kuwait Action Plan Region and the Implications of Selected World-wide Oil Spills to the Region. Combating Oil Pollution

in the Kuwait Action Plan Region. UNEP Regional Seas Rept. Studies, No. 44.

Greenpeace, 1992. The Environmental Legacy of the Gulf war. Greenpeace Report, 41 pp.

Gundlach, E.R., Boehm, P.D., Marchand, M., Atlas, R.M., Ward, D.M., and Wolfe, D.A. 1983. The fate of Amoco Cadiz oil. Sci. 221:122–129.

Gundlach, E.R. and Hayes, M.O. 1977. The Urquiola Oil Spill, La Coruña, Spain: Case history and discussion of methods of control and clean-up. Mar. Poll. Bull. 8:132–136.

Guzmán, H.M., Jackson, J.B.C., and Weil, E. 1991. Short-term ecological consequences of a major oil spill on Panamanian subtidal reef corals. Coral Reefs 10:1–12.

Habashi, B. 1991. Effect of the environmental disaster in Kuwait on the quality of coastal water. Paper presented at Kuwait Seminar on Environmental and Health Impacts of the Torched Wells and Iraqi Invasion of Kuwait. 9–11 November, Kuwait.

Harbard, C. and Wolstencroft, J. 1992. The ICBP/NCWCD Waterbird Survey of the Gulf Coast of Saudi Arabia during November–December 1991. International Council for Bird Preservation.

Hawley, T.M. 1992. Against the Fires of Hell: The Environmental Disaster of the Gulf war. Harcourt Brace Jovanovich Publ., New York.

Hellou, J. and Pyne, J.F. 1987. Assessment of contamination of petroleum – a role for bile metabolites. Environ. Toxicol. Chem. 6:857.

Heneman, B. 1991. The Gulf War Oil Spills: The ICBP Survey of the Saudi Arabian Gulf Coast, March, 1991. Report to the International Council for Bird Preservation and the National Commission on Wildlife Conservation and Development.

Hobbs, P.V. and Radke, L.F. 1992. Airborne studies of the smoke from the Kuwait Oil Fires. Sci. 256:987–991.

Horgan, J. 1991. The muddled cleanup in the Persian Gulf. Sci. American, October, 1991, p. 106–110

Jackson, J.B.C., Cubit, J.D., Keller, B.D., Batista, V., Burns, K., Caffey, H.M., Caldwell, R.L., Garrity, S.D., Getter, C.D., Gonzalez, C., Guzman, H.M., Kaufmann, K.W., Knap, H.W., Levings, S.C., Marshall, M.J., Steger, R., Thompson, R.C., and Weil, E. 1989. Ecological effects of a major oil spill on Panamanian coastal marine communities. Sci. 243:37–44.

Jahns, H.O., Bragg, J.R., and Dash, L.C. 1991. Natural cleaning of shorelines following the Exxon Valdez spill. Proc. 1991 Oil Spill Conf., Sponsored by United States Coast Guard, American Petroleum Institute, and United States Environmental Protection Agency:167–176.

Jones, D.A. and Richmond, M.D. 1992. Intertidal and subtidal marine habitat surveys, pp. 1324–160. In: Final Report of Phase I, Establishment of a Marine Habitat and Wildlife Sanctuary for the Gulf Region. National Commission for Wildlife Conservation and Development and the European Community Commission.

Kawash, N. 1992. Gulf patrols fail to curb dumping in world's most polluted waters. Arab News, November 11, 1992, p. 10.

KFUPM/RI, 1983a. Assessment of Potential Pollutants in the Arabian Gulf. The Research Institute, King Fahd University of Petroleum and Minerals, Dhahran, Saudi Arabia.

KFUPM/RI, 1983b. An Investigation of the Potential Health Effects of the Nowruz Oil Spill on Saudi Arabian Fisheries. Phase II – Monitoring Program. The Research Institute, King Fahd University of Petroleum and Minerals, Dhahran, Saudi Arabia.

KFUPM/RI, 1987. Final Report. Northern Area Biological Study, 1985/86. Prepared for the Arabian American Oil Company by Water Resources and Environment Division, Research Institute, King Fahd University of Petroleum and Minerals, Dhahran, Saudi Arabia, 166 pp. and Appendices.

KFUPM/RI, 1988a. Final Report – Assessment of Toxicants in Selected Commercial Species of Fishes from the Arabian Gulf. Water Resources and Environment Div., Research Institute, King Fahd University of Petroleum and Minerals, Dhahran.

KFUPM/RI, 1988b. Assessment of Toxicants in Commercial Species of Fish in the Arabian Gulf. The Research Institute, King Fahd University of Petroleum and Minerals, Dhahran, Saudi Arabia.

KFUPM/RI, 1990. Final Report: Aramco Sustaining Research Project – Ennvironmental Studies,

Vol. II, Tarut Bay Biotopes Investigation - Seagrass and Sand Benthos. Prepared for Saudi Aramco by the Water Resources and Environment Division, Research Institute, King Fahd University Petroleum and Minerals, Dhahran, Saudi Arabia.

KFUPM/RI, 1991a. Evaluation of Shoreline Cleanup Methodology: Musallamiyah Test Site. Prepared for the Meteorology and Environmental Protection Administration by the Research Institute, King Fahd University of Petroleum and Minerals, Dhahran, Saudi Arabia, Technical Progress Report No. 47.

KFUPM/RI, 1991b. Evaluation of Shoreline Cleanup Methodology: Sulayq Test Site. Prepared for the Meteorology and Environmental Protection Administration by the Research Institute, King Fahd University of Petroleum and Minerals, Dhahran, Saudi Arabia, Technical Progress Report No. 48.

KFUPM/RI, 1991c. Interim Report: Gulf War Oil Spill 1991 - Arabian Gulf War Oil Spill Research Program. prepared for the Meteorology and Environmental Protection Administration by the Water Resources and Environment Division, Research Institute, King Fahd University Petroleum and Minerals, Dhahran, Saudi Arabia.

KFUPM/RI, 1991d. Sustaining Research Project - Marine Environmental Studies. Annual Progress Report - 1991. Prepared for Saudi Aramco by the Water Resources and Environment Division, Research Institute, King Fahd University Petroleum and Minerals, Dhahran, Saudi Arabia.

KFUPM/RI, 1991e. Evaluation of the Effectiveness of the BioSea Bioremediation Agent for Degradation of the Arabian Gulf War Oil Spill. Water Resources and Environment Division, Research Institute, King Fahd University Petroleum and Minerals, Dhahran, Saudi Arabia, Oil Spill Technical Progress Report No. 38.

KFUPM/RI, 1991f. Karan Island - Assessment Report and Remediation Proposal. Water Resources and Environment Division, Research Institute, King Fahd University Petroleum and Minerals, Dhahran, Saudi Arabia, Oil Spill Technical Progress Report No. 26.

KFUPM/RI, 1991g. Physical Properties and Metal Concentrations in Oil Slick Samples. Third Progress Report. The Research Institute, King Fahd University of Petroleum and Minerals, Dhahran, Saudi Arabia.

KFUPM/RI, 1991h. Physical Properties of Oil Slick and Metal Concentrations in Oil Slick Samples and Seawater Samples. The nineth Progress Report, The Research Institute, King Fahd University of Petroleum and Minerals, Dhahran, Saudi Arabia.

KFUPM/RI, 1991i. Gulf Oil Spill 1991. Arabian Gulf Atmospheric Pollution Research Program. 1st Interim Report. The Research Institute, King Fahd University of Petroleum and Minerals, Dhahran, Saudi Arabia.

KFUPM/RI, 1992. Sustaining Research Project - Marine Environmental Studies. Annual Progress Report - 1992. Prepared for Saudi Aramco by the Water Resources and Environment Division, Research Institute, King Fahd University Petroleum and Minerals, Dhahran, Saudi Arabia.

Kinsman, D.J.J. 1964. Reef coral tolerance of high temperatures and salinities. Nature 202:1280–1282.

Krahn, M.M. and Malins, D.C. 1982. Gas chromatographic mass spectrometric determination of aromatic hydrocarbon metabolites from the liver of fish exposed to fuel oil. Journ. Chromatography, pp. 248–299.

Lardner, R.W. and Das, S. 1991. On the computation of flows driven by density gradient: Residual currents in the Arabian Gulf. Appl. Math. Model. 15:282–294.

Lehr, W. and Cekirge, H.M. 1979. GULFSLIK I, A Computer Simulation of Oil Spill Trajectories in the Arabian Gulf. Research Institute, King Fahd University of Petroleum and Minerals, Dhahran, Internal Rept. no. 25 .

Lehr, W.J., Cekirge, H.M., Fraga, R.J., and Belen, M.S. 1983. Empirical studies of the spreading of oil spills. Oil Petrochemical Poll. 2:7–11.

Linden, O., Abdulraheem, M.Y.,Gerges, M.A., Alam, I., Behbehani, B., Borhan, M.A., and Al

Kassab, L.F. 1990. State of the Marine Environment in the ROPME Sea Area. UNEP Regional Seas Rept. Studies, No. 112, 33 pp.

Luten, B.J., Bouquet, W., Riekwel-Booy, W., Rauchbaar, A.B., and Schlote, M.W.M. 1987. Mercury in flounder, *Platichtys flesus*, cod, *Gadus morhua*, and perch, *Perca fluviatilis,* in relation to their length and environment. Bull. Environ. Contam. Toxicol. 38:318–323.

Maki, A.W. 1991. The Exxon Valdez oil spill: Initial environmental impact assessment. Environ. Sci. Technol. 25:24–29.

Mathews, C.P., Kedidi, S., Fita, N., Al-Yahaya, A. and Al-Rasheed, K. 1992. Appropriate Managerial Response to the Critical Condition of Saudi Prawn Stocks following the Oil Fires and Oil Slick Originating in Kuwait in February 1991. Saudi Fisheries Company.

McCain, J.C. 1984a. Marine Ecology of Saudi Arabia. The intertidal infauna of the Sand Beaches in the Northern Area, Arabian Gulf, Saudi Arabia. Fauna of Saudi Arabia, 6:53–78.

McCain, J.C. 1984b. Marine Ecology of Saudi Arabia. The nearshore soft-bottom benthic communities of the Northern Area, Arabian Gulf, Saudi Arabia. Fauna of Saudi Arabia, 6:79–101.

McCain, J.C., Sadiq, M. Al-Rabeh, A., Fayad, N., and Fadlallah, Y.H. 1992. The impact of the Gulf war Oil Spill on the coastline of Saudi Arabia. Paper presented at the Fourth General Conference, Third World Academy of Sciences, Kuwait, 23–26 November.

McCain, J.C. and Hassan, M. (in press). The effects of the Gulf war on ichthyoplankton in Saudi Arabian Gulf waters. Arabian Journ. Sci. Eng. (accepted)

McCain, J.C., Beard, D.W. and Fadlallah, Y.H. (in press). The influence of the Kuwaiti Oil Well Fires on seawater temperature in the western Arabian Gulf. Mar. Pollution Bull. (accepted)

McCain, J.C., Qurban, M., Lennon, D.J. and Chesterton, N.J. (in press). Remediation of salt marshes after the Gulf War Oil Spill. Arabian Journ. Sci. Engin.(accepted)

MEPA, 1987. An Assessment of Biotopes and Coastal Zone Management Requirements for the Arabian Gulf. Meteorology and Environmental Protection Administration, Coastal and Marine Management Ser., Techn. Rept. No. 5.

MEPA, 1991a. Shoreline Assessment and Cleanup Report. Prepared by the Meteorology and Environmental Protection Administration, 1.

MEPA, 1991b. Gulf War Oil Spill: Shoreline Cleanup Plan. Shoreline Assessment. Meteorology and Environmental Protection Administration, v. 2.

MEPA, 1991c. Gulf War Oil Spill: Shoreline Cleanup Plan. Shoreline Cleanup Plan and Cost Estimate. Meteorology and Environmental Protection Administration, v. 3.

MEPA, 1991d. Survey of Coastal Jetties and Dikes. Prepared by the Meteorology and Environmental Protection Administration.

MEPA, 1992. Arabian Gulf. Saudi Arabia: An Assessment of Biotopes and Coastal Zone Management Requirements for the Arabian Gulf. Meteorology and Environmental Protection Administration, Coastal and Marine Management Series, Rept. 5., 248 pp.

Mielke, J. 1990. Oil in the Ocean: The Short and Long-Term Impacts of a Spill. Congressional Res. Ser., July 24, 1990.

Miller, J.D. 1989. An assessment of the Conservation Status of Marine Turtles in Saudi Arabia. Meteorology and Environmental Protection Administration, Coastal and Marine Management Series, Technical Rept. 9, 209 pp.

Murty, T. and El-Sabh, M.I. 1985. Modeling the movement of oil slicks in the inner Gulf of the Kuwait Action Plan region during stormy periods: Application to the Nowruz oil spill. UNEP Regional Seas Reports and Studies, no. 70, pp. 279–298.

Neff, J.M. 1987. Biological effects of oil in the marine environment. Chem. Eng. Progr., November 1987:27–33.

NERC, 1991. Oil in the Arabian Gulf – Its Behaviour and Effects. Prepared for the Department of the Environment by the Natural Environment Research Council, United Kingdom.

Al Nokhathah, M. 1992. Shrimp season ends on a depressing note. Al Nokhathah, Saudi Fisheries Company, no. 14, December 1992.

Nomani, A.A., Thankachan, T.S., Al Sulami, S., and Kutty, P.C.M. 1992. Monitoring of Kuwait oil spill contaminants in water samples from SWCC seawater MSF plants in the eastern coast of Saudi Arabia. pp. 1178–1193. Proc. Int. Conf. on Chemistry and Industry, November 14–16, Bahrain.

Al Obayyed, S.A. and Khan, N.A. 1992. Oil spill – Its cleanup operation and immediate impact on the environment in Jubail. Proc. 1st Bahrain Int. Conf. on Environment, February 24–26, 1992.

OGJ, 1991. Saudis map $450 million Gulf spill cleanup. Oil Gas Journ., November 18, 1991 pp. 116–117.

OGJ, 1992. Survey to assess Persian Gulf spill effects. Oil Gas Journ., February 10, 1992, p. 38.

Al Omran, L. 1992. Pilot survey to assess the total PAH's level in coastal and marine environment of Kuwait. Poster Paper presented at the Fourth General Conference, Third World Academy of Sciences, Kuwait, 23–26 November.

OSIR, 1992. One year after the Persian Gulf oil spill, agencies still assessing amount of oil spilled and environmental effects. Oil Spill Intelligence Rept., 15(5):1–7.

Pellew, R. 1991. Caught in the crossfire. IUCN Bull. 22 (3):17–18.

Perttila, M., Tervo, V. and Parmanne, R. 1982. Age dependence of the concentrations of harmful substances in Baltic herring. Chemosphere 11:1019–1026.

Preen, A. 1989. The Status and Conservation of Dugongs in the Arabian Region. Meteorology and Environmental Protection Administration, Coastal and Marine Management Series, Technical Rept. 10, 200 pp.

Preen, A. 1991. Report on the Die-off of Marine Mammals associated with the Gulf War Oil Spill. Report for National Commission for Wildlife Conservation and Development.

Prena, J. 1992. The intertidal softbottom macrofauna in Dawhat al-Musallamiya and Dawhat ad-Dafi (Western Arabian Gulf), pp. 161–170. In: Final Report of Phase I, Establishment of a Marine Habitat and Wildlife Sanctuary for the Gulf Region. National Commission for Wildlife Conservation and Development and the European Community Commission.

Price, A.R.G. 1979a. Saudi Arabian Gulf shrimp resources. Proc. Saudi Biol. Soc. 3:293–302.

Price, A. R. G. 1979b. Temporal variations in abundance of penaeid shrimp larvae and oceanographic conditions off Ras Tanura, western Arabian Gulf. Est. Coast. Mar. Sci. 9:451–465.

Price, A.R.G. and Jones, D.A. 1975. Commercial and biological aspects of the Saudi Arabian Gulf shrimp fishery. Bull. Mar. Res. Centre, Saudi Arabia, 6, 24 pp.

Al-Rabeh, A.H., Cekirge, H.M., and Gunay, N. 1989. Stochastic model of oil spill fate and transport. Applied Math. Model. 13:322–329.

Al-Rabeh, A.H., Gunay, N., and Cekirge, H.M. 1990. A hydrodynamic model for wind driven and tidal circulation in the Arabian Gulf. Appl. Math. Model. 14:410–419.

Al-Rabeh, A.H., Gunay, N., and Cekirge, H.M. 1992. Modeling the fate and transport of Al-Ahmadi oil spill. Water Air Soil Poll., 63.

RDA, 1992. Final Report. Study on Fisheries Industry. Prepared for the Bahrain Ministry of Finance and National Economy by RDA International, Inc., Placerville, CA.

Readman, J.W., Fowler, S.W., Villeneuve, J.P., Cattini, C., Oregioni, B. and Mee, L.D. 1992. Oil and combustion-product contamination of the Gulf marine environment following the war. Nature 358:682–684.

ROPME, 1986. Evaluation of Present Situation on Land-Based Sources of Pollution with KAP Member States. Report of the Regional Organization for the Protection of the Marine Environment.

Sadiq, M. 1992. Marine Chemistry of Toxic Metal in Marine Environments. Marcel Dekker Inc. New York.

Sadiq, M. (in press). Effect of the 1991 Gulf Oil Spill on metal concentrations in seawater. Mar. Pollut. Bull (accepted).

Sadiq, M., Zaidi, T.H. and Al Muhanna, H. 1992. Sample weight and digestion temperature critical factors in mercury determination in fish. Bull. Environ. Contam. Toxicol. 47:335–341.

SAFISH, 1992. Pollution reduces shrimp production in the Gulf by 99 percent. Al-Majallah Mag. [in Arabic], no. 647, p.70.

Shaw, W.S. 1992. Smoke at Bahrain during the Kuwait oil field fires. Weather 47(6):220–226.

Sheppard, C. and Price, A.R.G. 1991. Will marine life survive the Gulf War? New Scientist, 1759:36–40.

Sheppard, C.. Price, A.R.G. and Roberts, C. 1992. Marine Ecology of the Arabian Region. Academic Press, London, 359 pp.

Shinn, E.A. (1976. Coral reef recovery in Florida and the Persian Gulf. Env. Geol. 1:241–254.

Sleeter, T.D., Butler, J.N. and Barbash, J.E. 1980. Hydrocarbons in sediments of Bermuda Region: Lagoonal to abyssal depths. In: Petrakis, L. and F. T. Weiss (eds.), Petroleum in the Marine Environment, American Chemical Society, Advances in Chemistry, Series 185.

Spaulding, M., Anderson, E., Tatsu, I., and Howlett, J. 1992. (in press). Simulation of oil trajectory and fate in the Arabian Gulf from Mina Al-Ahmadi. Oil Chemi. Poll.

Summers, R.W., Underhill, L.G., Pearson, D.J., and Scott, D.A. 1987. Wader migration systems in southern and eastern Africa and western Asia. Wader Study Group Bull. 49:15–34.

Tawfiq, N.I. 1991. Response by Saudi Arabia to the environmental crisis caused by the Gulf war, pp. 40–43. In: Al Shatti, A.K.S. and J. M. Harrington, The Environmental and Health Impact of the Kuwaiti Oil Fires, Proc. Int. Symp., Univ. Birmingham, 17 October 1991.

Varanasi, U., Stein, J.E., and Nishimoto, M. 1989. Chapter 4. In: Varanasi, U. (ed), Metabolism of Polycyclic Aromatic Hydrocarbons in thhe Aquatic Environment, CRC Press Inc.

Youngblood, W. Blumer, W.M., Guillard, R.L., and Fiore, F. 1971. Saturated and unsaturated hydrocarbons in marine benthic algae. Mar. Biol. 8: 190–201.

Walsh, J. 1992. Saddam's War on Wildlife . Nova Television Documentary.

WCMC, 1991. Gulf War Environmental Information Service: Impact on the Marine Environment. World Conservation Monitoring Center, Cambridge.

Zwarts, L., Felemban, H., and Price, A.R.G. 1991. Wader counts along the Saudi Arabian Gulf coast suggest that the Gulf harbours millions of waders. Wader Study Group Bull. 63:25–32.

CHAPTER 3

Environmental impact of Kuwait oil fires: upper atmosphere

3.1. Summary

- The burning oil wells in Kuwait, no doubt, created the world's largest oil field fire in history. The environmental implications of the oil fires generated tremendous media and scientific interest.
- Pre-war predictions of the environmental impact of Kuwait oil fires were extremely gloomy. It was suspected that a large quantity of soot particles would be produced by the fires that will remain suspended in the stratosphere for several weeks and eventually moved up to the troposphere, thus enhancing the greenhouse effect. Atmospheric warming and modification of the earth's radiation balance could change the global climate, weakening the Indian monsoon, and adversely affecting other climatological sensitive regions of the world. Some ecologists warned that there could be "petroleum winter effect", if the Kuwait oil fields were ignited. Using mathematical models, many scientists discredited the "petroleum winter effect" theory.
- Early estimates of number of damaged oil wells ranged between 550 to 800. After extinguishing the fires, the Kuwait Oil Company stated that 613 oil wells were burning, 76 were gushing only, and 99 were damaged (no fire and no gushing). A total of 788 Kuwait oil wells sustained damages as a result of Iraqi sabotage.
- In general, the numerical calculations predicted that the smoke plume would rise only to an altitude of 1 to 4 km, with an insignificant contribution from the self-lofting phenomenon, Smoke residence time was predicted to be short, generally a matter of days. Dry deposition processes would play a major role in the removal of soot from the atmosphere. Kuwait oil fires would significantly affect local/regional climate but there would be insignificant impact on a global scale. Airborne observations supported the above predictions.
- Remote sensing, especially satellite imagery, were successfully used in determining the number of oil wells and dispersion and composition of

smoke from Kuwait oil fires. In general, there was good agreement between the satellite images and results of ground and airborne investigations.

- Estimates of the amount of oil burning in the Kuwait oil fires were plentiful, varying between 1.2 and 7.5 million barrels per day. An oil burning rate of 6 million barrels per day was established by the Kuwait Oil Company which has been adopted in this book. A total emission of about 1.12 billion barrels of crude oil during the whole episode of Kuwait oil fires was calculated based on the chronological information of fire fighting and other pertinent data.
- Soot emission rates used in all mathematical modeling studies (5–10 percent) were almost double than those determined during airborne investigations (0.3 to 3.8 percent). Soot/smoke emission was dependent on the type of fire. For example, white smoke plumes were rich in salts, whereas the black smoke plumes contained larger portions of soot/smoke particles. Lake fires produced a more sooty mass than did the oil well fires.
- Several different shapes of flames were noticed during burning of the Kuwait oil fields. The smoke produced by Kuwait oil fires varied from gray-white to dense dark in color. The whitish tinge to the smoke was imparted by high concentrations of inorganic salts, especially that of sodium chloride. Physical and chemical characteristics of different types of flames and smoke plumes are reviewed.
- In general, smoke particles were found to be spherical with a diameter from about 0.1 to 0.3 μm. Their concentrations decreased with an increase in distance from the fire source. Smoke particles clustered to form a chain aggregate commonly referred as soot. The proportion of aggregate increased with the age of the plume. In contrast to the earlier common belief of smoke being hydrophobic, the majority of the smoke particles were found to be hydrophilic. Smoke particles collected from the super-composite plume (smoke plumes from individual oil wells combined together to form a single plume) consisted of about 30% salt, 15–20% soot, 8% sulfate crystals, and about 30% organic chemicals.
- Carbon dioxide accounted for over 83% and CO was about 0.5% of the emission products from the Kuwait oil fires. The total quantities of CO_2, CO, and elemental C (largely soot) produced by the Kuwait oil fires were estimated at 133, 0.8, and 8 million metric tonnes, respectively. Total carbon dioxide emission was estimated to be <2% of the global emission from the burning of fossil fuel and biomass. It is, therefore, expected that the effects of burning oil wells in Kuwait may be significant on a regional scale. Global effects including the greenhouse warming are likely to be insignificant.
- Total emission of SO_2 from Kuwait oil fires was estimated at 2.3 million metric tonnes. It is believed that SO_2 emitted by the fires was rapidly oxidized to SO_4. This phenomenon decreased ambient concentrations of SO_2 to almost background levels within 4 to 5 hours after emission. The area of high SO_2 concentrations was estimated to be within 100 km downwind of the source. Over 80% of SO_2 that was estimated to be converted to SO_4 was removed within 100 km of the fire source. Dry deposition of soot particles

played a major role in SO_4 removal from the atmosphere. All indications suggest that SO_2 emissions from the Kuwait oil fires would have insignificant impact on the global or regional environments. Hydrogen sulfide concentrations, even close to the source, were found to be near background levels which suggests that it was efficiently burned in the fires.

- Concentrations of NO_x in the plume was found to be low to decrease to background levels within about 100 km from the fire source. A total of 80,000 tonnes of NO_x are estimated to emitted from the burning oil wells in Kuwait. Although there was evidence of ozone depletion near the fire source, NO_x emissions from the Kuwait fires pose no global or even regional environmental threat.
- Ozone was found to be depleted near the fire source, primarily due to its reaction with NO_x and secondarily due to its reaction with hydrocarbons. Ozone depletion was more pronounced in the upper layers as compared with the lower portion of the smoke plume. Real-time airborne measurements suggested that solar radiation was a critical factor in the depletion of O_3 by NO_x. Because of the inability of the plume to rise to the stratosphere, the depletion of O_3 by photochemical reaction with NO_x emitted by the Kuwait oil fires poses no global environmental problem.
- Measurements in the Kuwait smoke plume indicated that ethane and propane were the most abundant non-methane hydrocarbons (NMHC). These chemicals decreased rapidly with distance from the fire source. Concentrations of NMHC found in the smoke plume were comparable with clean to moderately polluted atmosphere. Concentrations of methane (CH_4) were found to be slightly elevated in the smoke plumes. It was anticipated that quantities of NMHC and methane emitted by the Kuwait oil fires were not sufficiently high to induce a global or regional environmental impact.

3.2. Introduction

In the very early days of the Gulf conflict, the Iraqi regime made it clear that it intended to use oil as weapon, if needed. To prepare for this eventuality, six oil wells were ignited by Iraqis in December, 1990, as an experiment to determine the most efficient method of well destruction. Considering the results of this experiment, 30 to 40 pounds of military explosive type C4 with sandbag damping was attached to each oil well. The government of Iraq mobilized about 1000 military explosive experts and 30–40 engineers, together with the forced labor who worked in the oil fields, to sabotage and set fire to about 1000 oil wells in Kuwait (Garwin and Kendall, 1991). When the air war began on January 16, 1991, about 60 oil wells were blown up by the Iraqis and probably another 34 might have been destroyed by the allied bombing (Riva, 1991). During the last days of the Gulf war, sabotage attempts were made on over 800 additional Kuwait oil wells. It was reported that about two-thirds of the well heads were blown off and the rest badly damaged. Many wells were gushing (flowing without a blaze) oil at typical rates of one or two thousand cubic meters per day (Garwin and Kendall, 1991).

It was widely accepted that, in a desperate move, Iraq might sabotage Kuwait oil wells. Therefore, satellite imagery was used to detect fires in Kuwait during the Gulf war. Smoke plumes from the oil well fires were observed for the first time on a System Probatoire d'Observation de la Terre (SPOT) image on February 8, 1991. It was further confirmed on 15 February when LANDSAT Thematic Mapper (LANDSAT TM) imagery depicted a smoke plume in Kuwait. The smoke plume was also detected on February 21, 1991 on National Oceanic and Atmospheric Administration (NOAA) Advanced Very High Resolution Radiometer (AVHRR) imagery (Williams et al., 1991). The Research Institute of King Fahd University of Petroleum and Minerals, Dhahran (Saudi Arabia) utilized NOAA-AVHRR satellite images and observed the first fire on January 24, 1991. Kwarteng and Bader (1992) noticed a gradual spread of fires which peaked during the last week of February, 1991.

The world was shocked on the first confirmatory news that Kuwait oil wells had been deliberately ignited. The burning oil wells in Kuwait, no doubt, created the world largest oil field fire in history. Public and media interest in the fires was tremendous. Many articles were written by reporters, scientists, and politicians alike, even before the start of fires in Kuwait (Bakan et al., 1991; Browning et al., 1991; Small, 1991; Sandia, 1991; Turco et al., 1990). The international community was alarmed to think about the ultimate environmental impact of the oil fires in Kuwait (Barnaby, 1991; Browniing et al., 1991; El-Baz, 1992; Cook, 1991; Emerson, 1991; Hahn, 1991; Kelly, 1991; KFUPM/RI, 1991a, 1992; Linden, 1991; Marshall, 1991; Peck, 1991; Pope, 1991; Schwartz et al. 1991; Warner, 1991; World Oil, May 1991). As a result of international interest, many scientific teams attempted to understand the nature and extent of environmental crisis caused by the Kuwait oil fires. The results of some of the efforts related to the impact on the upper atmosphere are reviewed here.

3.3. Pre-war predictions of environmental impact of Kuwait oil fires

Soon after the Iraqi occupation on August 2, 1990, the international community started demanding that Saddam Hussain get out of Kuwait and threatened to evict him forcefully if their demand fell on deaf ears. Proud to be aggressors, the Iraqi leaders warned of a scorched earth policy in retaliation for the forced eviction from Kuwait. They planned to damage Kuwait's oil industry by igniting oil storage facilities and blowing up all oil wells in Kuwait oil fields. It seems that the blueprint of Saddam's plan was intentionally leaked to the world. The objective was to scare the world by threatening the environmental disaster that would follow military action which would ultimately prove to be the doomsday for the world in general and the Gulf region in particular. This tactic was meant to move the common people to force the political leaders not to evict Iraq forcefully so that the world could gradually forget Kuwait as a sovereign country.

Unfortunately, the world scientific community became part to the political saga of the Gulf conflict. In January, 1991 in a meeting in London, many scientists predicted dreadfully pessimistic projections regarding the environmental consequences of would-be-oil fires. Notable of those were Jordanian Royal Science Advisor Dr. Abdullah Touchan (National Review, 1991), Carl Sagan - an astronomer at Cornell University, Richard P. Turco of the University of California, and John Cox, a consultant in the Gulf region. Touchan predicted that oil fires would consume oxygen within several km and every living creature in that radius would die instantly. Turco et al. (1990) noted a large scale hemispheric effects. They calculated that a major injection of soot in the stratosphere would disrupt monsoon precipitation and cause severe depletion of stratospheric ozone over the northern Hemisphere. They predicted that black soot particles would remain suspended for weeks, absorb the sun's radiation, and could eventually move up to the troposphere where they would remain for months, if not years. John Cox proposed that the smoke cloud from 300 burning oil wells could cool India enough to prevent the summer monsoon rainfall. Such a loss in rainfall and associated reduction in the solar radiation could adversely effect crop productivity in the Indian Subcontinent and starve several hundred million people worldwide (Raloff and Monastersky, 1991). Rao (1991), using satellite imageries from INSAT-1B, NOAA-11 and IRS-1A, concluded that prolonged smoke from the burning oil wells would have a weakening effect on the Indian monsoon. It was cautioned that a large quantity of soot particles would be produced from Kuwaiti oil fields that would enhance the greenhouse effect (Aldhous, 1991). He speculated that warming of the upper atmosphere and modifications in the earth's radiation balance could change the global climate, weakening the Indian monsoon and adversely affecting other climatological sensitive regions of the world. In summary, many of the scientist warned of a "petroleum winter effect", if the Kuwait oil fields were to be ignited.

The above doomsday predictions were based on the following generalized reasoning. It was assumed that open burning of crude oil would produce large quantities of smoke/soot particles. These particles would be transported to the stratosphere by self-lofting where they would remain suspended for extended periods. The stratospheric smoke particles would absorb solar radiation and, therefore, reduced sunlight falling on the earth's surface. The absorption of solar radiation would lead to the heating of atmosphere in the smoke/soot layers and a reduction of solar heat on the earth's surface. The resultant change in the solar radiation balance would have dire and long-term ecological implications. Some scientists, referring to the horrible consequences of the nuclear winter effect in case of a nuclear war, termed the environmental consequences of the Kuwait oil fires as the "petroleum winter effect". The question of the smoke from the burning oil wells in Kuwait presenting a serious threat to the global climate was the subject of considerable scientific debate and widespread public concern.

In a reaction to the doomsday theory, many scientists and government agencies undertook research to predict the effect of oil fires in Kuwait. For example, before the start of the Gulf war, the United States Department of Energy asked Sandia National Laboratories to determine the potential military, environmental, and Kuwaiti oil reservoir depletion from Iraqi sabotage of the Kuwaiti oil fields. Sandia (1991) estimated that if several well heads were blown up, these well heads could spew out five to six million barrels crude oil daily, but oil reservoirs would not be severely damaged. Bakan et al. (1991) investigated the impact of smoke from the burning oil wells on global climate by numerical experiments using a coupled atmospheric-oceanic general circulation model with an interactive soot transport model and extended radiation scheme. Their results showed a decrease of approximately 4°C in the ambient temperature of the Gulf region. Outside this region, the predicted changes were small and statistically insignificant. The insignificant global impact of the oil fires was expected because of the low height of the smoke plume. The low injection levels enables the soot particles to be washed out or removed by dry deposition much more quickly than the periods assumed by the "doomsday" scientists. Bakan et al. (1991) predicted a typical smoke/soot life time of 21 days (i.e. the smoke will settle out of the atmosphere in about 21 days). They also predicted that only 0.3% of the soot would be transported to the stratosphere, where it may remain suspended for several years. In general, Bakan et al. (1991) discredited the "petroleum winter effect" theory.

Similarly, Browning et al. (1991) estimated that the smoke plume is very unlikely to reach the stratosphere, and that the bulk of the smoke would remain within the lower troposphere before being deposited on the ground within a week or so of its emission. Consequently, the changes in the Asian monsoon resulting from the smoke are likely to be small compared with the natural interannual variability. Their calculations showed that, beneath the plume within about 200 km of the source, it is likely that there will be a reduction of day-time temperature of about 10°C. They predicted that the plume might remain visible for up to a few thousand km (horizontal distance). Episodes of severe acid rain, black snow, and photochemical smog, comparable with the highest concentrations around major industrial areas, were predicted to occur out to a distance of a thousand km or more downwind. Carbon dioxide release, estimated to be about 1% of current global emission, would have an insignificant effect of global warming. Small (1991) stressed that the environmental impact of well fires will depend mainly on the amount of smoke and the injection height. From the available information, he suggested a rise of smoke to an altitude of about 1 to 3 km.

Smoke is a strong absorber of solar radiation. A large smoke mass in the atmosphere can lead to a reduction in solar radiation. Global climate calculations of Small (1991) showed only minimal climate changes for a small absorption optical depth. The atmosphere can be affected on a local scale (i.e., about 1000 × 1000 km^2 area). As the smoke spread over the regional (10^{13} m^2) area and on global scale (10^{14} m^2), the smoke clouds will be thinned but the weather can possibly be influenced. From his calculations, Small (1991) concluded that his estimates of smoke production from the burring oil fires in Kuwait do not support any of the purported climate impacts. He noted that the smoke will not be injected high enough to spread over a large area of the northern hemisphere, at least not sufficient to produce a measurable drop in the ambient temperature or cause a failure of the monsoon in the Indian Subcontinent. A small increase in the global CO_2 budget was expected due to Kuwait oil fires. Reviewing the published information, Hahn (1991) concluded that mathematical calculations suggest that the effect of the smoke emission in Kuwait on the Asian summer monsoon would be small.

Some of the earliest predictions regarding the environmental impacts of Kuwait oil fires were gloomy. However, many of numerical calculations showed that smoke/soot would not be injected high enough to reach stratosphere but would remain suspended within 1 to 5 km altitude. The amount of CO_2 produced would not contribute significantly to the greenhouse effect. The atmospheric residence time of smoke was expected to be short – about 20 days. The environmental effects would be largely local and insignificant impact at the global scale was expected. From a global perspective, solar radiation and ambient temperature would not be altered enough to adversely affect monsoon intensity and distribution.

3.4. Burning oil wells in Kuwait

At the beginning of 1990, Kuwait's total proven oil reserves were officially estimated at 94.5 billion barrels, and proven gas reserves at 1400 billion m^3r (Arab Oil and Gas Directory, 1990). The oil and gas reserves are found in eleven onshore and three offshore fields (Figure 3.1, Table 3.1). The bulk of Kuwait's reserves are confined to the Greater Burgan Oil Field and are estimated at 50 billion barrels of oil. The Greater Burgan Oil Field, which is Kuwait's largest producer, is the second largest field in the world after Saudi Arabia's Ghawar Oil Field. Raudhatain, the second largest onshore oil field in Kuwait, has proven reserves estimated at 6 billion barrels. The Sabriya Oil Field has reserves of approximately 3.8 billion barrels and Minagish Oil Field has 1.8 billion barrels (Arab Oil and Gas Directory, 1990).

Figure 3.1 shows the geographical locations of some major oil fields in Kuwait. The oil fields exist in two main sectors to the north and south of Kuwait City. The southern sector contains the Greater Burgan Oil Field, the world's second largest oil field. The Greater Burgan is comprised of the Magwa, Ahmadi, and Burgan oil fields. In addition to the Greater Burgan, this sector also contains the Menagiesh, Umm Gahdair and Wafra Oil Fields. The Wafra Oil Field is equally shared by Saudi Arabia and Kuwait. The northern sector consists of Bahara, Sabria, Rawdatain, and Ratqa Oil Fields. The Bahara Oil Field is the oldest in Kuwait (Al Besharah, 1991).

In the very initial days of the Gulf conflict, the Iraqi leader Saddam Hussain

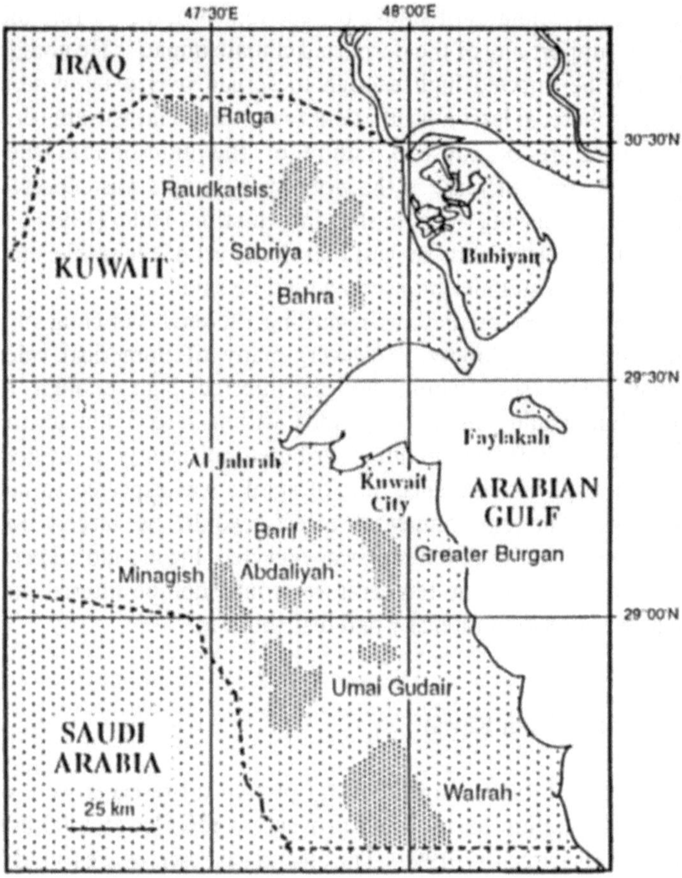

Figure 3.1. Kuwait oil fields (from World Oil, 1991).

shocked the world by announcing his intentions to use oil as a weapon. It was known that when oil became a weapon, the Gulf environment would certainly be a victim. In a deliberate sabotage attempt, the Iraqis attached 13 to 18 kg of military explosive type C4 to about 1000 oil wells. During the last days the Gulf war, attempts were made to blow these wells. The first sign of oil fires was pick by SPOT imagery on February 8, 1991. It was later confirmed on 15 February when a Landsat TM images depicted a the smoke plume in Kuwait. Further evidence of oil fires was provided on February 21, 1991 by NOAA AVHRR imagery (Williams et al., 1991). Kwarteng and Bader (1992, 1993) detected the first fire on January 24, 1991 using NOAA-AVHRR images and followed their progress. They reported that the number of oil fires reached its climax in the third week of February and counted a total of 625 fires on a Landsat TM image taken on March 23, 1991. The smoke from Kuwait oil fires was visible over Dhahran, Saudi Arabia, more than 300 km south of Kuwait, on February 23, 1991.

Table 3.1. Estimates of the number damaged oil wells in Kuwait.

Reporter	Burning	Gushing	Damaged*	Total
Kuhn (1991)	550	-	-	-
Hobbs and Radke (1992)	-	-	-	749
Barnaby (1991)	580–600	100	-	700
Johnson et al. (1991)	600	-	-	-
Garwin and Kendall (1991)	690	42	-	732
Kramer (1991)	-	-	-	732
World Oil (1991)	-	-	-	600–700
Touby (1991)	670	62	-	732
Al-Besharah (1992)	613	76	99	788
Al Hassan (1992)	730	-	-	-
Tawfiq (1992)	613	76	99	788
EPC (1991)**	613	76	99	788
Cutter (1991)	700	-	-	-
Schwartz et al. (1991)	>500	-	-	-
Kwarteng and Bader (1992)	625	-	-	-
Cook (1991)	500	-	-	-
Linden (1991)	600	-	-	-
El-Baz (1992)	732	-	-	-
Mage et al. (1991)	628	77	122	827
Kuwait Oil Company@	613	76	99	788
NOAA (1992)	611	79	108	798
Gerges (1992)	613	76	-	-
Lameloise et al. (1991)	550	-	-	-

* Number of damaged oil wells consists of all wells on fire, gushing oil but not on fire, and damaged but neither on fire nor gushing oil.

** Value adopted as reference in this book.

@ Cited by EPC (1991) and Al-Besharah (1991).

To predict the environmental impact of the oil fires, mathematical models required the input of the amount of oil being burned. The amount of oil being burned was dependent on the number of damaged oil wells. During the early stages of this catastrophe, the number of burning wells could only be guessed. There was a considerable confusion on how many of the wells were burning. In this respect, the closeness of wells to each other in many fields and the intermingling of individual smoke plumes cast doubt on the results of satellite imagery. The burning oil well estimates varied between more than 550 and less than 800 damaged oil wells. The information is summarized in Table 3.1.

Mathematical calculation to assess environmental impact required a knowledge of the exact location of the burning oil wells. Some estimates were available on the state of damage in each oil field. Some details concerning the burning oil wells in Kuwait are given in Table 3.2.

In summary, Iraqi troop set fire to oil wells even during the air campaign. The first fire was detected on January 24, 1991. The oil fires gradually increased in their intensity and number during the war and peaked during the last week of

Table 3.2. Number of burning oil wells in each Kuwait oil field.

Oil Field	Wells Before Invasion[a]	Wells On Fire[b]	Wells On Fire[c]	Wells Gushing[c]	Wells Damaged[c]
Burgan	433		291	24	28
Magwa	148	*350	98	6	21
Ahmadi	89		60	3	17
Kashman	7		0	0	1
Dharif	4				
Abdulia	5				
Minagish	96	50	27		7
Umm Gudair	45	50	26	2	11
Bahrah	8	25	3	2	
Sabriya	72		39	4	9
Raudhatain	84	50	62	2	5
Ratqa	144		1		
Wafra	–	25	6	33	
Romeilla	–	75			
Total	1135	625	613	76	99

[a] World Oil (1991)
[b] Kwarteng and Bader (1992)
[c] Al-Besharah (1991) and EPC (1991)
* Represents Greater Burgan Oil Field that includes Magwa, Ahmadi, and Burgan Oil Fields

February, 1991. Satellite imagery was successfully used to determine the number of burning oil wells.

3.5. Atmospheric dispersion of smoke/soot

Atmospheric dispersion of smoke/soot, both vertically and horizontally, is very critical in determining its impact on global and local environments. Smoke dispersion from the burning oil wells in Kuwait was dependent on smoke injection heights, wind speed and direction, emission velocity of gases within the plume, temperature and humidity, emission products and characteristics, atmospheric stability layers, and radiation disruption in the area. Information on many of these factors was not available at the beginning of the Gulf war. Therefore, early dispersion modeling efforts made several assumptions. The outcome of model computations was only as good as those assumptions and the quality of input data. The dispersion of smoke was investigated using mathematical models, satellite imagery data, aircraft measurements, and monitoring activities. A brief review of some of these efforts is presented herein.

3.5.1. Dispersion of smoke using numerical models

Atmospheric modeling scientists were the first to address the environmental impact of the Kuwaiti oil fires. For example, before the start of the Gulf war, the United States Department of Energy contracted Sandia National Laboratories to determine the potential military, environmental, and Kuwaiti oil reservoir depletion possible due to Iraqi sabotage of the Kuwaiti oil fields. Later many other scientists and organizations predicted smoke/soot dispersion using a variety of numerical models. A comprehensive review of these numerical models is beyond the scope of this book. An excellent review of the modeling activities was given by WMO (1991, 1992). A brief discussion, largely extracted from WMO (1991, 1992), of some of these numerical models is given below.

- TASC (Technical Analytical Services) developed and ran near real-time meteorological and air pollution impact assessment models. These models were coupled with an improved plume model. TASC compared its model outputs with satellite visualization and reported to be in general agreement (Chase and Bass, 1991).
- Lawrance Livermore National Laboratory applied the Atmospheric Release Advisory Capability (ARAC) model to map optical density in Kuwait. This model provides near real-time assessment of concentration, dose, and deposition of soot. An improved plume rise modeling feature was added to ARAC capabilities (Sullivan, 1991).
- The Air Research Laboratory (ARL) of National Oceanic and Atmospheric Administration (NOAA) developed trajectory modeling capabilities and mesoscale meteorological models (e.g., CSO-RAMS Model). NOAA was very active in supporting operations in the Kuwait region with its modeling tools (Heffter, 1991). The results of these models were validated using NOAA AVHR imagery. The plume position showed reasonable agreement in most cases when a model resolution of 80 km was used. However, the coarse resolution of the above model (100 km) prevented more thorough verification with these data (WMO, 1992).
- The Meteorologie Nationale (France) developed two numerical models: Emeraude – a global model, and Peridot – a limited area model. During the Gulf war, the Peridot model was applied to Kuwait oil well fires. Meteorologie Nationale also developed an Eulerian model of atmospheric dispersion. These capabilities were applied to the Kuwait region (Bompay, 1991).
- The Meteorological Office (UK) used a 3-D Monte Carlo model to simulate the transport of the plume and to assess dry and wet deposition, surface layer time-integrated concentrations, and episodes of high acid rain. The model plume was compared with visible plume satellite and airborne pictures and was found to be in general agreement (Smith, 1991).
- Max-Planck-Institut für Chemie, Germany, suggested two numerical models; (1) the local scale model CALGRID – an urban air pollution model,

and (2) the Regional Scale Model (Stem-II) – a long range dispersion model that could be used to study dispersion of emission products from the burning oil wells in Kuwait (Carmichael, 1991).

- The Cologne University of Germany preferred a regional scale numerical model called EURAD. This model was basically developed for Europe to study acid rain. It was modified for the Gulf application (Jakobs et al., 1991).
- The United States Army Atmospheric Sciences Laboratory (ASL) developed a model of Atmospheric Chemical Hazards (MACH). It is a real time computing facility. This modeling capability was used to predict downwind concentrations, doses, and deposition of emission products over Kuwait and the Gulf region (Meyers, 1991).
- The effects of smoke on the local, and regional meteorological conditions and short-term climate perturbation was investigated by the Atmospheric Environment Services (Canada) using the Canadian Climate Center General Circulation Model (CCC/GCM) (Blanchet and Vupputuri, 1991).
- Penner et al. (1991) developed the LLNL global climate/chemistry model that uses the winds and precipitation fields from the National Center for Atmospheric Research (NCAR) Community Climate Model (CCM) to simulate monthly averaged surface concentrations of soot (Penner, 1991). The model calculations showed general good agreement with satellite observations, particularly during quasi persistent meteorological conditions. This model missed the local mesoscale flow perturbations (WMO, 1992).
- The Royal Netherlands Meteorological Institute, in collaboration with other scientists, developed and validated a real-time puff dispersion model for application to the Gulf crisis. It uses meteorological and atmospheric boundary layers from the European ECMWF prognostic model. This model was modified to the Gulf region (Verver, 1991).
- The Building and Fire Research Laboratory of National Institute for Science and Technology (NIST) of the United States developed advanced fire modeling capabilities which were applied to the Kuwait oil fire situation (Snell, 1991).
- The Institute for Atmospheric Physics in Germany employed a non-hydrostatic solution of Navier-Stokes Equation and developed the MESOSCOP model which was applicable to oil fires and air pollution (Schumann, 1991).
- The Research Institute of King Fahd University of Petroleum and Minerals, Dhahran, Saudi Arabia used the RAM Model within the User's Network for Applied Modeling (UNAMAP). This model was used to assess the impact of burning oil wells in Kuwait within a radius of 150 km from the fire source. The Air Resources Laboratory's Air Transport and Dispersion Model (ARL-ATADM) was adopted to determine the movement of smoke for regional impact assessment.

WMO (1992) concluded that the horizontal resolution of the operational regional scale model (80 km) was too coarse to resolve important mesoscale

features. It pointed out that many important processes were not addressed in many of the above simulation models. These processes are:
- Thermally induced circulation pattern in Kuwait and other Gulf countries
- Lack of adequate vertical resolution in the numerical models
- Particles' radiative effects on the atmospheric stability layers over Kuwait.
- Majority of the regional models ignored both dry and wet deposition of soot. Dry deposition was found to play an important role in scavenging smoke/soot from atmosphere.
- Limited accuracy of meteorological forecasting used in the models. The accuracy and frequency of the historical meteorological data in the Gulf region were below the standards required for many of the above numerical models.
- Lack of data on source characterization, such as the number of fires and their distribution, amount of soot produced, combustion efficiency, CO_2 and other gases emission rates, etc.
- Inaccurate plume height specifications in the model resulting from lack of understanding of stability layers and other meteorological phenomena.

From the foregoing, it was concluded that mathematical calculations of the dispersion of smoke/soot had many serious limitations. Mathematical modeling can prove a good prediction tool provided that they simulate the real time situation closely and the input data are more reliable and frequent. The results of mathematical modeling efforts with respect to the Kuwait oil fires should, therefore, only be considered in very broad general terms.

3.5.2. An overview of the results of numerical modeling

Soon after the occupation of Kuwait, the Iraqi Government threatened the world with dire environmental consequences if forcefully evicted ["We swear ....that we will make the Gulf a graveyard for all those who think of committing aggression starting with these cowardly American Navies" Radio Baghdad]. Anticipating the environmental catastrophe that might befall the Kuwait region, many scientists aired their concerns. Numerical models were applied to different scenarios – from the most optimistic to the most pessimistic situations. The numerical modeling efforts heightened as hostilities in the region escalated and continued throughout and even well beyond the Gulf war. Some salient conclusions and recommendations of some of these investigations are reviewed here.

Browning et al. (1991) applied a simple plume rise model and the British Meteorological Office's mesoscale weather prediction model to estimate the behavior of the smoke plume. Their calculations showed that the plume was not likely to rise to the stratosphere. They predicted that the bulk of the smoke/soot would remain into the lower troposphere. They estimated soot and smoke lifetime in terms of weeks. In general, they concluded, that the effects of Kuwait oil fires would be significant in the Gulf region only but insignificant at a global scale. The model predicted a the smoke plume rise of about 2 km in winter and 3 km in summer. Their calculations showed that initially the plume would rise quickly by mainly conventional forces and later more slowly by self-lofting through solar heating. The mesoscale model calculations, under different scenarios showed that the smoke plume will always fail to reach stratosphere (about 15 km over Kuwait). Therefore, the atmospheric lifetime of smoke was expected to be short – in terms of weeks. Browning et al. (1991) argued that analogies of nuclear winter and large volcanic eruption to the Kuwait oil fires were inappropriate. They used the British Meteorological Office's long range dispersion model to investigate long-term atmospheric distribution of the smoke. Close to the fires, the results were similar to the mesoscale model. With the passage of time, the plume was shown to spread rapidly broadening towards the south. The model predicted a lifetime for the smoke and soot between 5 to 15 days. Dry deposition proved to be the principal scavenging mechanism for sulfur dioxide and soot particulates.

Graf (1991) utilized the CALGRID (a local scale) and STEM II (a regional scale model) to study dispersion of emission products from the burning oil wells in Kuwait. The model predicted a significant cooling effect of about 4°C for the summer of 1991 over the Gulf region (about $500 \times 500\ km^2$). On a global scale, the model predicted insignificant climatic changes. Graf predicted that the environmental impact of the burning oil wells will disappear soon after their extinction due to the relatively short residence time for the soot. The model also predicted a weakening of monsoon in the Indian Subcontinent (WMO, 1991).

Blanchet and Vupputuri (1991) reported the results of the Canadian Climate Center General Circulation Model (CCC/GCM) as it was applied to the investigation of the effects of smoke on local and regional meteorological conditions and to short-term climate perturbations. It showed that, on a small scale and near source, the smoke plume would produce cooling due to reduced radiation. At the regional scale with smoke located in lower stratosphere, the model predicted a warming effect. It also showed that the magnitude of warming was sensitive to the height of smoke layer. For the long-term, the model predicted a cooling of 5 to 10°C in the Kuwait area. Beyond 100 km from the source, natural variability in radiation dominated. Unlike Graf's model (1991), this model predicted an increase in the monsoon over India due to enhanced convection and moist air flow at lower levels.

Penner et al. (1991) developed the LLNL global climate/chemistry model that uses the winds and precipitation fields from the National Center for Atmospheric Research (NCAR) Community Climate Model (CCM) to simulate monthly averaged surface concentrations of soot. They predicted that, in March 1991, high winds would carry soot as far as Mauna Loa. The soot emissions were predicted to be retain close to the source in July, but that average concentrations of more than $100\ ng/m^3$ could be found throughout the Indian Subcontinent and parts of southwest Asia. This concentration contour also encompasses much of northeastern Africa and extends northwest to the Black Sea. Penner (1991) concluded that the burning oil wells in Kuwait appeared to enhance optical depth from soot over a large area surrounding the source. The model also predicted a widespread surface concentration of sulfate and confinement of H_2S within a local region (WMO, 1991).

Using the CSIRO climate models, Manins and Bouma (1991) estimated the effects of burning oil fields in Kuwait on the Middle East. They predicted that the smoke plume would rise initially to 3 km. They suspected that solar heating would further elevate the smoke but only to the lower atmosphere and suggested that because of this the smoke will be removed rapidly. Simulating continuous emission into the lower and middle atmosphere during typical January and February conditions, Manins and Bouma (1991) noted that smoke would not be transported to the Southern Hemisphere. The pollutants were predicted to disperse rather rapidly over a large area in the Northern Hemisphere. They also noted that dry deposition of soot would be appreciable. Similarly, simulation for typical July conditions suggested an absence of leakage of smoke to the Southern Hemisphere. They concluded that Kuwait oil fires would significantly affect climatological conditions at a regional scale but not at either a hemispheric or global scale.

Schumann (1991) used a mesoscale model to study the vertical rise of the smoke plume. His calculations showed a plume rise to 3.5 km, then an eastward drift. He observed that, at 20 km downwind from the fires, the plume would extend from 2 to 3 km in altitude, and spread laterally and quickly.

Small (1991) calculated smoke production as 0.016 million tonnes per day if oil emission is assumed to be 1.6 million barrels per day. He stressed that the environmental impact of the well fires would depend mainly on the amount of smoke and injection height. From the available information, he suggested a rise of the smoke plume to 1 to 3 km. From the calculations, Small (1991) concluded that smoke will not be injected high enough to spread over a large area of the Northern Hemisphere.

The numerical calculations of Hahn (1991) suggested that the smoke would not rise higher than the planetary boundary layer, about 3300 m at a maximum over Kuwait. He concluded that the environmental consequences would be significant only in the Gulf region. He showed that most of the smoke mass would be transported through the 2000–3000 meters layer. After aerial transport of about 1.5 to 2 km from the source, the smoke cloud would be thinned to such an extent that it would become non-detectable. He suggested a smoke particulate residence time of 5 days. Results of model calculations suggested that the major smoke deposition would occur over Kuwait and large parts of Iran, Saudi Arabia, and other Gulf Sates. He estimated that about 20% of the emitted smoke will be deposited within a 50 km radius of the source.

Bakan et al. (1991) investigated the impact of the smoke plume from the burning oil wells in Kuwait on the global climate by numerical experiments using a coupled atmospheric-oceanic general circulation model with an interactive soot transport model and extended radiation scheme. In this study, they assumed a constant burning rate of 3.1 million barrel per day, that oil burning started on 15 February 1991 and continued to through January, 1992, and that 10% of the oil (mass basis) would be converted to soot. Their calculations indicated that, for the ranges of atmospheric static stabilities found over the Gulf region, the soot injection would not normally exceed 2 to 4 km in height. Their calculations suggested a 20 day residence time for soot and smoke in the troposphere. The highest altitude of smoke rise would be attained during summer when increased solar radiation would enhance convective activities. Self-lofting of smoke would also be highest in summer. They found negligible soot injection (about 0.3 percent) into the stratosphere. The model calculations suggested only regional impact of the burning oil wells in Kuwait.

In general, the numerical calculations predicted that the smoke plume would rise to 1 to 4 km in altitude. Self-lofting was predicted to have insignificant effect on smoke rise. The residence time of the smoke particles was estimated to be about 15 days and dry deposition was thought be the major process in the removal of soot from the atmosphere. The Kuwait oil fires were believed to have a significant effect on local/regional climate but have an insignificant impact on global scale.

3.5.3. Satellite imagery to study Kuwait oil fires

Satellite images can be used to detect fires and smoke dispersion. Some of the characteristics of the smoke, such as particle size distribution, can also be determined using satellite imagery. For example, LANDSAT™ images, visible in band 5, 6, and 7, were used to determine the number of fires. A comparison between band 6 and 7 images can revealed fires even through the smoke (Williams et al., 1991). National Oceanic and Atmospheric Administration (NOAA) Advanced Very High Resolution Radiometer (AVHRR), System Probatoire d'Observation de la Terre (SPOT) imageries, and many other satellite systems can also be used to detect the glow of fires.

During the early days of Kuwait oil fires, many efforts were made to determine the number of burning oil wells and the extent of both horizontal and longitudinal dispersion of the smoke. Kwarteng and Bader (1992, 1993) detected the first fire on January 24, 1991 using NOAA-AVHRR images. They observed large scale fires first at the Wafra Oil Field and later that the fires were spread gradually and systematically to other Kuwait oil fields. The number of oil fires reached its maximum in the fourth week of February 1991. Kwarteng and Bader (1992, 1993) counted a total of 625 fires on LANDSAT™ images on March 23, 1991. They observed 29 fire sources in Umm Gudair, 365 in the Greater Burgan, 23 in the Wafra Oil Field. Figure 3.2 shows fires in the Greater Burgan and the Umm Gudair Oil Fields, a part of the work done by Kwarteng and Bader (1992, 1993). Individual fires can clearly be seen on this LANDSAT™ images. Using the LANDSAT™ imagery, Williams et al. (1991) counted a total of 529 oil fires in late April 1991.

Satellite imagery was also used to determine the day to day distribution of smoke plumes from the Kuwait oil fires. In the satellite imagery investigations, more prominence was given to the horizontal smoke dispersion and less to the vertical rise. Smoke from the burning wells was visible in early February 1991.

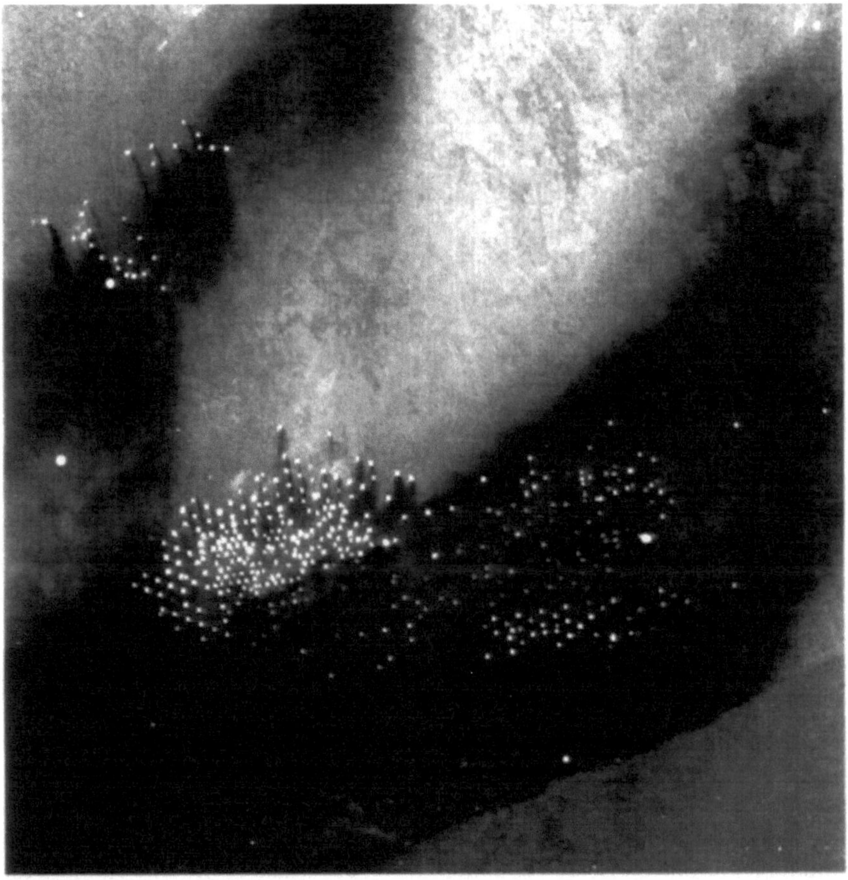

Figure 3.2. Satellite images of oil fires in the Greater Burgan and Umm Gadair Oil Field (courtesy of Dr. Kwarteng of KFUPM/RI).

LANDSAT™ images taken on February 15 and 16, 1991, showed many smoke plumes from individual oil wells extending to the north and south of Kuwait City. Many more oil wells were ignited by Iraqis in anticipation of the ground attack. These fires were visible on all the remote sensing images acquired during the third week of February (Williams et al., 1991). During the winter and spring, the wind carried the smoke over central and southern Saudi Arabia and the Arabian Gulf. A NOAA-AVHRR image on February 21, 1991, showed the plume extending about 300 km to the south of Kuwait (Horgan, 1991b). Smoke dispersed over an area of about 7000 km² was noticed on a NOAA-AVHRR image taken on March 4, 1991. A thin veil of smoke over an area of about 180,000 km² was visible in a NOAA-AVHRR image taken on March 18 (Limaye et al., 1991). In METEOSTAT™ images, thin smoke was observed as far from the fire source as Pakistan in the east and Turkey in the north (Williams et al., 1991).

Satellite imagery was also used to determine the height of the smoke plume. In NOAA-AVHRR images, clouds could be seen at 7 to 9 km altitude that cast a shadow on the smoke plume at a height of about 3 to 5 km below (Limaye et al., 1991). Space Shuttle imagery was used to study the smoke palls inducted by the Kuwait oil fires and found no self-lofting (Lulla and Helfert, 1991). NOAA-AVHRR imagery, i.e., opacity of the smoke at different wavelengths, was used to determine smoke particle size. Most of the particles were below 1 μm size (Limaye et al., 1991). Many such examples can be cited with reference to the use of satellite imagery in investigating the Kuwait oil fires.

Some of the most commonly used satellite images are discussed above. There are many more satellites that could be used to study the environmental crisis resulting from the Kuwait oil fires. Some of these imageries are:

System Probatoire d'Observation de la Terre (SPOT)
LANDSAT™ 4 and LANDSAT™ 5
Polar-Orbiting Environmental Satellites (NOAA-10 and NOAA-1)1
The United States Defense Meteorological Satellites (DMSP)
Meteosat Satellites of the European Space Agency
Space Shuttle Missions

In summary, remote sensing was successfully used for collecting information on the Kuwait oil fires. In general, there was a good agreement between the satellite images and the ground and airborne investigations.

3.5.4. Airborne investigations of smoke dispersion

After the cessation of hostilities in the Gulf region, the world scientific community initiated research and monitoring programs of the Kuwait oil fires, concentrating on plume characterization. For this purpose, several teams of foreign scientists undertook daring airborne missions in and around the fires going through and around the smoke plume. A brief chronological summary of these missions is given in Table 3.3.

The airborne investigations mainly examined the horizontal rise of plume

Table 3.3. A summary of aircraft flights over Kuwait oil fires.

Flight Date 1991	Aircraft Type	Major Organizations
March 23–31	C–130 Hercules	British Meteorological Office
May 19–June 14	Lockheed Electra	The United States National Center for Atmospheric Research (NCAR)
May 15–June 15	Convair C–131	University of Washington (USA)
May 20–June 6	Piper Navajo	Umwelt Data
July 31–August 8	Bell 206 Helicopter	Saudi Arabian Air Force/National Aeronautical and Space Administration /United States Environmental Protection Agency
July 30–August 17	Gruman G–1	The United States Department of Energy

and characterization of smoke, rather than the horizontal dispersion of smoke. Johnson et al. (1991) of the British Meteorological Office recorded airborne observations of the smoke plume in March 1991. They noted that smoke from individual well fires combined, rose, and the plume top reached about 4.5 km in altitude. On a few occasions, the team found a detached smoke mass rising above the well defined smoke top. Based on the visual airborne observations, they concluded that the smoke was mainly confined to the lower half of troposphere and was not lofted to the stratosphere in significant amounts. They found a marked vertical wind shear leading to a differential advection of the plume, e.g., the lower part of the plume transported southwards down the Arabian Gulf while the upper part was moved over Iran.

Hobbs and Radke (1992) reported visual observations of the smoke plume recorded during airborne investigations by the Washington State University. Based on these observations, it was reported that the team never detected the smoke plume above 6 km in altitude. The plume from the northern oil fields merged with that from the southern fields to form a super composite of the smoke plume which was about 40 km wide near the sources. The top of the plume was flat, occasionally with wavy features. The team also noted small masses of breakaway smoke. The base of the plume (lower side of the plume) was generally also flat and confined to about 0.5 to 2 km in altitude. Individual fires produced different plumes, ranging from black to whitish gray. A few fires, presumably from natural gas, produced no visible plume. Hobbs and Radke (1992) found an inverse relation between the distance from the fire sources and smoke concentrations. They found that smoke concentrations beyond 1000 km from Kuwait almost approached background levels. Daum et al. (1992) studied the Kuwait oil fire plume during mid-summer 1991. They found the smoke plume well above the ground, in some cases 1–2 km, and it was never observed above 5 km in altitude. They found that the majority of the plume particles were about 2–3 μm in diameter.

In summary, most of the airborne investigations found that the smoke plume rose to a maximum altitude of 5 km, and had a flat top with occasional small breakaway patches from the smoke plume. The lower boundary of the smoke was also found to be well defined and flat at about 1 km in altitude. The self-lofting phenomenon did not significantly effect the smoke plume rise.

Results of numerical calculations, satellite imagery investigations, and airborne studies consistently showed that the environmental effects of Kuwait oil fires were significant only in the Gulf region and insignificant at a global scale. The smoke was emitted from an area of about 40 $\times$ 40 km; the individual plume rose, spread, and merged with other plumes to form a large cloud which, depending on the wind direction and speed, fanned out into predominantly to the north, east, and south. Most of the smoke was transported within an altitude range of 2 to 5 km. After aerial transport of about 1000–2000 km from the source, the smoke cloud thinned to such an extent that it became undetectable, very close to the background levels. Hahn (1991) estimated that, on the average, the smoke cloud covered about 6 $\times$ 10^{11} m^2. During an airborne investigation, Hobbs and Radke (1992) found that smoke concentrations beyond 1000 km from Kuwait almost approached background levels. Processing the data collected by NOAA-AVHRR, Limaye et al. (1991b) showed that the smoke plume extended over an area of about 7000 km² on March 4, 1991, and about 180,000 km² on March 18, 1991.

KFUPM/RI (1991a, 1992) used the ARL-ATAD model (Air Resources Laboratory Regional Transport and Dispersion Model) to determine smoke dispersion in the Gulf region. To run this model they retrieved upper atmospheric data from the synoptic weather charts (700 mb and 850 mb pressure level) that were updated twice daily by MEPA. Data on wind speed, wind direction, temperature, pressure, and stability height were estimated from the weather chart. They used prevailing meteorological conditions rather than

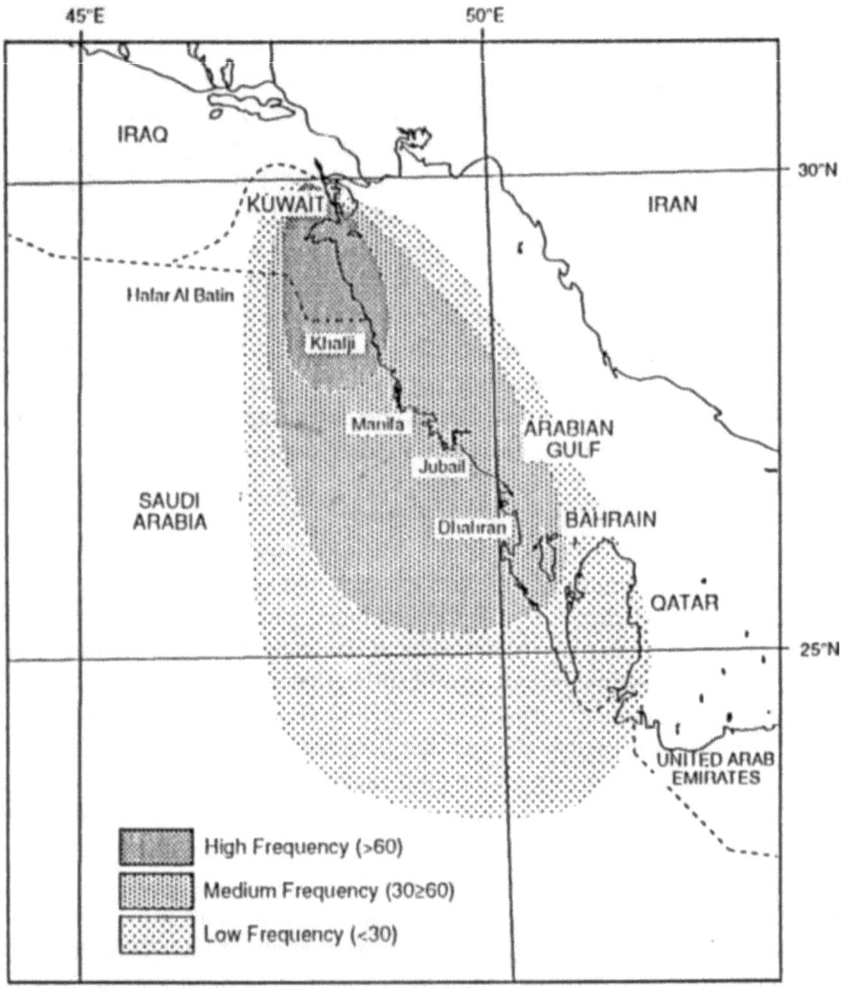

Figure 3.3. Smoke dispersion in the Gulf region during March–May 1991.

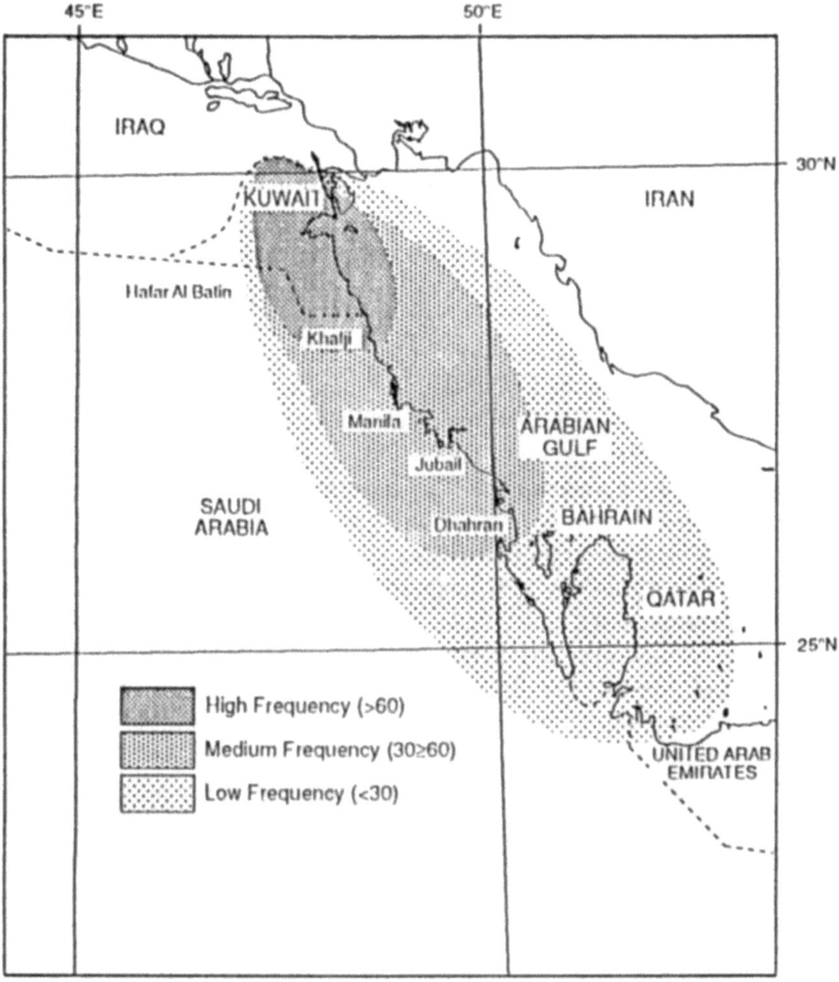

Figure 3.4. Smoke dispersion in the Gulf region during June–August 1991.

historical data to predict smoke dispersion on a daily basis. They also simulated monthly smoke dispersion characteristics from March to October 1991. The most probable smoke dispersion estimates for March through July (when the wind was mostly northerly and smoke was frequently observed in Dhahran) are reconstructed in Figures 3.3 and 3.4. It is evident from this figure that Kuwait, and a large portion of Saudi Arabia, Bahrain, and Qatar were covered with smoke during this period. Many hundreds of fires were extinguished by the time the predominant wind direction changed in August. Therefore, newly impacted areas, largely in Iraq, Iran, and Turkey, were not affected as much as the above areas. It is, therefore, suspected that alteration in the meteorological conditions due to Kuwait oil fires was most pronounced in the area under smoke during the period February through August, in Kuwait, northeastern Saudi Arabia, Bahrain, and Qatar.

3.6. Ground monitoring of smoke/soot

Air pollution laboratories throughout the world are monitoring atmospheric air particulates loads. Many of these reported an enhancement of smoke or soot concentrations in the air at their monitoring locations. Some of the reports are reviewed herein.

Linden (1991) reported that black rain related to smoke from the Kuwait oil fires was found as far as Qatar, 645 km to the south of Kuwait. There were press reports of black snow in the Indian held Kashmir, about 2600 km to the east of Kuwait (Linden, 1991). The Gulf region experienced unusual weather patterns during March, 1991. By mid-March 1991, heavy rain was frequent and southerly winds were predominant. This unusual weather with southerly winds transported smoke over Iraq, Turkey, and Iran. This weather pattern seems to be responsible for black snow in Pakistan and India too. In addition, airborne investigations in April–May 1991, identified scattered patches of smoke rising above the general smoke body. These "drifters" could have been responsible for the black snow over the Himalayas ranges (Hobbs and Radke, 1992; Johnson et al., 1991). The British Meteorological Office, using numerical models, had predicted that blackened snow could be observed over the Himalayas (Warner, 1991).

The soot from Kuwait oil fires was also observed over Japan. Okada et al. (1992) examined air particulates collected during April 1991 at an altitude of 7.5 km over Japan. They found an unusually high portion (53 percent) of soot containing particles in the sub micrometer-size range (0.15–1 μm radius). Relatively high vanadium concentrations were detected in some samples of air particulates. They noticed that such a high concentrations of soot particles in the troposphere had not been reported previously. From the data, Okada et al. (1992) suspected that these soot particles were probably emitted by the burning oil wells in Kuwait. Numerical modeling had predicted that an insignificant fraction of smoke (about 0.3 percent) might be self-lofted to the troposphere, and if this occurred, the smoke could be transported over long distances. The soot from the Kuwait oil fires that was reported over Japan by Okada et al. (1992) might be due to this phenomenon. Smoke has been reported as far north as Russia, and as far east as Pakistan, well over 20,000 km from Kuwait (Horgan, 1991a; 1991b). There were also some reports that smoke from burning oil wells in Kuwait might have reached Hawaii, USA.

3.7. Estimates of burning crude oil

One of the important parameters used in assessing the potential environmental impact of the Kuwait oil fires using numerical models is the amount of oil being burned each day. Before any further discussion, we wish to differentiate between oil emission rate and oil burning rate. Oil emission rate is defined as the quantity of oil per unit time that is flowing out of the ground. This oil may or may not be burned. The portion of the emitted oil that is burnt as it comes out of the ground is referred to as burning rate or combustion rate. The quantity of the emitted oil that is not burned is referred to as residual or residue oil.

In a free-flowing and free-burning situation, like the oil fires in Kuwait, it is extremely difficult to estimate each of the above types. The rate of oil flow from any well can easily be determined under controlled conditions; however, this becomes extremely difficult and highly speculative for headless or damaged head oil wells. The emission rate under free-flow conditions depends on the reservoir pressure, piping characteristics (diameter, length, and type), and the extent of damage to the oil well head and pipes. In addition, many of these conditions, especially reservoir pressure and extent of damage might change with time. It may be concluded that any accurate estimate of the oil flow from the damaged and burning oil wells in Kuwait can never be made. Therefore, it was expected that oil flow estimates would be as plentiful as the number of estimators.

Before the start of the Gulf war, the United States Department of Energy contracted Sandia National Laboratories to determine the potential military, environmental, and Kuwaiti oil reservoir depletion impacts which could arise from Iraqi sabotage of the Kuwait oil fields. Sandia (1991) estimated that if several well heads were blown up, these well heads could spew out 5–6 million barrels crude oil daily, but oil reservoirs would not be severely damaged. Manins and Bouma (1991) assumed 1.5 million barrels of oil emission per day. Based on the measurements taken by an aircraft, the British Meteorology Office estimated the emission quantity to vary between 3.3 and 5.5 million barrels per day (Johnson et al., 1991). The Ministry of Oil in Kuwait surveyed the burning, gushing, and damaged oil wells and the extent of the oil lakes formed on the ground near the damaged wells. Based on the information about the pressure at the well head, reservoir capacity and pressure, etc., the Ministry calculated an emission rate of 6 million barrels of oil and 100 million cubic meters of gas per day. Since the Kuwaiti Ministry of Oil has all necessary information on well capacity and reservoir pressure in each oil field, their estimates are expected to be reasonably close to the reality, though some believe that these estimates could be biased in favor of Kuwait due to the possible payment of reparations by Iraq in the future. The estimates of the Kuwaiti Ministry of Oil are also supported by many independent calculations (El Desouky and Abdulraheem, 1991).

Many environmental scientists have estimated the burning rate or combustion rate of oil from the burning oil wells in Kuwait. Bakan et al. (1991), in their study of the impact of smoke from the burning oil wells on global climate, assumed 3.1 million barrel per day of a constant burning rate. Assessing the environmental effects of the burning oil wells in Kuwait, Browning et al. (1991) assumed that about 1.5 million barrels of oil was burned every day. Similarly, Small (1991) assumed a combustion rate of 1.6 million barrel per day. Many other scientists also estimated oil emission from the burning and damaged oil wells in Kuwait. A brief summary of these estimates is given in Table 3.4.

Table 3.4. Estimates of oil emission by the Kuwait oil fires.

Source	Number of Oil Wells	Oil Emission (Million Barrels per Day)
Hahn (1991)	500–600	1.5–3.0
Johnson et al. (1991)		3.9
Small (1991)		1.2
Browning et al. (1991)		1.5
Bakan et al. (1991)		3.1
WMO (1991)	625	2.4
Manins and Bouma (1991)		1.5
Williams et al. (1991)	611	5.0–6.0
Horgan (1991a)	550	6.0
Barnaby (1991)	600	6.0
EPC (1991)	613	6.0
Johnson et al. (1991)		2.3–7.5
Schumann (1991)		7.5
KFUPM/RI (1992)	625	2.4
NOAA (1992)	611	6.2

Table 3.4 indicates that oil burning estimates vary widely, ranging between 1.2 and 7.5 million barrels per day. Lack of data on Kuwait oil industry, reservoir pressure, the nature and the extent of damage, free-flow behavior of the oil wells, reservoir pressure behavior under free-flow conditions, and many other factors contributed to the confusion regarding oil emission estimates. The best estimate of oil burning rate was probably provided by the Kuwait Ministry of Oil (i.e., 613 burning oil wells and an oil emission of 6 million barrels per day). For this book, we have adopted the estimate of the Kuwait Ministry of Oil of about 6 million barrels per day of oil being emitted by the burning and gushing oil wells in Kuwait.

Soon after the cessation of hostilities, the Kuwait Government contracted international organizations and companies to extinguish the burning oil wells. During the early days of this effort, some fire fighting experts estimated that the oil fires would take years to control. However, all burning wells were extinguished by November 6, 1991. From February to November 1991, the number of burning wells decreased consistently and as a consequence of this, the oil emission rate also decreased. NOAA (1992) estimated that the oil burning rate in mid-March was 6.2 million barrel per day, which decreased to 5.1 millions by early June, 3.3 millions by mid-August, and 1.8 millions barrel/day by late September 1991. These estimates are higher than those proposed by the Kuwait Ministry of Oil. On November 6, 1991, the Emir of Kuwait ceremonially extinguished the last burning oil wells thus ending one element of the greatest environmental disaster in human history.

The burn rates, reported by NOAA (1992) were calculated by the Analytical Science Corporation (TASC) under a NOAA contract (WMO, 1992). Saudi Aramco in Dhahran, Saudi Arabia also collected similar information. The data collected by the Saudi Aramco were used to establish a chronology of fire extinguishing efforts and the temporal burning rate for each field as listed in Table 3.5. Total oil burning rate for the Kuwait oil fires was calculated from the information available and is plotted in Figure 3.5. This figure shows an initial gradual decrease in the burning rate and then a sharp decrease. At the beginning of the extinguishing effort, fire fighters were faced with many technical and logistic problems that were overcome later. Moreover, the Government of Kuwait became more effective and efficient with time and increased fire fighting teams from four (in the beginning) to over 22 (at the peak of the fire fighting operation). The sharp decrease in the oil burning rate shown in Figure 3.5 actually reflects the results of these efforts. The data reported in Table 3.4

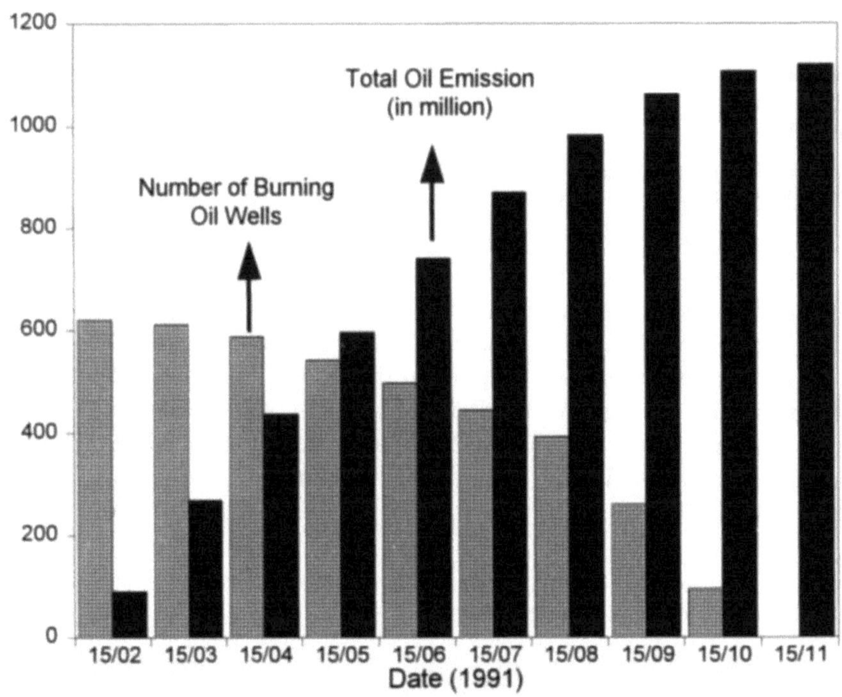

Figure 3.5. Temporal distribution of burning oil wells and total oil emissions from the Kuwait oil fires.

and 3.5 on oil emission from each oil field, oil well pressure, and probable oil emission were used to calculate the total amount of oil emitted by the damaged oil wells in Kuwait. These calculations showed a total of 1.12 billion barrels of crude was emitted during the Kuwait oil fire episode. This value will be used as a reference in this book.

Table 3.5. Temporal progress in extinguishing the burning oil wells in Kuwait.

Date	Oil Wells										
	Rau	Sab	Bah	Ahm	Mag	Waf	Min	Umm	Bur	Total Oil Wells	Flow Rate@
February*	62	39	3	60	98	6	27	26	291	613	6.0
March 15	62	38	3	60	98	6	27	26	290	610	5.7
April 15	62	38	3	58	87	0	27	26	285	586	5.5
May 15	62	38	3	55	50	0	27	26	280	541	5.0
June 15	62	38	3	35	40	0	27	26	265	496	4.6
July 15	62	38	3	18	23	0	26	26	248	444	4.1
August 15	62	38	3	0	0	0	25	26	238	392	3.3
September 15	54	33	3	0	0	0	9	26	135	260	2.2
October 15	20	19	1	0	0	0	0	0	55	95	0.8
November 15	0	0	0	0	0	0	0	0	0	0	0.0

Oil Field Abbreviations:

Rau - Raudhatain
Sab - Sabriya
Bah - Bahrah
Ahm - Ahmadi
Mag - Magwa
Waf - Wafra
Min - Minagish
Umm - Umm Gudair
Bur - Burgan

* Information from Al Besharah (1991)
@ Flow is in million barrel per day

From the foregoing, it can be seen that oil burning estimates were as numerous as those who wished to make them. The estimates varied widely from 1.2 to 7.5 million barrels per day. The estimate of Kuwait Oil Company (EPC, 1991) of 6 million barrels per day was adopted in this book. From the chronological information on fire fighting activities, data on reservoir pressure, well flow, etc., an estimated total amount of 1.12 billion barrels of crude oil was released during the whole Kuwait oil fires episode. This estimate of total oil emission will be used in this book.

3.8. Estimates of soot/smoke production

The quantity of smoke produced is important because it influences many aspects of the environment. Very large amounts can induce effects with global implications such as a reduction in air and sea temperatures, the amount of solar radiation reaching the earth's surfaces, upper atmosphere perturbation resulting in reduced monsoon in impacted areas, etc. Manins and Bouma (1991) assumed that the 10,000 tonnes of soot emitted by the fires would remain loft for at least 24 hours and would be distributed over a distance of 700 km from the fire source covering a 100,000 km^2 area. They further suggested that about 70% of sunlight under the plume will be absorbed, thus causing a decrease in ambient temperature equivalent to 10°C. Moreover, soot is a good absorbent of organic chemicals. Inhalation of smoke loaded with organic chemicals may cause serious health problems. It is, therefore, important to collect information on the oil fire emission factors or rates of smoke production.

The efficiency of combustion effects the type and quantity of emission products. In a 100% complete combustion situation, only water and carbon dioxide are produced. However, incomplete combustion is the rule of nature and thus a number of emission products are generated during combustion, with soot/smoke as one of the major components. Because of its importance, many investigators estimated combustion efficiency or soot production. Hobbs and Radke (1992), from their airborne observations through the smoke plume, estimated a combustion rate (soot production) of 1.6 to 2.8% for pool fires and 0.3 to 0.6% for the composite fire plume in Kuwait. They noticed that soot production in Kuwaiti fires was closely associated with flame appearance. They observed very little soot (about 4% by mass) was present in white smoke whereas black smoke contained about 20 to 25% soot and variable amounts of salts. Fires associated with the oil pools produced the blackest smoke, which contained up to 48% soot by mass. The mass of the smoke particles in the super composite plume was estimated to be about 30% salt, 15–20% soot, 8% sulfate, and about 30% organics. The Union of Concerned Scientists estimated that the fires in Kuwait burned with a poor 70 to 90% efficiency, and suspected that a stew of poison gases and solids was being shot in to the atmosphere (Linden, 1991). Some of the estimates of soot/smoke production rates are listed in Table 3.6.

The data reported in Table 3.6 can be grouped into two categories: those reporting soot emission factors below 3% and those reporting emission factors >5 percent. Another observation is that all reported values which were based on actual and/or airborne measurements report soot emission factors of less than 3 percent. Most of the numerical model used an assumed smoke emission factor of 5% or higher. Investigators using numerical models adopted data that were reported with reference to nuclear winter episodes. It may be pointed out that about 25 to 40% of the soot generated by the Kuwait oil fires was inorganic salts (Hobbs and Radke, 1992). We feel that airborne observations are more accurate and should be adopted in assessing the environmental impact of the Kuwait oil fires. Considering the data in Table 3.6, a soot/smoke production rate of 2% has been adopted for this book. This rate is estimated from the observations of Daum et al. (1992) and Hobbs and Radke (1992). From the estimates of a total crude oil emission of 1.12 billion barrels and assuming that 90% was burning, a total amount of about 2.5 million tonnes of soot/smoke would have been produced during the Kuwait oil fires episode.

Table 3.6. Summary of the combustion rate estimates for the Kuwait oil fires.

Investigator	Burn Rate (%)
Hahn (1991)	5–10
Hobbs and Radke (1992), Oil lake fires	1.6–2.8
Hobbs and Radke (1992), Oil well fires	0.3–0.6
Laursen et al. (1991 cited by WMO, 1992)	0.3
Bakan et al. (1991)	10.0
Small (1991)	8.6
Manins and Bouma (1991)	5
Snell (1991)	5–15
Penner (1991)	5
Schumann (1991)	10
EPC (1991)	5–15
Browning et al. (1991)	6
Daum et al. (1992) (measured in July–August 1991)	3.3–3.8

From the foregoing it was deduced that the soot emission rate used in the numerical calculations was at least double that estimated during the airborne investigations. Soot production was dependent on the type of fire and was reflected by the color of smoke. White smoke plumes were rich in inorganic salts whereas black smoke plumes contained a larger portion of soot/smoke particles. The lake fires produced denser smoke and, therefore, higher soot concentrations than in the smoke produced by the oil well fires. About 2.5 million tonnes of soot/smoke was produced from Kuwait oil fires.

3.9. Morphological characteristics of Kuwait oil fires

The morphological characteristics of Kuwait oil fires were reported to be variable. The media (television, newspapers, magazines, etc.) showed different shapes of flames. In addition, the color of the smoke plume varied between grayish white to dark black. These variations are known to effect emission products. Therefore, knowledge of the morphology of the Kuwait oil fires is desirable. Information available on the morphological characteristics of the Kuwait oil fires are reviewed. Two main characteristics are considered here; the type of flame and the color of the smoke plume.

3.9.1. Morphology of fire flames

The type of flame contributes to plume rise and emission products. A knowledge of the physical appearance of the flame may help in the assessment of the environmental implications of the Kuwait oil fires. Some important types of flame are briefly reviewed. Four common type of oil fires were observed in Kuwait (Al Besharah, 1991) as follows:

Jet Flame The Jet Flame was long, cylindrical and often resembled a long cone. This was a very common type of flame in the Kuwait oil fields. Crude was released at with very high pressure under free-flow conditions as a jet spray. Usually in such cases, the well head was completely destroyed, and only the lower flange of the well head was left in place. Smoke plumes from such flames rose with very high speed, dessipated quickly, and were considered to be of high combustion efficiency. This means that there would have been a relatively high production rate for carbon dioxide and other gases with lesser amounts of soot. Since the crude was emitted at high speed under high pressure, a rain of oil mist might be expected in the vicinity of the well head. No oil lake was formed at the base of a well head with this type of fires.

Ball Flame The Ball Flame height was smaller but wider than the Jet Flame. The well head was completely blown out. Crude oil came out at a lesser pressure than that of the Jet Flame. The smoke plume rise was slower and to lower altitudes than the Jet Flame. In many cases oil lakes were formed. In a few instances, ground accumulation of oil caused a flickering of the flame. Combustion was expected to be incomplete and less efficient than that of Jet Flame, especially in the inner part of the flame. Thus the Ball Flame produced more soot and less gaseous pollutants as compared with the Jet Flame.

Moustache Flame The Moustache Flame was horizontal jet type flame (i.e., from both side of well head). This probably resulted due to the formation of a cake or other obstruction at the partially damaged well head. Oil lakes were commonly formed around the well head and smoke was dissipated near the ground. Combustion was thought to be incomplete. The smoke plume was dense. The soot content of the emission was higher than the above two types.

Lake Fires Lake fires resembled a typical bush fire and, therefore, the smoke plume remained relatively low. The speed of plume rise was the slowest of the various types of flames and combustion was incomplete. Smoke from these fires became denser and black with time as the light crude components were either evaporated or burned. The quantity of smoke produced by Lake Fires was estimated to be five times higher than those of the well fires (Al Besharah, 1991). Smoke intensity increased with the weathering of the spilled crude. Because of incomplete combustion, high quantities of soot were produced.

The above mentioned types of flame are shown schematically in Figure 3.6.

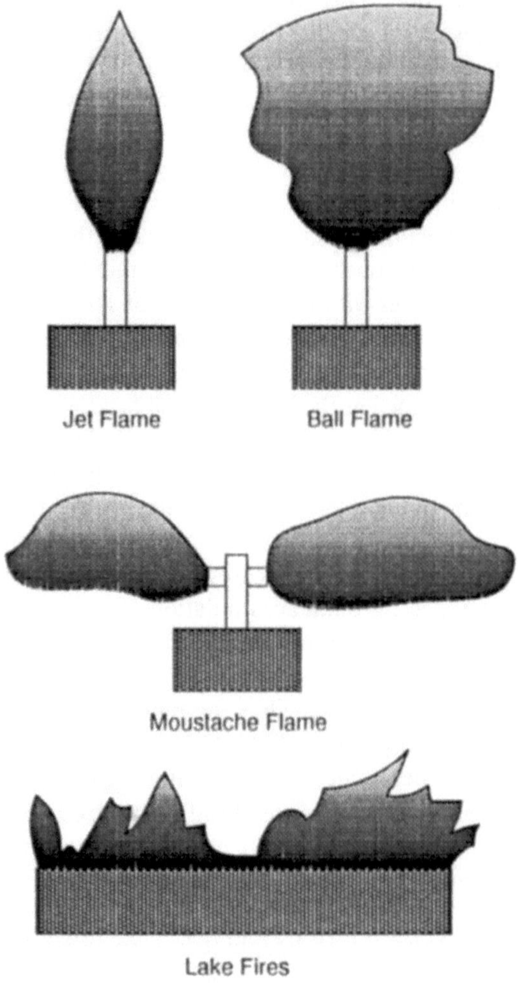

Figure 3.6. Common flame types in burning the Kuwait oil fields.

3.9.2. Morphology of the smoke plume

The press, scientific media, and television graphically presented the smoke plume as different shades of color varying from black to white. Similar observations were reported by airborne investigations. Considering the published reports, Al Besharah (1991) concluded that smoke color from the Kuwait oil fires was dependent on the following:

a. Type of Crude Burned: Kuwait has many crude types varying from light to very heavy (Table 3.9). Light crude is burned more efficiently and thus would produce less smoke/soot. The resultant color of the smoke would be relatively less black than that of heavier crude oils. It has been reported that lake fires produced dense black smoke and soot production was five times higher than that of the oil well fires (Al Besharah, 1991). This was partially attributed to the increased proportion of the heavy fractions of crude in the oil lakes.

b. Type of Damage Sustained Oil Well Head: Damage to the well Head determined the type of flame and combustion efficiency. These factors influenced the density and blackness of the smoke plume.
c. Extent of Combustion: Carbon dioxide and water are the two products emitted with the complete combustion of fossil fuels. The amounts of smoke/soot, organic chemicals, and gases produced are directly related to the degree of combustion. Smoke from the Kuwait oil fires would have been more dense black in cases where incomplete combustion was predominant. For example, combustion was relatively incomplete in Ball and Moustache Flames as compared to the Jet Flame and the resultant smoke plumes were more black with higher concentrations of soot particles.
d. Amount of Water in Crude: It was reported that white tinge of smoke increased with an increase in the water content of the crude. This will be discussed later.

Individual fires produced different plumes ranging from dense black to white in appearance. A few fires, presumably natural gas, produced no visible plume. It was a common belief that the whitish color tinge was dependent on the amount of water vapor in the plume (Al Besharah, 1991; Johnson et al., 1991). Hobbs and Radke (1992) discounted this theory. They reported that dew point measurements within the whitish gray of the smoke plume and outside in the ambient air were similar and relative humidity in the plume was also low. These plumes were found to contain considerable amounts of salts which scattered light efficiently to produce the white appearance. They found that more than 80% of the mass of the particles in one of the white plume was salt, mainly sodium chloride. Similarly, Daum et al. (1992) reported high concentrations of salts in the aerosol. They observed high concentrations of calcium sulfate and sodium chloride. They suggested that the out-of-plume aerosol may have had an important soil-derived component to enhance salt, and particularly calcium, concentrations. Observations by a joint team of NASA/USEPA/Saudi Air Force scientists also found high concentrations of inorganic salts in the whitish smoke as compared with the black the smoke plume (Cofer, 1991). It is believed that the smoke plume color is dependent on the amount of salts and not water although there is a good correlation between the amount of brine water and concentrations of salts in a crude (high water content - high salt concentrations). From the foregoing, it was concluded that the white smoke produced less soot than the black smoke. A comparison of white and black smoke is given in Table 3.7.

The different types of smoke varied in their chemical composition. Results of carbon partitioning and other chemical determinations conducted by the scientific team from the Washington State University are given in Table 3.8.

The table shows that:

1. Concentrations of carbon monoxide were higher in lake fires and the smoke plume was dense black.
2. Carbon dioxide concentrations seem to be similar in both black and white smoke plumes but higher than the lake fires. This suggests that oil lake fires were burning with less efficiency.
3. Methane and non-methane hydrocarbons were higher in the lake fires. This may be related to inefficient combustion.
4. Nitrogen oxides were found to be similar in all types of smokes.
5. Concentrations of different carbon fractions varied widely in the pool fires.
6. Combustion is relatively incomplete in the lake fires.
7. Concentrations of inorganic constituents were higher in the whitish the smoke plume as compared with that of black or mixed plume.

The above data suggest that chemistry of a smoke plume is affected by many inter-dependent variables. It is expected that the chemistry of smoke produced by the oil lake fires is complex and the smoke contains relatively more high carbon organic compounds than the smoke produced by the oil well fires. Among the oil fires, the chemical composition of black smoke seemed more complex than that of the whitish smoke.

3.10. Chemistry of the smoke plume of Kuwait oil fires

The chemistry of the smoke plume or the quantities of emission products largely depends on the chemical composition of the crude oil which is being burned. For example, crude high in sulfur will be expected to produce higher concentrations of sulfur dioxide and maybe hydrogen sulfide. Heavy crude will generate more soot than light crude and so on. When many of the Kuwait oil fields were burned, there were many different kinds of crude oils being burned simultaneously. To give an idea of the extent of this variability, information on the crude oils of the Arabian Gulf region was collected and is summarized in Table 3.9. Total sulfur content varied between 0.6 to 4.55% in these crude grades. The probable Kuwaiti crude oils involved in the fire episode contained total sulfur from 2.52 to 4.55% – a relatively high sulfur concentration. Almost all investigators have assumed total sulfur as 3% in the crude oil burned in Kuwait, which seems to a reasonable assumption. The crude oils from Kuwait also vary in hydrogen sulfide concentration. These concentrations range between 9.8 to 315 mg/ml. The Kuwaiti crude oils contain appreciable concentrations of toxic metals such as nickel and vanadium. Table 3.9 lists various components of Kuwait crude oils.

In a broad sense, there were two main components of the smoke plume; solid particulates and gases. The solid particulates fraction largely contains inorganic salt crystals, organic chemicals, and soot particles. The gaseous fraction largely consists of carbon dioxide and carbon monoxide. The minor gases in the fraction include sulfur dioxide, hydrogen sulfide, nitrogen oxides, methane, ozone, and aromatic hydrocarbons. The chemical behavior of these constituents is discussed below.

3.10.1. Smoke/soot particles in the plume

All airborne studies of the Kuwait oil fires (British Meteorological Office, National Center for Atmospheric Research-NCAR, University of Washington, German G-1, and Department of Energy) investigated the morphology of smoke particulates. In general, the smoke particles were spherical with a diameter of about 0.1 to 0.3 μm (Johnson et al., 1991; Hobbs and Radke, 1992). Smoke particles combined to form a chain-aggregate commonly referred to as soot. The morphological characteristics of smoke particles seemed to be altered by the type of smoke and distance (age) from the source. Johnson et al. (1991) observed that the individual particles in the far-field (1000 km from Kuwait) were largely spherical and more closely packed into near-spherical clusters as compared with those collected from the near-field (182 km from the fires). Hobbs and Radke (1992) reported that some of the soot particles in the black plume formed chained aggregates within a few seconds of combustion. In comparison with the black smoke plumes, fewer chain-aggregates were detected in the white plume during this study. The aged black smoke also contained more non-spherical particles as compared with the mixed or white the smoke plume. In contrast to the earlier, commonly held belief that smoke was hydrophobic, the majority of smoke particles were found to be hydrophilic. This was partially caused by inorganic salts adsorbed on the smoke particles. These smoke/soot particles served as cloud condensation nuclei in cumuli-form or strati-form clouds (Hobbs and Radke, 1992).

As stated earlier, individual fires produced different colors of plumes ranging from dark black to the whitish gray in appearance. It was common belief that this whitish color was dependent on the amount of water vapors in the plume (Al Besharah, 1991; Johnson et al., 1991). Contrary to this, Hobbs and Radke (1992) emphasized that water content in the Kuwait oil fire plumes and the atmosphere outside the plume was similar. They found that white smoke plumes contained considerably higher amounts of inorganic salts, especially sodium chloride, as compared with black smoke plumes. The inorganic salt crystal was believed to scatter light so efficiently as to produce the white appearance. The presence of higher concentrations of inorganic smoke in the white the smoke plume was further confirmed by the airborne investigations of Daum et al. (1992).

As expected, the distribution of smoke particles in the plume changed with time and dispersion. A few airborne investigations determined the number of smoke particles in the plume. In general, total particle concentrations within a few kilometers were around 10^5 per cm^3. At a distance of about 300 km from the fires, concentrations of smoke particles were decreased to 5000–15,000 per cm^3 and further decreased to almost background levels (300–500 per cm^3) when the distance from the fires was increased to beyond 1000 km (Hobbs and Radke, 1992). We have already discussed the dispersion of smoke in the atmosphere and have indicated the most probable impacted areas in Figures 3.3. Therefore, smoke dispersion is not discussed further.

Table 3.7. Characteristics of white and black smoke.

Black Smoke	White Smoke
Quantity of inorganic salts low.	Quantity of inorganic salts, particularly of sodium chloride, very high.
Concentration of water vapor low.	Amount of water vapor high.
Produces high soot concentrations (about 20–25% soot).	Produces low soot concentrations (about 4% soot).
Burning efficiency low.	Burning efficiency relatively high.
More soot particles formed chain-aggregates within seconds of combustion.	Fewer soot particle formed chain-aggregates within seconds of combustion.
Aged-smoke contained non-spherical particles.	Aged-smoke contained non-spherical particles but less than the black smoke.
Relatively high concentrations of CO in the smoke plume.	Very low concentrations of CO in the smoke plume.

Information on the chemical composition of soot particles is limited. Hobbs and Radke (1992) reported that the smoke particles collected from the super composite plume (the smoke plume where fires from the individual wells combine together to form one plume) consisted of about 30% salt, 15–20% soot, 8% sulfate crystals, and about 30% organic chemicals. Sodium chloride was the major component of salt fraction. Smoke particles from individual fires varied widely. For example, the particles from the fire producing white smoke contained more salts and less elemental carbon. A sample of particles from a white smoke plume was found to contain 80% inorganic salt (Hobbs and Radke, 1992). The particle sample collected from the black smoke plume contained higher concentrations of elemental carbon and less salts. The smoke produced from the lake fires was very thick and contained very high concentrations of soot particulates. Differences between the white and black smoke have already been discussed and are presented in Tables 3.7 and 3.8.

Table 3.8. Approximate concentrations of chemical parameters in different types of the smoke plumes (data from Cofer, 1991).

Parameter	Black Smoke	White Smoke	Oil Lake Smoke	Mixed Smoke
Carbon dioxide (ppm by volume)	375–446	377–442	393–429	317–402
Carbon monoxide (ppm by volume)	0.58–3.35	0.25–0.58	1.51–18	0.51–1.41
Methane (ppm by volume)	1.79–1.88	1.78–1.88	1.83–5.61	1.78–1.86
Non-methane hydrocarbons (ppm by volume)	0.57–2.93	0.89–1.69	1.15–15.4	0.57–1.02
Nitrogen oxides (ppm by volume)	0.32–0.32	0.32–0.32	0.32–0.33	0.32–0.32
Chloride ($\mu g/m^3$)	15–182	3220	–	9–10
Sulfate ($\mu g/m^3$)	9–177	630	–	10–18
Calcium ($\mu g/m^3$)	6–84	415	–	1–8

– No information available

Table 3.9. Physico-chemical properties of some crude oils from the Arabian Gulf region (from HPI Consult, 1987).

Crude oil type	Specific Gravity API	Total S (%)	Total N (%)	H_2S (mg/ml)	Ni (μg/ml)	V (μg/ml)
Abu Al Bu Khoosh (UAE)	0.8670	2.00		6.70	12.00	4.3
Murban (UAE)	0.8223	0.78	0.047	2.70	1.00	1.35
Umm Shaif (UAE)	0.8372	1.51	0.061	3.60	3.00	2.20
Zakum, lower (UAE)	0.8216	1.05	0.050	2.90	0.36	1.56
Zakum, upper (UAE)	0.8591	2.00	0.097		8.77	4.42
Mubarek (UAE)	0.8392	0.62		3.20	1.23	
Aboozar (Iran)	0.8927	2.48	0.160	18.30	71.00	
Bahrgansar (Iran)	0.8916	2.45	0.260	20.40	72.00	7.90
Dorrood (Iran)	0.8565	2.35	0.054	5.65	23.00	5.00
Foroozan (Iran)	0.8686	2.50	0.067		36.00	6.50
Iranian Heavy (Iran)	0.8707	1.73	0.230	9.40	116.00	5.25
Iranian Light (Iran)	0.8554	1.35	0.170	6.40	35.00	3.50
Rostam (Iran)	0.8447	1.55	0.070	3.30	16.50	2.84
Salmon (Iran)	0.8549	1.91	0.087	5.60	11.60	4.10
Soroosh (Iran)	0.9452	3.30	0.260		101.00	12.00
Basrah Heavy (Iraq)	0.9052	3.50		4.00	90.00	9.73
Basrah Light (Iraq)	0.8559	1.95	0.095	6.50	18.00	4.18
Basrah Medium (Iraq)	0.8696	2.58		12.60	50.00	5.40
Kirkuk (Iraq)	0.8487	1.97	0.120	4.61	29.00	3.80
North Rumaila (Iraq)	0.8559	1.98		7.50	27.00	4.80
Kuwait Export (Kuwait)	0.8680	2.52	0.120	9.78	30.00	5.30
Burgan (Kuwait)	0.9134	3.37		42.20	34.00	7.70
Rataawi (Kuwait)	0.9094	4.24	0.240	80.00		39.00
Eocene-Wafra (Kuwait/Saudi)	0.9420	4.55		315.00	59.20	8.90
Hout (Saudi Arabia)	0.8606	1.91		6.03	28.00	5.01
Khafji (Kuwait/Saudi)	0.8838	2.85		17.50	55.00	8.04
Ratwai (Kuwait/Saudi)	0.9123	4.07		36.50	55.00	9.50
Oman Export (Oman)	0.8427	0.79		8.20	9.50	3.35
Dukhan (Qatar)	0.8203	1.27		5.28	2.26	1.70
Qatar Marine (Qatar)	0.8442	1.42		5.93	7.00	2.72
Arab Light (Saudi Arabia)	0.8575	1.79	0.098	6.25	20.50	3.58
Arab Extra Light (Saudi Arabia)	0.8352	1.19	0.042	3.86	1.10	1.83
Arab Medium - Kharasaniah (Saudi Arabia)	0.8712	2.40	0.137	9.41	22.00	5.44
Arab Medium - Zulf/Marjan (Saudi Arabia)	0.8696	2.48	0.137	11.01	43.00	5.05
Arab Heavy (Saudi Arabia)	0.8871	2.85	0.168	19.00	57.00	7.9
Mubarek (UAE, Sharjah)	0.8392	0.62		3.20	1.23	

S - Sulfur
N - Nitrogen
H_2S - hydrogen sulfide gas
Ni - Nickel
V - Vanadium

3.10.2. Carbon emission (CO_2/CO/C) in the smoke plume

One of the major components of smoke and a combustion product is carbon dioxide (CO_2). When crude oil is burned, it is largely converted to carbon dioxide and water. Therefore, the emission of carbon dioxide may effect combustion efficiency. In addition, CO_2 is the major contributor to global warming, commonly known as the greenhouse effect. Because of these reasons, the emission of CO_2 from the oil fires in Kuwait was of particular concern to environmental scientists. As a matter of fact, pre-war doomsday predictions of the Kuwait oil fires were based on the projected impact of the emission of very large quantities of smoke and CO_2 that might be produced from this catastrophe and injected high enough to the stratosphere to seriously effect global warming. During the early days of the Gulf conflict, numerical models were applied to assess the production of CO_2 from the Kuwait oil fires.

Carbon monoxide (CO) is a minor component of total carbon emission in the Kuwait oil fires. The fraction of CO will increase under limited air or oxygen conditions. Carbon monoxide will ultimately be converted to CO_2. Appreciable quantities of elemental carbon or soot were expected to be produced from the Kuwait oil fires. The rate of soot production has already been discussed in Section 3.8. Many investigators estimated total carbon emissions from the Kuwait oil well fires. Some of these estimates are listed in Table 3.10. According to this table, the estimates of soot production vary widely. The data listed in Table 3.10 were extracted from numerical investigations. Different investigators assumed different daily burn rates and combustion efficiencies for the fires. Data on oil burn rates and soot production have already been discussed in Section 3.8.

Many investigators estimated carbon emission rates based on airborne observations of the smoke plume. Hobbs and Radke (1992) measured

Table 3.10. Emission estimate of burning oil wells in Kuwait.

Reference	Burn Rate (mbd)	Total C (ttpd)	Soot (ttpD)	Total S (ttpd)	Total NO_X (ttpd)
Johnson et al. (1991)	3.9	442	17.5	16.7	1.2
Browning et al. (1991)	1.5	164	13.7	5.5	1.4
Small (1991)	1.6	173	15.9		
Bakan et al. (1991)	3.1		44.1		
Hahn (1991)	1.5–3.0		1–22	11–21	3.5–7
WMO (1991)	2.4		22.4	0.54	
Horgan (1991a)	6.0	100	50		
Barnaby (1991)	4.3	328	25	10.2	2.55
El Desouky and Abdulraheem	6.0			55–65	0.5–3.0

mbd – Million barrels per day
S – Sulfur
ttpd – Thousand tonnes per day
C – Carbon
NO_X – Nitrogen oxide gases

concentrations of CO in the smoke plume about 80 km from the Kuwait City and found them to be below the typical urban levels in the United States. They reported concentrations of CO and CO_2 as 0.5 and 83 percent, respectively, in the super composite (where smoke from all oil fields mix) plume. They calculated an emission rate for CO_2 and CO as:

CO_2 Emission (tonnes per day)	
Greater Burgan Oil Field	1.7 million
Northern Oil Fields	0.13 million
Total Emission	1.8 million
CO Emission (tonnes per day)	
Greater Burgan Oil Field	0.0089 million
Northern Oil Fields	0.0012 million
Total Emission	0.0103 million

Based on the airborne measurements of the British Meteorological Office, Jenkins et al. (1991) calculated a total CO_2 emission of 161 million tonnes per year, however, the uncertainty in this figure was large, i.e. about 40 percent. The estimates of Jenkins et al. (1991) is lower than that of Hobbs and Radke (1992). The above estimates were based on data collected during airborne investigations that were undertaken during March–April 1991. At that time, accurate estimates of crude oil emission and burn rates were not established. Later most of this information was provided by the Kuwait Oil Company (EPC, 1991; El Desouky and Abdulraheem, 1991). From the chronological data of fire fighting and data related to crude oil emissions, the total crude oil discharge was estimated at 1.12 billion barrels. Using this estimate of crude oil emission and burning emission rates from Hobbs and Radke (1992), the following amounts of CO_2, CO, and elemental carbon emissions were calculated.

Total emission of CO_2	133 million metric tonnes
Total emission of CO	0.8 million metric tonnes
Total emission of elemental carbon	8 million metric tonnes

The quantities of CO_2 and CO emitted by the Kuwait oil fires were about 2% of the global emission from fossil fuel and biomass burning and 0.1% of the worldwide emission from all sources. Similar conclusions on the global contribution of CO_2 and CO were reached by Johnson et al. (1991). They speculated that CO_2 produced by the Kuwait oil well fires would only be about 3% of the total annual fossil fuel emissions. From the foregoing, it is concluded that:

1. Carbon dioxide accounted for over 83% of the emission products of Kuwait oil fires. Concentrations of CO were more than 150 times lower than those of CO_2 suggesting good overall burning efficiency.

2. The total quantities of CO_2, CO, and elemental C (largely soot) produced by the Kuwait oil fires were 133, 0.8, and 8 million metric tonnes, respectively. These calculations are based on a total oil discharge of 1.12 billion barrels and CO_2, CO and elemental C emission rates at 83, 0.5 and 5% of the oil burned, respectively.
3. The total emission estimates calculated above are lower than those estimated by the numerical models and real-time measurements. This is largely because of the differences in the amount of oil discharge assumed by the various investigators.
4. Total emission estimates suggest that CO_2 emission from the Kuwait oil fires would have a negligible effect on global warming.

Concentrations of different organic compounds in the smoke plume were dependent on many factors. As expected, concentrations of CO_2 and CO in the plume decreased with an increase in distance from the fire source. A typical distribution of these and other gases as a function of distance is shown in Figure 3.7. The decrease in emission rates was more prominent for CO than for CO_2. Daum et al. (1992) determined organic constituents in the aerosol near the fire source and aged smoke. They found that about two-third of this fraction was elemental carbon, and the remaining one-third was other organic compounds. Concentrations of carbonate-carbon was found to be negligible (Daum et al., 1992). In addition to distance, concentrations of organic compounds were also effected by their source of origin. For example, comparing the smoke plumes of

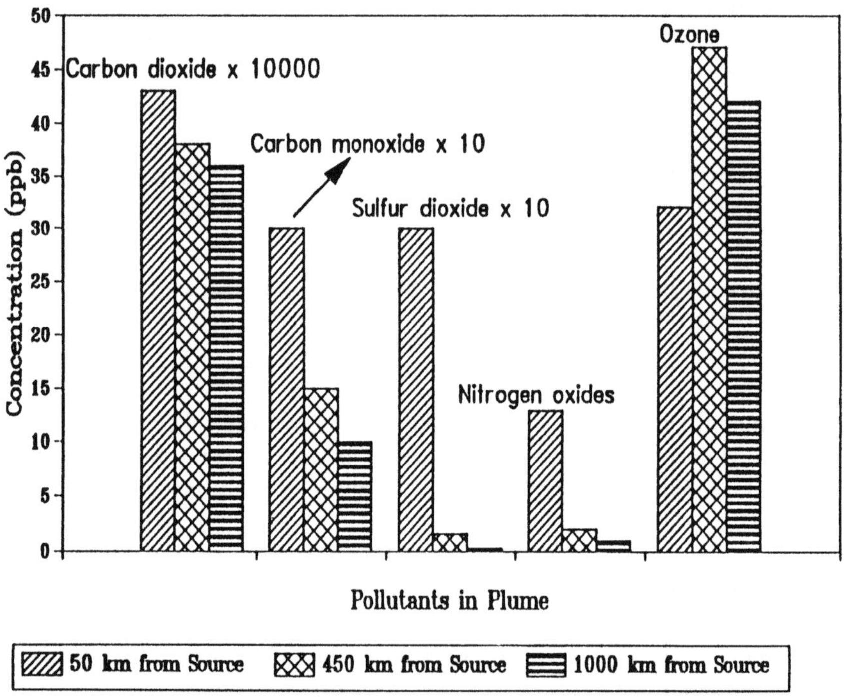

Figure 3.7. Concentrations of gaseous pollutants in smoke plume of the Kuwait oil fires.

different origin, Cofer (1991) found that oil lake fire smoke contained the lowest concentrations of CO_2 but that CO levels were the highest. He also found higher concentrations of methane and non-methane hydrocarbons in the lake fires the smoke plume. He concluded that combustion in the lake fires was not efficient.

Total CO_2 emission was estimated to be <2% of the global emission from the fossil fuel and burning biomass. It is, therefore, expected that the effects of burning oil wells in Kuwait may be significant on regional scale but those on a global scale, including the greenhouse effects, are likely to be insignificant.

3.10.3. Sulfur emissions ($SO_2/H_2S/SO_4$) in the smoke plume

Sulfur is present in almost all crude oil. Its emission rate in the smoke plume is largely controlled by its concentration in the burning crude. Information on sulfur concentrations in the Arabian Gulf crude oils is summarized in Table 2.10. Total sulfur content varied between 0.6 to 4.55% in these crude oils. Kuwait crude oil contained total sulfur concentrations from 2.52 to 4.55%, a relative high sulfur concentration. Crude oils from Kuwait also vary in hydrogen sulfide concentration. Concentrations range between 9.8 to 315 mg/ml (parts per million). Therefore, large quantities of sulfur were expected to be emitted from the Kuwait oil fires.

In developing a post-war scenario of the probable environmental impact of the Kuwait oil fires, many investigators calculated sulfur emissions (Table 3.10). This table also contains some of the results of airborne measurements over Kuwait. A review of the data in Table 3.10 indicates that sulfur emission estimates vary widely. Each investigator calculated emission rates using different assumptions. The British Meteorological Office collected data on the real-time concentrations of SO_2 and H_2S in the smoke plume of Kuwait oil fires in March 1991 (Johnson et al., 1991). From these measurements, Johnson et al. (1991) estimated a sulfur emission of 5.7 million tonnes per year. Hobbs and Radke (1992) reported an average concentrations of SO_2 in the plume 80 km downwind of Kuwait City on May 30, 1991, as 106 ppb (parts per billion) by volume, slightly above the background concentrations at that time. Concentrations of H_2S were slightly elevated above background levels because of its efficient combustion in the plume. Hobbs and Radke (1992) reported that SO_2 emissions were about 1.6% of the fuel burned. On May 30, 1991, Hobbs and Radke (1992) estimated SO_2 emissions as:

SO_2 emission (metric tonnes per day)	
The Greater Burgan Oil Field	19,000
Northern Oil Fields	800
Total Emission (daily) from all fires	20,000
Total emission from all fires for all time	2.3 million tonnes@

@This value was based on the emission of 1.12 billion barrels total crude oil spillage during the Kuwait oil fire episode. Kuwaiti crude oils were assumed to contain an average of 3% sulfur.

Calculations of a sulfur budget for the Kuwait oil fires, based on SO_2 concentrations, showed that a large portion of the sulfur in the crude oil was still unaccounted. This suggests that probably a large portion of sulfur was oxidized and ultimately scavenged from the atmosphere. From the airborne data, Hobbs and Radke (1992) found that concentrations of SO_2 decreased rapidly with time and distance from the fire source. This behavior of SO_2 is shown in Figure 3.7. They concluded that depletion of SO_2 in the plume was due to its transformation to SO_4 rather than dilution. They proposed that oxidation of SO_2 was rapid. The rate of oxidation of SO_2 during the first 3–4 hours of travel in the atmosphere was reported to be about 50% of the total emissions. Similarly, Johnson et al. (1991) also hinted of the possible early oxidation of SO_2 to SO_4. It was generally suspected that SO_2 would probably convert to SO_4 due to oxidation during combustion. The above assessment of SO_2 oxidation was vividly refuted by later plume measurements by a team of scientists from the United States Department of Energy, Brookhaven National Laboratory (USA), Battelle Pacific Northwest Laboratory (USA), and Saudi Aramco Oil Company (Saudi Arabia). They investigated the oxidation of SO_2 in the plume about 100 km from the source by using chloride as a tracer (Daum et al., 1992). They determined the chemical composition of particulates collected in and out of the flame and used these data to conclude that the oxidation rate of SO_2 was very slow, less than 1% per hour as compared to about 50% as proposed by Hobbs and Radke (1992). In their discussion, they argued that chloride was not reactive and could be used as a tracer to allow the calculations of the relative abundance of sulfate, calcium, SO_2, and elemental carbon (with respect to chloride). Based on these calculations, they argued that most of sulfate in the plume, especially that of aged smoke, was a result of mixing of background air with higher concentrations of sulfate. They concluded that the oxidation of SO_2 to SO_4 made a negligible contribution to total sulfate in the plume, especially in the super composite and aged smoke plumes. However, Daum et al. (1992) did not propose a convincing mechanism to account for the large missing amounts of would-be SO_2. With the objective to resolve this controversy we present our explanation. In attempting this, background information will be reviewed first and the question will then be addressed.

It was reported that the color of the smoke plume from Kuwait oil fires varied from dark black to whitish-gray. It was also agreed (Daum et al., 1992; Hobbs and Radke, 1992) that plume color was dependent on the content of salts and brine in it, an increase in mixing of brine from the oil fields with the emitted crude enhancing the whitish tinge of the plume. Data on the chemical composition of brine will help to critically evaluate some of the data reported by Daum et al. (1992). Unfortunately, information on the chemical composition of brine in Kuwait oil fields could not be found. Assuming that the majority of brine follows a general pattern in chemical composition, though it may vary in specifics, concentrations of selected cations and anions in the Arab-D brine were estimated as follows:

Cations (mg/l brine or ppm)@	
Sodium	38690
Calcium	15620
Magnesium	1710
Potassium	1864
Strontium	838
Anions (mg/l brine or ppm)@	
Chloride	94660
Sulfate	290
Carbonate	138
Bromide	904

@ Data taken from KFUPM/RI (1991b)

These data show that concentrations of chloride and sodium are several times higher than those of sulfate and calcium. We discussed the question of chemical composition of brine with many petroleum engineers and geologists. All of them stated that, in general,

1. Brine contained very high concentrations of chloride – several times higher than sulfate;
2. Concentrations of sodium are several times higher than those of calcium;
3. In brine, combined concentrations of sodium and chloride are always many times higher than those of calcium and sulfate.

Information on metal concentrations in Kuwaiti crude oils were also collected from the literature and these data are shown in Table 3.11.

Table 3.11. Metal content (mg/kg) of typical Kuwait crude oils@.

Metal	Rataawi	Burgan	Kuwait Export
Vanadium	39	31	32
Nickel	19	10	6
Potassium	2	2	1
Sodium	1	1	2
Copper	0.1	0.2	0.2

@ Data taken from Al Besharah (1992)

Calcium concentrations in the crude oils oils are probably low, and therefore have not been reported. Cahill (1992) also found minor quantities of sodium chloride, but not any calcium salts, in laboratory combustion experiments conducted on eight samples of Kuwait export crude oil taken several years before the Gulf crisis. It can be concluded that concentrations of sodium in the Kuwait crude oils are higher than those of calcium. If the oil reservoir is calcareous, calcium may be found in appreciable concentrations, but certainly not enough to counter balance the concentrations of sodium coming from the brine in the case of the Kuwait oil fires.

An additional factor to consider was the nature of the fire fighting operation. Huge amounts of salty water (sabkha water, groundwater, or seawater) were pumped over the well head to cool it. During this operation, an appreciable amount of salty water would have evaporated thus incorporating inorganic salts into the solid fraction of the smoke plume. These waters contained very high concentrations of chloride and sodium as compared with calcium.

From the foregoing, it is concluded that inputs of chloride and sodium to the Kuwait oil fires were several times higher than those of calcium and sulfate. Relatively higher concentrations of sodium and chloride should, therefore, be expected in the emission products. Daum et al. (1992) collected plume particles on filters from the northern border of Saudi Arabia to Oman. Concentrations of chloride, sodium, calcium and sulfate were determined along with other constituents. Surprisingly, and contrary to all expectations, they found calcium concentrations almost double those of sodium. The sulfate concentrations in the plume near the source and aged smoke were as high as those of chloride. The unexpected analytical results of Daum et al (1992) on which they based all their calculations and later argument leave only two possibilities: either the analytical data were wrong or sodium and chloride have been scavenged from the plume much earlier than the start of their plume particulate sampling. We do not question the integrity of analytical data of Daum et al. (1992); however, sodium and chloride might have been deposited by dry deposition processes.

There are reports suggesting that soot particles in the plume were coated with sodium chloride salt (Hobbs and Radke, 1992; Cofer et al. 1992; Daum et al. 1992; Parungo, 1992; Cahill, 1992; Johnson et al., 1992). In general, total particle concentrations within a few km of the oil fires were around 10^5 per cm^3 up to 5000–15,000 per cm^3 at a distance of about 300 km. From these reports and data listed above, it can be assumed that:

1. Chloride would be the most abundant element, followed by sodium, calcium, and sulfate. Contrary to this logic, Daum et al. (1992) found major constituents including calcium, sulfate ~ chloride, and sodium in the aerosol samples collected near Kuwait (over 100 km away from the fire source) and in the aged smoke.
2. Soot particles were mainly coated with sodium chloride although they contained about 8% of sulfate.
3. Soot particles were scavenged heavily through dry deposition within a few kilometers of the fires. Airborne investigation reports suggest that over 85% of the particles were removed within 300 km and about 50% within 10 kilometer of the fires.

From the foregoing, it seems that both groups (Hobbs and Daum and their respective associates) were right. Hobbs and Radke (1992) and Cahill (1992) made their aerosol observations closer to the fire sources than those of Daum et al. (1992), who took their measurements between about 150 km to 800 km from the fire sources in Kuwait. Assuming an average speed of 4 meters per second (a predominant wind speed during August), the plume would have taken about 7 hours to reach the nearest location sampled by Daum et al. (1992). Hobbs and

Radke's (1992) observations were the closest possible to the source. They proposed that the majority of sulfur oxidation would have taken place in the first few hours after combustion. We believe that the oxidation of SO_2 to SO_4, as suggested by Hobbs and Radke (1992), was indeed rapid. Most sodium, chloride, calcium, and sulfate reacted to form sodium chloride and calcium sulfate. Relatively high temperatures near the fire source might have catalyzed the formation of these salts on soot surfaces. As mentioned above, a good portion of the soot would have been dry deposited in the vicinity of the fires. This phenomenon of rapid oxidation of SO_2 to SO_4 and the subsequent removal from the plume through dry deposition results in the elimination of most of sodium, chloride, calcium, and sulfate from the plume that was originally present in the crude. This is probably why most of the airborne investigations failed to measure high concentrations of SO_2 in the plume. Early removal of sulfate due to dry deposition resulted in lower estimates of sulfur emission by airborne investigations as compared with those predicted by the numerical models. As the plume was transported a greater distance, certainly there was mixing of background air with the very high concentrations of calcium carbonate. These processes (mixing and decreased sulfur oxidation) seem responsible as emphasized by Daum et al. (1992), i.e., the oxidation of SO_2 was not rapid in the plume of Kuwait oil fire and the enhancement in SO_4 concentrations in the plume particles can be explained based on the mixing of background soil.

The following points are based on the foregoing:

- Total emissions of SO_2 from Kuwait oil fires is estimated to be 2.3 million tonnes. This calculation was based on a total oil emission of 1.12 billion barrels and an SO_2 emission rate of 1.6% of the oil burned as calculated by Hobbs and Radke (1992).
- Sulfur dioxide gas emitted by the fires was rapidly oxidized to SO_4. This decreased ambient concentrations of SO_2 to almost background levels within 4 to 5 hours after its emission. The area of high SO_2 concentrations is estimated to be within 100 km downwind of the sources. Measurements made by Daum et al. (1992) between about 150 and 800 km away from the Kuwait oil fires were probably not suitable to indicate rapid oxidation of sulfur to sulfate.
- Hydrogen sulfide concentrations even close to source were found to be near background levels which suggest that it was efficiently burned in the fires.
- Concentrations of SO_2 in the smoke plume were not sufficiently large to account for SO_4 oxidation. However, SO_4 concentrations were enhanced in the dust particulates. It is suspected that oxidized SO_4 might have been adsorbed by the dust particles and ultimately was scavenged and dry deposited. Considering the composition of brine and the major inorganic constituents one would expect in the plume, it is proposed that most of sulfur as sulfate, sodium, and chloride were scavenged from the plume in the vicinity of the source and deposited to the ground.

3.10.4. Nitrogen oxide emission (NOx) in the smoke plume

As shown in Table 3.9, crude oils from the Gulf region contained appreciable concentrations of nitrogen (a mean of about 0.2 percent). Therefore, NO_x could be expected as a combustion product of Kuwait oil fires. Because of potential of NO_x to contribute to the depletion of ozone in the stratosphere, there was considerable interest in the estimation of NO_x emission rates for the Kuwait oil fires.

Both model calculations and real-time measurements were carried out to assess NO_x emission from the Kuwait oil fires. A brief summary of estimated emission rates derived from the numerical calculations is given in Table 3.10. These estimates vary widely from 500 to 7000 tonnes per day. The basic assumption regarding rates of oil emission seems to be responsible for the discrepancies among the various models.

Starting in March 1991, several aircraft flights were undertaken to study the physical and chemical characteristics of the smoke plume from the Kuwait oil well fires. On May 30, 1991, the scientific team from the University of Washington (NCAR) measured 9.1 parts per billion (by volume) of NO_x in the plume about 80 km downwind from the sources. This level was slightly above background concentration (about 1 ppb) in the atmosphere at that time. Hobbs and Radke (1992) estimated the NO_x emission rate at 0.05% of the fuel burned. Their observations also suggested the rapid depletion of NO_x in the plume at a rate of about 60% per hour. The implication of these observations was that the elevated levels of NO_x could only be expected in a small area around the fire sources. Moreover, NO_x seems to have reacted with other components of the plume. Hobbs and Radke (1992) provided evidence of near-source ozone depletion and suggested that a chemical reaction between NO_x and ozone was responsible for this depletion. The depletion of NO_x as a function of distance is shown in Figure 3.7. As discussed above, concentrations of NO_x decreased rapidly with an increase in distance from the source. From the airborne measurements, Johnson et al. (1991) observed increased activity of photochemical oxidation and thus, relatively lower emissions of NO_x than those predicted by Browning et al. (1991). Because of the rapid removal of NO_x via photochemical reactions, Johnson et al. (1991) argued that NO_x emission from the Kuwait oil fires is unlikely to influence the global NO_x background or global troposphere ozone.

A NO_x emission rate of 0.05% of the burned fuel is adopted as a reference in this book. Considering a total oil emission during the course of Kuwait oil fires of 1.12 billion barrels and the above NO_x emission rate, a total of 80,000 metric tonnes of NO_x was possibly released to the atmosphere due to the burning oil wells in Kuwait. This estimate is much lower than those proposed in the numerical models (Table 3.10).

From the foregoing, it is concluded that concentrations of NO_x in the plume were very low and comparable to background levels about 80 km downwind from the fire sources. It seems that NO_x was involved in the depletion of ozone in the vicinity of the fire source. A total of 80,000 metric tonnes of NO_x has been estimated to have been emitted from the burning oil wells in Kuwait. It is anticipated that NO_x emission from the Kuwait fires poses no global or even regional environmental threat.

3.10.5. Ozone (O_3) depletion in the smoke plume

Global ozone depletion in the stratosphere is one of the most pressing environmental issues of today. It was reported that crude oils from Kuwait contain about 0.2% nitrogen. During the early days of the Gulf crisis, it was feared that emission of large quantities of NO_x from the Kuwait oil fires might be lofted to the stratosphere along with the smoke plume where, in the presence of solar radiation, it would react with ozone, thus destroying the already depleted ozone layer in the stratosphere. This situation could have long-term global implications. Fortunately, in general, the plume remained below 5 km altitude and was never carried to stratosphere (about 13 km altitude) to any appreciable extent. Thus, the threat of global depletion of ozone was averted.

It was suspected that the depletion of ozone in the atmosphere and troposphere may induce regional environmental consequences. To assess this threat, concentrations of ozone in the plume were measured by the airborne research team from the University of Washington, USA. After analysis of these data, Hobbs and Radke (1992) reported that, near the fire sources ozone was found to be depleted probably due to its reaction with NO_x within a few kilometers of the source. Farther downwind, ozone was further depleted probably because of its reaction with alkanes and possible adsorption on the soot particles. In the diffused region of the smoke, where ultraviolet radiation was greater, ozone was regenerated. Hobbs and Radke (1992) reported an average rate of ozone production for the first 3.6 hours of plume transport to be 1.8 parts per billion (by volume) per hour. During a research flight by the British Meteorological Office, Johnson et al. (1991) investigated the vertical and horizontal distribution of ozone in the smoke plume. They found that in the lower layers of smoke, NO_x and O_3 were negatively correlated, indicating gas phase titration of O_3 with NO_x, but little horizontal variation in O_3 and NO_x. Conversion of O_3 to NO_x is largely a photochemical reaction. Because of the smoke cloud, radiation reaching the lower layers of smoke was probably not sufficient to initiate oxidation of NO_x by O_3. This may be largely responsible for the observed profile of NO_x and O_3 in the lower layers of smoke. In the upper layers of the smoke plume, O_3 and NO_x were positively correlated, suggesting oxidation of NO_x by O_3. These observations suggest the high degree of dependency of the O_3 reaction with NO_x. Enhanced concentrations of O_3 were found in the far-field (i.e., about 1000 km from the sources). Similar observations were noted by Daum et al. (1992).

From the foregoing, it is concluded that :

1. An initial depletion of O_3 in the near-source region was primarily due to its reaction with NO_x and secondarily to its reaction with hydrocarbons.

2. Because of the inability of plume to rise to stratosphere, the depletion of O_3 by photochemical reaction with NO_x emitted by the Kuwait oil fires poses no global environmental problem.
3. Real-time measurements suggest that radiation was a critical factor in the depletion of O_3 by NO_x.
4. No regional or local environmental effect is expected from O_3 depletion as a result of Kuwait oil fires.

3.10.6. Hydrocarbon emissions from the Kuwait oil fires

Hydrocarbons constitute a major proportion of crude oil. With complete combustion, these hydrocarbons are converted to CO_2 and water. A portion of the parent hydrocarbons in the crude oil or their intermediary combustion product may be released during incomplete combustion. The emission rate of hydrocarbons is, therefore, controlled by combustion efficiency, since efficient fires emit less of hydrocarbons. In general, the combustion efficiency of Kuwait oil fires was good (over 83 percent). However, there was still the potential to release large quantities of volatile and non-volatile organic hydrocarbons. Concentrations of hydrocarbons in the plume were determined during airborne investigations of the smoke plume. Johnson et al. (1991) measured concentrations of non-methane hydrocarbons (NMHC) in three smoke samples taken from the thickest part of the plume on March 30, 1991. They observed a decrease in concentrations of all seven hydrocarbons analyzed (ethane, ethene, propane, propene, benzene, toluene, n-hexane, and NMHC). Ethane and propane were found at the highest concentrations in the plume. These data are plotted in Figure 3.8. The airborne research team of the University of Washington also measured NMHC in the plume. Hobbs and Radke (1992) reported that the concentrations NMHC in the plumes ranged from 55 to 827 ppb of carbon. This is typical of a clean to moderately polluted atmosphere. The NMHCs were dominated by alkanes (65 to 82% of NMHC on a per carbon basis) followed in descending order by alkenes, or olefins (made up to 7 to 23% of NMHC), aromatic hydrocarbons (concentrations range between 5 to 19% of NMHC on a per carbon basis) and alkynes (about 1 to 7% of NMHC).

Concentrations of methane (CH_4) were slightly elevated in the smoke plume about 80 km downwind of Kuwait City on 30 May 1991 (Hobbs and Radke, 1992). An emission factor of 0.16% of the oil burned was reported by Hobbs and Radke (1992). Using this emission factor, we calculated a total emission of about 0.3 million tonnes of methane from the burning oil wells in Kuwait. The emission of methane is not sufficiently large so as to induce a global, regional, or local environmental impact. Moreover, its effect on global warming is expected to be negligible.

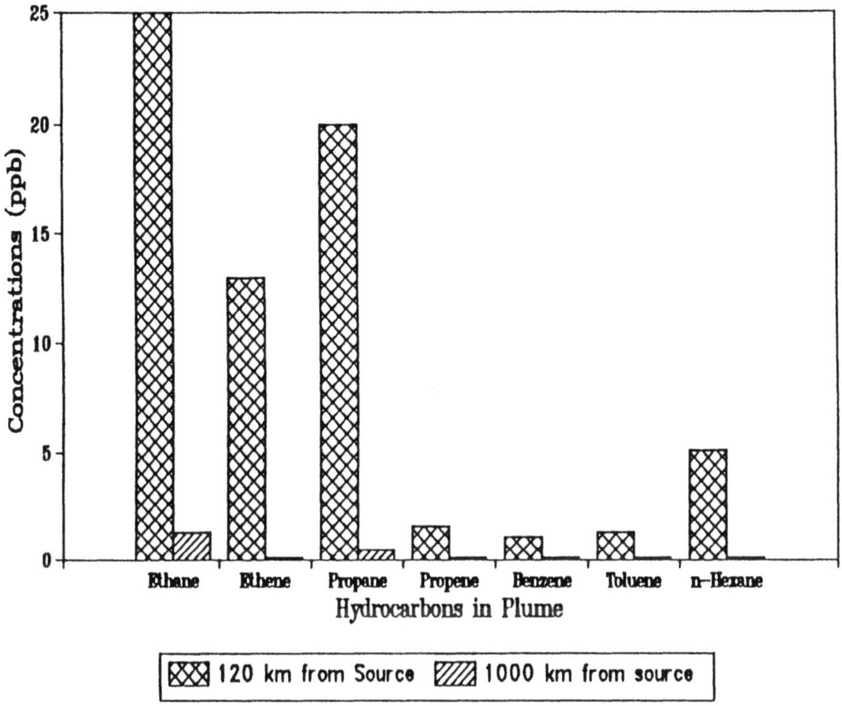

Figure 3.8. Concentrations of volatile hydrocarbons in smoke plume of the Kuwait oil fires.

References

Al Besharah, J. 1991. The Kuwaiti oil fires and oil lakes – facts and numbers. In Environmental and Health Impact of the Kuwaiti Oil Fires. Proc. of an International Symp. held at the University of Birmingham, 17th October, pp. 12–15.

Al Hassan, J.M. 1991. The Iraqi Invasion – Environmental Catastrophe. In Environmental and Health Impact of the Kuwaiti Oil Fires. Proc. of an International Symp. held at the University of Birmingham, 17th October, pp. 6–11.

Arab Oil and Gas Directory, 1990. Kuwait: Arab Petroleum. Research Center, Paris France, pp. 201–232.

Bakan, S., Chlond, A. , Cubasch, U., Feichter, J., Graf, H., Grassel, H., Hasselman, K., Kirchner, I., Latif, M. Roeckner, E., Sausen, R., Schlese, U. Schriever, D. Schult, I., Schumann, U., Sielmann, F., and Welke, W. 1991. Climate response to smoke from the burning of oil wells in Kuwait. Nature 351:367–371.

Barnaby, F. 1991. The environmental impact of the Gulf war. Ecologist 21:166–172.

Blanchet, J.P. and Vupputuri, R.K.R. 1991. Assessment of short term climate effects of Kuwait oil fire smoke using numerical models. World Meteorological Organization. Global Atmosphere Watch. No 72, Draft Report of the WMO Meeting of Experts on the Atmospheric Part of the Joint UN Response to the Kuwait Oil field Fires, Geneva, April 27–30, pp. 44–46.

Bompay, f. 1991. Operational capabilities of METEO France for the emergency response to the Kuwait oil field fires. World Meteorological Organization. Global Atmosphere Watch. No 72, Draft Report of the WMO Meeting of Experts on the Atmospheric Part of the Joint UN Response to the Kuwait Oil field Fires, Geneva, April 27–30, pp. 37.

Browning, K.A. , Allam, R.J., Ballard, S.P., Barnes, R.T.H., Bennetts, D.A. , Maryon, R.H.,

Mason, P.J., McKenna, D., Mitchell, J.F.B., Senior, C.A. , Slingo, A., and Smith, F.B. 1991. Environmental effects from the burning oil wells in Kuwait. Nature 351:363–367.

Cahill, T.A. 1992. Aerosol measurements in the Kuwaiti oil fire plume. pp. A14–A15. Paper presented at the WMO Meeting of Experts on the Atmospheric Part of the Joint UN Response to the Kuwait Oil field Fires, Geneva, April May 25–29, Final Report on the WMO meeting of experts on the atmospheric part of the Joint UN. Response to the Kuwait Oil field Fires. Geneva, April. 27–30.

Carmichael, G. 1991. Local and regional air quality modeling tools. World Meteorological Organization. Global Atmosphere Watch. No 72, Draft Report of the WMO Meeting of Experts on the Atmospheric Part of the Joint UN Response to the Kuwait Oil field Fires, Geneva, April 27–30, pp. 65–66.

Chase, R.R.P. and Bass, A. 1991. Regional models of the health and environmental impact of the Kuwaiti oil fires. World Meteorological Organization. Global Atmosphere Watch. No 72, Draft Report of the WMO Meeting of Experts on the Atmospheric Part of the Joint UN Response to the Kuwait Oil field Fires, Geneva, April 27–30, pp. 38.

Cofer, W.R. III, (1991). NASA/EPA helicopter operation: Kuwait oil fires. An Internal Rept. Atmospheric Science Division, NASA-Langley Research Center, Verginia, USA.

Cofer, W.R. III, Stevens, R., Winstead, K., Pinto, J., Sebaacher, D., Abdulraheem, M., Al-Shafi, M., Mazurek, M., Rasmussen, R., Cahoon, D., and Levine, J.S. 1992. Kuwait oil fires: composition of source smoke. Cited by Daum et al. (1992).

Cook, W. 1991. Putting out the fires: battle under the sand. US News and World Report, March 11.

Cutter, S.L. Ecocide in Babylonia. Summer 41:26–31.

Daum, P.H., Al-Sunaid, A. , Busness, K.M., Hales, J.M., and Mazurek, M. 1992. Studies of the Kuwait oil-fire plume during mid-summer 1991. Final Report, US. Department of Energy, Brookhaven National Laboratory, Battelle Pacific Northwest Laboratory, and Saudi Aramco Oil Company, Saudi Arabia.

El Desouky, M and Abdulraheem, M.Y. 1991. The impact of oil well fires on the air quality in Kuwait. In Environmental and Health Impact of the Kuwaiti Oil Fires. Proc. of an International Symp. held at the University of Birmingham, 17th October, pp. 16–26.

El-Baz, F. 1992. The war for oil: effects on land, air and sea. Geotimes 37(5):13–15.

Emerson, S. 1991. When earth takes the hit. International Wildlife, July–August, pp 38–41.

EPC, 1991. State of the Environment Report. A Case Study of Crimes Against the Environment. Environmental Protection Council, Kuwait, pp. 1–49.

Garwin, R.L. and Kendall, H.W. 1991. Quenching the wild wells of Kuwait. Nature 354:11–14.

Gerges, M.A. 1992. Preliminary results of observations on atmospheric pollution resulting from Kuwait oil well fires, collected during the UN Inter-Agency Plan of Action (April–August 1991), United Nations Environment Program, May 15, pp. 1–12.

Graf, H., Appendix X, pp.19–21, Cited by WMO (1991).

Hahn, J. 1991. Environmental effects of the Kuwaiti oil field fires. Environ. Sci. Technol. 25:1530–1532.

Hobbs, P.V. and Radke, L.F. 1992. Airborne studies of the smoke from the Kuwait Oil Fires. Sci. 256:987–991.

Horgan, J. 1991a. Up in flames. Kuwait burning oil wells are a sad test of theories. Scientific Am. May, pp. 7–9.

Horgan, J. 1991b. Burning question. Scientists launch studies of Kuwait oil fires. Scientific Am. 265:17–24.

HPI Consultants Inc. 1987. Crude Oil Assay Handbook. First Edition, Austin, Texas, USA.

Jakobs, H.J., Ebel, A. , Hass, H., and Memmesheimer, M. 1991. Regional model simulation of the long range transport of atmospheric pollutants from the Kuwait oil fires. World Meteorological Organization. Global Atmosphere Watch. No 72, Draft Report of the WMO Meeting of Experts on the Atmospheric Part of the Joint UN Response to the Kuwait Oilfield Fires, Geneva, April 27–30, pp. 68–69.

Jenkins, G.J., Johnson, D.W., McKenna, D.S., and Saunders, R.W. 1992. Aircraft measurements of the Gulf the smoke plume. Weather 47(6):212–220.

Johnson, D.W., Kilsby, C.G., McKenna, D.S., Saunders, R.W., Jenkins, G.J., Smith, F.B., and Foot, J.S. 1991. Airborne observations of the physical and chemical characteristics of the Kuwait oil the smoke plume. Nature 353:617–621.

Kelly, J.M. 1991. Fires over Kuwait. Popular Sci., September, pp. 62–65.

KFUPM/RI, 1992. Gulf Oil Spill 1991. Arabian Gulf Atmospheric Pollution Research Program. 2nd Interim Report. The Research Institute, King Fahd University of Petroleum and Minerals, Dhahran, Saudi Arabia.

KFUPM/RI, 1991a. Gulf Oil Spill 1991. Arabian Gulf Atmospheric Pollution Research Program. 1st Interim Report. The Research Institute, King Fahd University of Petroleum and Minerals, Dhahran, Saudi Arabia.

KFUPM/RI, 1991b. Fundamental Study of Oil field Scaling. Final Report, Volume 5, Solubility Studies. The Research Institute, King Fahd University of Petroleum and Minerals, Dhahran, Saudi Arabia.

Kramer, M. 1991. Kuwait: back to the Past. Time, August 5, 12–18.

Kwarteng, A.Y. and Bader, T. 1992. Monitoring the Kuwait oil fires using satellite data. Proc. Proc. Ist Bahrain International Conference on Environment, 24–26 February, pp. 293–307.

Kwarteng, A.Y. and Bader, T. 1993. Monitoring the Kuwait oil fires using satellite data. Arabian J. Sci. Eng. (accepted).

Lagrone, S.F. 1991. Potential community exposure to toxic chemicals. Environ. Sci. Technol. 25:366–368.

Lameloise, P., Thibaut, G., Masse, R., Lafuma, J., Person, A., and Pasquereau, M. 1991. Measurement Campaign of the Regional Mobile Laboratory for Measurements of Air Quality in Kuwait. Final Report, March 27 to April 4, Paris, France.

Lane, C. 1991. What just lies ahead. pp. 28–29. Newsweek, March 4.

Limaye, S.S., Suomi, V.E., Valden, C., and Tripoli, G. 1991. Satellite observations of smoke from oil fires in Kuwait. Sci. 252:1536–1539.

Linden, E. 1991. Getting blacker everyday. Time, May 27.

Lulla, K. and Helfert, M. 1991. Smoke palls introduced by Kuwaiti oil filed fires mapped from the space shuttle imagery. Geocarto International 6(2):71.

Mage, D.T., El Desouky, M., Lameliose, P., and Thibaut, G. 1991. Pollution de l'air par les puits de petrole en feu au Koweit. Medicine and Hygiene 49:3445–3448.

Manins, P. and Bouma, W. 1991. Burning oil fields in the Middle East. Clean Air 25:47–48.

Marshall, 1991. Nuclear winter from Gulf war discounted. Sci. 251:372.

Meyers, R.E. 1991. Regional atmospheric modeling review and recommendations. World Meteorological Organization. Global Atmosphere Watch. No 72, Draft Report of the WMO Meeting of Experts on the Atmospheric Part of the Joint UN Response to the Kuwait Oilfield Fires, Geneva, April 27–30, pp. 70–72.

National Review, 1991. Black mischief. National Review, February 25.

NOAA , 1992.Kuwait oil fires extinguishing chronology (unpublished). Cited by WMO, 1992.

Okada, K., Ikegami, M. Uchino, O., Nikaidou, Y., Zaizen, Y., Tsutsumi, Y. and Makino, Y. 1992. Kuwait soot over Japan. Nature 355:120.

Parungo, F. 1992. Aerosol particles in Kuwait oil fires. pp. A–16. Paper presented at the WMO Meeting of Experts on the Atmospheric Part of the Joint UN Response to the Kuwait Oilfield Fires, Geneva, April May 25–29, Final Report on the WMO meeting of experts on the atmospheric part of the Joint UN. Response to the Kuwait Oilfield Fires. Geneva, April 27–30.

Peck, L. 1991. The spoils of War. Amicus J. 13:6–8.

Penner, J.E., Atherton, C.S., Dignon, J. Ghan, S.J., Walton, J.J., and Hameed, S. 1991. Tropospheric nitrogen: A three-dimensional study of sources, distribution, and deposition. J. Geophys. Res. 96:959–990.

Penner, J.E. 1991. Kuwaiti oil fires. World Meteorological Organization. Global Atmosphere Watch. No 72, Draft Report of the WMO Meeting of Experts on the Atmospheric Part of the Joint UN Response to the Kuwait Oilfield Fires, Geneva, April 27–30, pp. 40–43.

Pope, C. 1991. War on earth. Sierra May/June, pp. 54–58.
Raloff, J. and Monastersky, R. 1991. Gulf oil threatens ecology, maybe climate. Sci. News 139:71–73.
Rao, U.R. 1991. Gulf il spill and the monsoon. Nature 351:704.
Riva, J.P. 1991. Kuwait oil well fires updated. May 8 CRS Report for Congress.
Sandia, 1991. Operation quick response predicting environmental effects of oil sabotage. Energy & Environment, July, Sandia National Laboratories.
Schumann, U. 1991. Objectives of measurements of soot plumes leeward of Kuwait. World Meteorological Organization. Global Atmosphere Watch. No 72, Draft Report of the WMO Meeting of Experts on the Atmospheric Part of the Joint UN Response to the Kuwait Oilfield Fires, Geneva, April 27–30, pp. 31–33.
Schwartz, J., Manegolb, C.S., Liu, M., Waller, D., and Barry, J. 1991. Blasting down to the wire. Newsweek March 4, 30–31.
Small, D. 1991. Environmental impact of fires in Kuwait. Nature 350:11–12.
Smith, F.B. 1991. Aircraft measurements of the Gulf by the UK-Hercules aircraft: Provisional data; and the application of a multi-particle 3-D Monte-Carlo Model. World Meteorological Organization. Global Atmosphere Watch. No 72, Draft Report of the WMO Meeting of Experts on the Atmospheric Part of the Joint UN Response to the Kuwait Oilfield Fires, Geneva, April 27--30, pp. 39.
Snell, J. 1991. Programs pertinent to Kuwait oil fires impacts. World Meteorological Organization. Global Atmosphere Watch. No 72, Draft Report of the WMO Meeting of Experts on the Atmospheric Part of the Joint UN Response to the Kuwait Oilfield Fires, Geneva, April 27–30, pp. 35–36.
Sullivan, T.J. 1991. Real-time global mutual aid for atmospheric release of pollutants is possible today. World Meteorological Organization. Global Atmosphere Watch. No 72, Draft Report of the WMO Meeting of Experts on the Atmospheric Part of the Joint UN Response to the Kuwait Oilfield Fires, Geneva, April 27–30, pp. 62–63.
Tawfiq, N.I. 1991. Response by Saudi Arabia to the environmental crisis caused by the Gulf war. In Environmental and Health Impact of the Kuwaiti Oil Fires. Proc. of an International Symp. held at the University of Birmingham, 17th October, pp. 40–43.
Touby, F. 1991. The end of a firestorm. MacLean's 104(11):46–47.
Turco, R.P., Toon, O.B., Ackerman, T.P., Pollack, J.B., and Sagan, C. 1990. Climate and smoke: an appraisal of nuclear winter. Sci. 247:166–174.
Verver, Ge 1991. The RIVM/KNMI Puff Dispersion Model. World Meteorological Organization. Global Atmosphere Watch. No 72, Draft Report of the WMO Meeting of Experts on the Atmospheric Part of the Joint UN Response to the Kuwait Oilfield Fires, Geneva, April 27–30, pp. 48.
Warner, S.F. 1991. The Environmental consequences of the Gulf War. Environment 33(5):7–9, 25–26.
Williams, R.S., Heckman, J. and Schneeberger, J. 1991. Environmental Consequences of the Persian Gulf War 1990–91. National Geographic Res. & Exploration 7(Special Issue):1–48.
WMO, 1991. World Meteorological Organization. Global Atmosphere Watch. No 72, Draft Report of the WMO Meeting of Experts on the Atmospheric Part of the Joint UN Response to the Kuwait Oilfield Fires, Geneva, April 27–30. Report on the WMO meeting of experts on the atmospheric part of the Joint UN. Response to the Kuwait Oilfield Fires. Geneva, April 27–30.
WMO, 1992. World Meteorological Organization. Global Atmosphere Watch. No 82, Draft Report of the WMO Meeting of Experts on the Atmospheric Part of the Joint UN Response to the Kuwait Oilfield Fires, Geneva, April May 25–29, Report on the WMO meeting of experts on the atmospheric part of the Joint UN. Response to the Kuwait Oilfield Fires. Geneva, April 27–30.
World Oil, 1991. Kuwait: the mother of all blowouts. World Oil, pp. 38–48.

CHAPTER 4

Air pollution from the Kuwait oil fires: ground level

4.1. Summary

- Kuwait oil well fires peaked during the last week of February 1991. Early numerical calculations indicated that large quantities of primary air pollutants would be produced by the Kuwait oil fires.
- When the Kuwait oil fields were ignited, many national and foreign organizations sent scientific teams to monitor air pollution generated by the burning oil wells.
- Many numerical investigations predicted and airborne investigations reported that the smoke plume was largely suspended between 2 to 5 km altitude and it rarely touched the ground.
- Ground level measurements depicted that the Kuwait oil fires enhanced the ambient concentrations of inhalable air particulates, generally referred to as PM10, in the Gulf region, particularly in Kuwait and northeastern part of Saudi Arabia. On many occasions, the daily mean concentration of PM10 was higher than the air quality guideline proposed by MEPA (1402 H). The excesses were more prevalent during summer (shamal season, a period of north winds) in Kuwait, northeastern Saudi Arabia, and Bahrain than during the preceding spring.
- Airborne observations suggested that a major portion (83 percent) of the emissions from the Kuwait oil fires was carbon dioxide (CO_2), whereas carbon monoxide (CO) was less than 1 percent. All ground level measurements in Kuwait and Saudi Arabia showed CO concentrations below the proposed air quality guidelines. Information on the composition of soot at ground level was lacking.
- It was estimated that about 2.3 million tonnes of sulfur dioxide (SO_2) would have been produced from the Kuwait oil fires. Many national and international organizations measured ground level concentrations of SO_2 in Kuwait, Saudi Arabia, Bahrain, and Iraq. All the measurements were below the air quality limit. Occasionally, deviations from of the air quality guidelines were observed in and around the burning oil fields. A comparison with historical data collected in Kuwait and Saudi Arabia indicated a

significant increase in the annual mean concentration of SO_2 during 1991. This increase was attributed to the Kuwait oil fires. The monthly mean concentrations of SO_2 in the Eastern Province of Saudi Arabia was enhanced during the shamal season. This was the time when smoke was most visible (over 80% of the days) in this area. Concentrations of sulfuric acid and hydrogen sulfide (H_2S) in Kuwait and Saudi Arabia were low and typical of what would be expected in an urban area.

- The data collected by national and international organizations on the ambient concentrations of nitrogen oxides (NO_x) in Kuwait and Saudi Arabia were below the proposed air quality limit. Comparing these observations with the available historical measurements at Dhahran indicated no increase in NO_x that could be attributed to the Kuwait oil fires. Rather the concentrations seem to be affected by local events such as traffic load, etc.
- Concentrations of ozone (O_3) were consistently within the air quality guideline in the Gulf region. The historical data collected by Saudi Aramco suggest no impact of the Kuwait oil fires on the concentrations of this pollutant.
- Analyses of both PM10 and total suspended air particles (TSP) indicated a contribution of the Kuwait oil fires to metal concentrations in air particulates. In general, concentrations of nickel, vanadium, cadmium, molybdenum, titanium, and chromium were significantly ($p < 0.05$) higher in air particulates during the months of May, June, and October at Dhahran during 1991 than those collected in 1983. However, concentrations of lead and barium were lower in the 1991 particulate samples than those collected in 1982. A comparison with data from other parts of the world reveals that there are relatively higher concentrations of nickel, vanadium, and chromium in the inhalable particulates from the Gulf region.
- Many volatile organic chemicals were detected in the ambient air of Kuwait and Saudi Arabia during the period of the Kuwait oil fires. These data suggested no serious atmospheric contamination from these compounds. As expected, volatile organics were higher in Kuwait as compared to Saudi Arabia. Concentrations of volatile hydrocarbons were comparable to the level in the typical urban centers.
- Concentrations of total polycyclic aromatic hydrocarbons (PAH) in PM10 air particulates in Kuwait, Saudi Arabia, and Bahrain ranged from 7 to 387 ng/m^3. The data on individual PAH compounds were highly variable. As expected, the Kuwait atmosphere contained significantly higher concentrations of PAHs as compared with those from Saudi Arabia and Bahrain. Information on PAH concentrations in Iraq, Iran, and other Gulf countries were not available. Concentrations of total PAHs found in many of the air particulate samples from Kuwait and Saudi Arabia were higher than the controversial proposed air quality guideline (10.9 ng/m^3).

4.2. Introduction

Early predictions on the environmental implications of the Kuwait oil fires indicated that large quantities of primary air quality pollutants would be produced as combustion products of the Kuwait oil fires. Initially, the combustion products were supposed to be ejected into the upper atmosphere due to pressure and convectional forces. However, it was feared that inversions might result in the touch down of smoke. This episode, if it were to occur, would contaminate ambient air and might bring about serious short- and long-term human health impacts. Therefore, considerable national and international interest was generated in trying to obtain ground measurements of primary air quality parameters.

Ground level air quality is very crucial to the terrestrial ecosystem, particularly to human health and vegetation. It is generally believed that the Gulf crisis adversely effected ambient air quality, not only in Kuwait but also far from its boundaries including parts of Saudi Arabia, Iran, Iraq, Bahrain, Qatar, and the United Arab Emirates. The Kuwait oil fires were the single most important event that could have caused the deterioration of the ambient air quality in the Gulf region. The effects of the Kuwait oil fires on the air quality of upper atmosphere have already been discussed in Chapter 3. Here, available data on ambient air quality at ground level are summarized and discussed.

4.3. Some facts regarding the Kuwait oil fires

The amount of emission products of the Kuwait oil fires could effect air quality at ground level in the Gulf region. Many aspects of the Kuwait oil fires have been discussed in Chapter 3. The data that would be helpful in understanding the ambient air pollution during the Kuwait oil fires episode are summarized herein.

1. A total of 1.12 billion barrels of crude oil were emitted by oil well fires in Kuwait starting in February 1991 and continuing until the fires were extinguished on November 6, 1991.
2. The total quantities of carbon dioxide (CO_2), carbon monoxide (CO), and elemental C (largely soot) produced by the Kuwait oil fires were 133, 0.8, and 8 million tonnes, respectively (for details see Chapter 3).
3. The total emission of sulfur dioxide (SO_2) from the Kuwait oil fires is estimated to be 2.56 million tonnes (for detail see Chapter 3).
4. The emission of nitrogen oxides (NO_x) was estimated at about 80,000 tonnes.
5. A total of about 0.3 million tonnes of methane was estimated (Chapter 3).

From the foregoing, it can be concluded that the Kuwait oil fires would have altered the ground level air quality in the Gulf region. At the initial stage of the Kuwait oil well fires, it was feared that deterioration of air quality might bring about catastrophic results with perhaps an extensive loss of human lives in the impact area. Available data on the air quality monitoring at the ground level were collected and are discussed here. Only ground measurements are considered in this chapter. Air contamination of upper atmosphere has already been discussed in Chapter 3.

4.4. Potential pollutants from the Kuwait oil fires

The primary air quality pollutants (sulfur dioxide, inhalable particulates, nitrogen oxides, ozone, carbon monoxide, and hydrogen sulfide) are known to induce acute health complications if humans are exposed to high concentrations for prolonged periods. For example, sulfur dioxide (SO_2) is absorbed through the nose and upper airway while breathing which, in turn, produces increased airway resistance in the lung. High doses of SO_2 (i.e., 2–5 ppm or 5000–13,000 $\mu g/m^3$ air) may cause severe impairment of lung functions. The effect of SO_2 at lower concentrations (i.e., 0.5 ppm or 1300 $\mu g/m^3$) include a distasteful odor and irritation leading to airway resistance. Of several oxides of nitrogen are present in ambient air, only two are known to effect human health. These are nitric oxide (NO) and nitrogen dioxide (NO_2). These oxides, if transferred across the lung-blood barrier, can produce inactive forms of hemoglobin. It is estimated that eye and nasal irritation will develop after exposure to about 15 ppm of nitrogen dioxide (NO_2) and pulmonary discomfort will commence after a brief exposure to 25 ppm. Upon entering the respiratory system, CO combines rapidly with blood hemoglobin to form carboxyhemoglobin (COHb) in the lungs. This reduces the capability of hemoglobin to carry oxygen to other body tissues. Sulfur dioxide and nitrogen oxides may oxidize in the presence of water to form sulfuric acid and nitric acid, respectively. Acid aerosol is known to cause respiratory diseases. Most of the above pollutants can be emitted by the Kuwait oil fires during combustion and, therefore, monitoring of these pollutants is important.

In addition to the primary pollutants, combustion can produce many unknown organics. It is well established that crude oil is a naturally occurring product of very complex and variable composition. It contains several hundreds of organic- and inorganic chemicals. The majority of these chemicals will produce new by-products during combustion. Emission products of crude oil combustion are expected to be complex and many of them have human health implications. Potential exposure risks come from smoke, sulfur dioxide, nitrogen oxides, carbon monoxide, aliphatic and polycyclic aromatic hydrocarbons (both volatile and non-volatile), and toxic metals. Air contamination with these chemicals can significantly effect human health. The data collected on the ambient concentrations of the above contaminants and many others will be reviewed herein.

4.5. National/international air quality monitoring programs

Because of the presence of many foreign nationals in the conflict area, the international community decisively moved to study, control, and minimize the environmental impacts of the Gulf war. International efforts were concentrated on fighting the oil spill in the Arabian Gulf and the Kuwait oil fires. The efforts on the Kuwait oil fires can be divided in two: those related to the smoke plume characterization in the upper atmosphere (these have been discussed in Chapter 3) and those related to the ground level measurements of the pollutants (these will be discussed here). The local organizations involved in air quality monitoring at ground level were the Environmental Protection Council of Kuwait (EPC), the Meteorology and Environmental Protection Administration (MEPA) of Saudi Arabia, the Research Institute of King Fahd University of Petroleum and Minerals (KFUPM/RI) in Dhahran, Saudi Arabia, Saudi Aramco at Dhahran, the Environmental Protection Committee of Bahrain (BEPC), and the Environmental Protection Committee of Qatar. The United States Environmental Protection Agency (USEPA), the United States National Oceanic and Atmospheric Administration (NOAA), World Meteorological Organization (WMO), and British Meteorological Office are but a few of the many international organizations who participated in the air quality monitoring campaign. Both the national and international efforts to monitor air quality in the Gulf region are summarized in Table 4.1.

Table 4.1. Air quality parameters that were monitored during the Kuwait oil fires.

Responsible Organization	Location Monitored	Measurement Date	Air Quality Parameters Monitored
EPC[a](Kuwait)	Mansoria Kuwait	April 1991– to date	CO, NO_x, methane, NHC H_2S, SO_2, total S, PM10, O_3, NO, meteorological parameters
EPC[a] (Kuwait)	Reqa Kuwait	May 1991– to date	CO, NO_x, methane, NHC H_2S, SO_2, total S, PM10, O_3, NO
EPC[a] (Kuwait)	Jahra Kuwait	–	CO, NO_x, methane, NHC H_2S, SO_2, PM10, O_3
EPC[a] (Kuwait)	Al Zoor Kuwait	–	CO, NO_x, methane, NHC H_2S, SO_2, PM10, O_3
NIST[b] (USA)	Kuwait City	15/5/1991	smoke, PAH, SO_2, CO, CO_2, particulates
USEPA[c] (USA)	Many sites in Kuwait	March 13– March 27	SO_2, PAH, inorganic acids, volatile organics, metals, PM10 (portable)
AIRPARIF[e] (France)	4 sites in Kuwait	March 27– April 4	SO_2, CO, O_3, NO, NO_2, VOC, PAH
Japanese Team[f] (Japan)	4 sites in Kuwait	April 28– May 5	NO_2, NO_x, particles, SO_2, carbon, metals, anions
NILU[g] (Norway)	Umm Qasr	June 5–12	PM10, PAH, SO_2, NO_x
MEPA[h] (Saudi Arabia)	Subaheea, Kuwait	April – June	PM10, NO, SO_2, H_2S, O_3, NO_x, NO_2, meteorological data

Table 4.1. Continued

Responsible Organization	Location Monitored	Measurement Date	Air Quality Parameters Monitored
MEPA[h] (Saudi Arabia)	Mansoria Kuwait	April May	PM10, NO, SO_2, H_2S, O_3, NO_x, NO_2, meteorological data
KFUPM/RI[d] (Saudi Arabia)	Dhahran	March - Continue	PM10, PAH, TSP, modeling, metals, PAH
MEPA[h] (Saudi Arabia)	Khafji	April- to date	PM10, CO, SO_2, H_2S, O_3, NO_x meteorological data
Saudi Aramco[i] (Saudi Arabia)	7 sites	1982- to date	SO_2, H_2S, NO_x, O_3, PM10 hydrocarbons, meteorological data, metals, solar radiation, etc.
Royal Commission[j] (Saudi Arabia)	Jubail	for several years	SO_2, H_2S, NO_2, O_3, PM10, NMOC, meteorological data
USEPA[c] (USA)	Few sites in Saudi Arabia	March 13- March 27	SO_2, PAH, inorganic acids, volatile organics, metals PM10 (portable)
EPCB[k] (Bahrain)	Bahrain	March- to date	PM10, TSP, SO, H_2S, PAH, metals, meteorological data

CO - Carbon monoxide
CO_2 - Carbon dioxide
NO_x - Nitrogen oxides
H_2S - Hydrogen sulfide
SO_2 - Sulfur dioxide
PAH - Polycyclic aromatic hydrocarbons
AH - Aliphatic hydrocarbons
PM10 - Inhalable air particulates
TSP - Total suspended air particles
NHC - Non-methane hydrocarbons
VOC - Volatile organic compounds
S - Sulfur

[a] EPC (1991): Environmental Protection Council of Kuwait.

[b] Mulholland et al. (1991); NIST = The United States National Institute for Standards and Technology.

[c] USEPA (1991): The United States Environmental Protection Agency. Monitoring locations in Kuwait were: Al Maqua, Al Ahmadi,, Al Burgan, Al Wafra, Al Dhuba, Mina Abdullah, Um Al Haiman, Freedom City, Kuwait City, Adnan Hospital, Mubareek Al Kabeer, Al Farwaniah, Al Jahra. The USEPA monitoring locations in Saudi Arabia were Dhahran and Riyadh,

[d] KFUPM/RI (1992): The Research Institute of King Fahd University of Petroleum and Minerals, Dhahran, Saudi Arabia, operated a monitoring station at Dhahran. In addition, air quality data from other organizations were collected and compiled.

[e] Lameloise et al. (1991) monitored Kuwait City, Al Rawdatayn, Al Ahmadi, and Al Magwa Oil Fields.

[f] Okita (1991): Ahmadi, Riggae, Mansouria, Jahra.

[g] Norwegian Institute for Air Research made measurements near the northern border of Kuwait and Umm Qasr, Iraq (Sivertsen and Berg, 1991). This was the only organization that measured air quality in Umm Qasr, Iraq. In Kuwait, monitoring stations were established in Mansoria, Jahra, Rabiah, and Riga.

[h] Meteorology and Environmental Protection Adminstration (MEPA) monitored air quality in Khafji, Hafr Al Batin, Tanajib, and Dhahran in Saudi Arabia and two locations in Kuwait (cited by KFUPM/RI, 1992b).

[i] Al Sunaid (1992a): Abqaiq, Dhahran, Rahimah, Tanajib, Shedgum, Tarut Island, and Udhailiyah for air quality monitoring and collecting meteorological data by the Saudi Aramco.

[j] Royal Commission for Jubail and Yanbu has established four air monitoring stations in and around the Industrial City of Jubail, Sauid Arabia (cited in KFUPM/RI, 1992b).

[k] Environmental Protection Committee of Bahrain (Raveendran and Al Mahmood, 1992).

4.6. Smoke dispersion in the Gulf region

One of the major emission products of Kuwait oil fire was large quantities of soot/smoke. Important associations between acute episodes of asthma and particulate matter have been shown in many epidemiological studies. Inhalable suspended particulates probably presented the most serious health hazard during the Gulf crisis, except for participation in the armed conflict. Their concentrations are already naturally high and the presence of soot in the atmosphere may have synergistic effects with other pollutants on human health. Soot/smoke is a good absorbent of organic combustion products. Therefore, the toxicity of soot particles depends on their chemical composition, especially on the content of polycyclic aromatic hydrocarbons and other organics. Particulates may cause physical irritation and may have serious effects on health since organic pollutants such as polycyclic aromatic hydrocarbons are absorbed onto the particulate surfaces. Smoke is important because it is associated with other pollutants emitted by the Kuwait oil fires. Soot particles can adsorb large quantities of inorganic and organic compounds (Hobbs and Radke, 1992). In addition, smoke also effects solar radiation reaching the ground and thus alters temperature at ground level (Shaw, 1992). The effects of smoke on ground surface temperatures will be discussed later.

Dispersion of smoke plume in the upper atmosphere was discussed in Chapter 3. Observations taken at the ground level are presented here. The extent of smoke at a given location was shown to be dependent on wind direction and speed. Downwind locations were expected to experience higher smoke levels. The first smoke from the Kuwait fires was seen at Dhahran on February 23, 1991. Smoke became a common feature with the onset of the northerly (shamal) winds. This situation persisted throughout July, improved in August when the predominant wind direction changed towards the south and completely vanished by the end of October when all oil well fires were extinguished. The data reported by Shaw (1992), indicated smoke levels in Bahrain, as plotted Figure 4.1. The increase in elevated smoke periods coincides well with the northerly shamal winds that transported smoke to the south of the sources. A similar situation was observed in all the Saudi coastal cities. The magnitude of smoky hours decreased in Qatar and the United Arab Emirates because smoke was thinned when transported for such long distances. There were several press reports that suggested that large parts of Iran, Iraq, and Turkey were occasionally covered with smoke (Linden, 1991; Warner, 1991). Based on the information discussed in Chapter 3, a generalized representation of smoke dispersion pattern in the Gulf Region is shown in Figure 4.2.

No measurements were taken of smoke concentrations at ground level. It was probable that the smoke plume very rarely touched the ground because of meteorological conditions. The smoke was largely suspended between 2 to 5 km altitude (Daum et al, 1992; Hobbs and Radke, 1992). The smoke, therefore, was of more importance to the meteorology of the region rather than to human health.

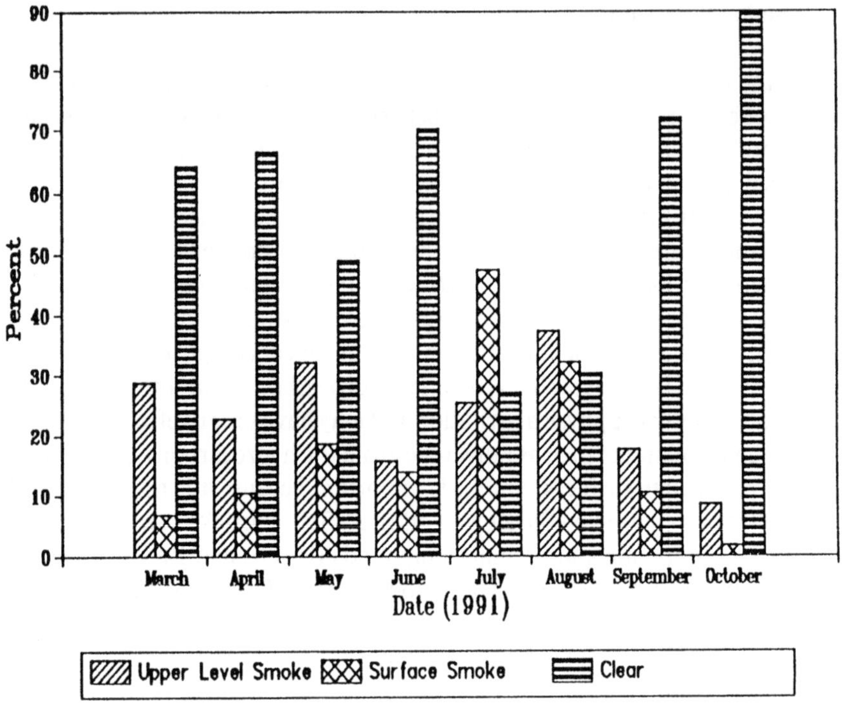

Figure 4.1. Prevalence of Kuwait oil fire smoke in Bahrain.

4.7. Concentrations of suspended air particulates

Suspended air particulates represent a broad class of chemically and physically diverse substances that exists as discrete particles in the atmosphere. The majority of these particles range between 0.005 to 100 μm in size. Originally, the USEPA based ambient measurements on particles larger than 25 μm in size and termed these "Total Suspended Particles", commonly referred to as TSP. In 1987, the USEPA replaced TSP with an indicator that includes only those particles that are smaller than 10 μm in size. Air particles less than 10 μm in size can reach the tracheobronchial and deeper regions of the lung and thus are a more suitable indicator the adverse health effects related to air particulates. The fraction of air particles of less than 10 μm in size are referred to as PM10. PM10 as an indicator of air quality has been adopted universally, i.e. as an indicator of inhalable fraction of air particulates. Measurements of ambient concentrations of PM10 are becoming a standard air monitoring practice worldwide.

The health effects of PM10 particles include adverse effects on the respiratory system, aggravation of any existing respiratory and cardiovascular complications, alteration in the body defensive systems, damage to lung tissues, and premature mortality. Because of the human health significance, air quality

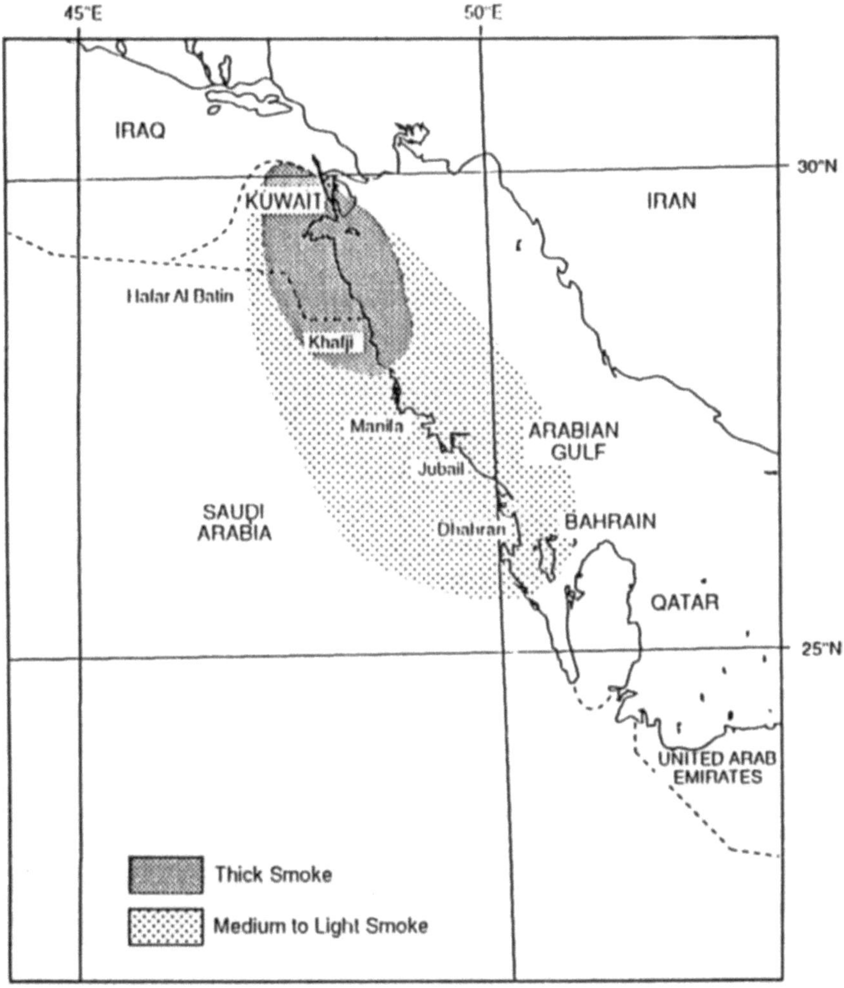

Figure 4.2. **General distribution of smoke from Kuwait oil fires in the Gulf region.**

guidelines with respect to particulate matter in the ambient air have been proposed worldwide. A review of these guidelines is beyond the scope of this book. However, for reference purposes, we list the air quality guidelines proposed by the USEPA (1978) and MEPA (1402 H) in Table 4.2. Guidelines for other primary air pollutants are also given in this table.

The Kuwait oil fires emitted large quantities of soot particles and inorganic compounds (Hobbs and Radke, 1992; Bakan et al., 1991; Penner, 1991; EPC, 1991; Browning et al., 1991; Cahill, 1992; Daum et al., 1992). Even at Dhahran, several hundreds km away from Kuwait, the skies were dark for much of the time from February through August 1991 (shamal season). In June and July, thick smoke clouds filled the sky for more than 80% of the days. To show the

Table 4.2. Air quality guidelines proposed by the USEPA (1978) and MEPA (1402 H).

Averaging Time	Maximum Concentration	Allowable Exceedances
Sulfur Dioxide (SO_2)		
1 hour	730 $\mu g/m^3$ (0.28 ppm)	twice a month
24 hours	365 $\mu g/m^3$ (0.14 ppm)	once a year
1 year	80$\mu g/m^3$ (0.03 ppm)	(none)
Inhaleable Particulate (PM10)		
24 hours	150 $\mu g/m^3$	once a year
1 year	50 $\mu g/m^3$	(none)
24 hours	340 $\mu g/m^3$	once a year
1 year	80 $\mu g/m^3$	(none)
Photochemical Oxidants (as Ozone, O_3)		
1 hour	235 $\mu g/m^3$ (0.12 ppm)	none
1 hour	295 $\mu g/m^3$ (0.15 ppm)	twice a month
Nitrogen Oxides (as Nitrogen Dioxide, NO_2)		
1 hour	660 $\mu g/m^3$ (0.35 ppm)	twice a month
1 year	100 $\mu g/m^3$ (0.05 ppm)	(none)
Carbon Monoxide (CO)		
1 hour	40 mg/m^3 (35 ppm)	once per year
8 hours	10 mg/m^3 (9 ppm)	once per year
Hydrogen Sulfide (H_2S)		
1 hour	195 $\mu g/m^3$ (0.14 ppm)	twice a month
24 hours	40 $\mu g/m^3$ (0.03 ppm)	once a year
Lead (Pb)		
Quarterly	1.5 $\mu g/m3$	none

Italicized guidelines were proposed only by Meteorology and Environmental Protection Administration of Saudi Arabia (MEPA, 1402 H). The remaining guidelines are common to both organizations.

gravity of smoke episode, PM10 filters collected under different weather conditions at KFUPM/RI sampling station in Dhahran, Saudi Arabia, are shown in Figure 4.3. On a smoky day, the filter paper was dark black. The intensity of darkness on the filters decreased with clearing sky (decrease in smoke concentrations at and above the ground). The relationship between different weather conditions and the smoke cloud is clearly demonstrated in Figure 4.3. One could easily imagine the quality of air humans were inhaling in Kuwait, northeastern Saudi Arabia, Bahrain, Qatar and parts of Iran and Iraq. As shown in Figure 4.2, the situation was more serious in Kuwait and northeastern Saudi Arabia because smoke covered these areas almost continuously from March to August 1991. The predominant northerly winds pushed smoke over these areas during this period. When the predominant wind direction changed, most of the oil fires were already extinguished and the smoke problem was improved considerably.

Because of their health significance and their potential for enhancement of the toxic effects of other air pollutants, many national and international organizations attempted to monitor ground level concentrations of air particulates in the Gulf region (Table 4.1). For obvious reasons, most of these

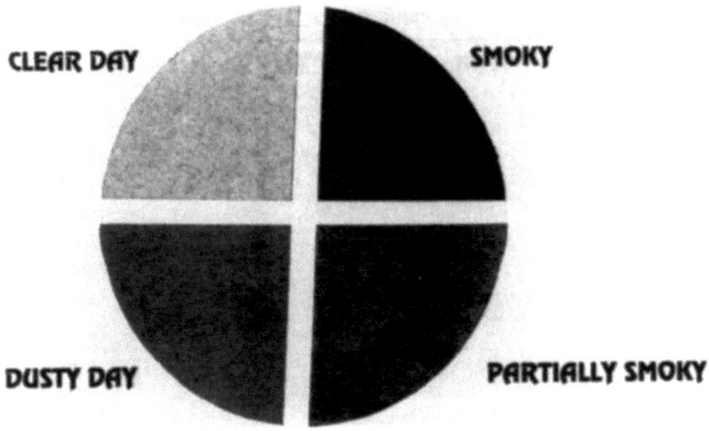

Figure 4.3. Typical air particle sampling filters collected under different weather conditions in Dhahran during Kuwait oil fires.

measurements concentrated in Kuwait, Saudi Arabia, and Bahrain. Information on air particulate loads in Iraq and Iran during the Gulf crisis is scarce. Since PM10 particulate correlate better with human health impact than do TSP data, only data on PM10 concentrations in the air are given in Table 4.3.

The monitoring campaign experienced many technical and practical problems while attempting to collect data on the ambient air quality. Because of the complete breakdown of governmental and scientific infrastructures in Kuwait, air quality monitoring could not be well planned and coordinated at the very early critical period of the Kuwait oil fires. In addition, many international teams were motivated by their national needs. For example, the USEPA concentrated its efforts where American troops were stationed. The NILU monitored air quality at Umm Al Qasr because Norwegian medical staff were located there. The problem of coordination between different foreign teams was further compounded by the lack of efficient scientific and governmental infrastructures in the region, particularly in Kuwait. As claimed by the Kuwaiti authorities, the majority of scientific equipment were either removed or damaged by the Iraqis. This situation delayed the start of air monitoring program after the cessation of hostilities in Kuwait. At best the data collected in Kuwait during fire episode are fragmented. It is difficult to gain a comprehensive picture of air quality in Kuwait during the war and the oil well fire crisis from these measurements. The collected data are best suited for the estimation of generalized trends. Some exceptions to these trends, especially in and around the burning oil fields are expected.

Systematic monitoring of PM10s in Kuwait was started in May 1991 by the Environmental Protection Council. Air quality monitoring was initiated at five locations: Salmiya, Fahaeel, Al Armada, Farwania, and Andalus. Mean concentration of PM10 at these locations during May through September 1991

Table 4.3. Concentrations of PM10 air particulates in the Gulf region during 1991.

Organization	Location	Measurement Date	PM10 concentration ($\mu g/m^3$)		
			Minimum	Maximum	Mean
EPC[a] (Kuwait)	Many sites in Kuwait	May–July	50	2030	–
	Ahmadi,[12] Kuwait	April–September	100	935	326
NIST [b] (USA)	Kuwait City	May 15	300	900	600
USEPA[c] (USA)	Many sites in Kuwait	March 13–March 27	10	5400	–
AIRPARIF[d] (France)		March 27–April 4	32	2030	–
Japanese Team[e] (Japan)	4 sites in Kuwait	April 28–May 5	–	1014	530
NILU[7] Norway	Kuwait	June 5–12	214	2228	636
NILU[f] (Norway)	Iraq	July	8	400	43
KFUPM/RI[g] (Saudi Arabia)	Dhahran	April 1991–March 1992	38	1461	–
Saudi Aramco[h] (Saudi Arabia)	Dhahran	April–November 1991	33	1117	–
	Abqaiq		28	1222	–
	Rahimah	-do-	33	1067	–
	Tanajib	-do-	27	1342	–
	Riyadh	-do-	97	1861	–
EPCB (Bahrain)[i]	Bahrain	July 31–August 4	139	673	199

[a] EPC (1991)
[b] Mulholland et al. (1991)
[c] USEPA (1991)
[d] Lameloise et al. (1991)
[e] Okita (1991)
[f] Sivertsen and Berg (1991)
[g] KFUPM/RI (1992)
[h] Al Sunaid (1992a)
[i] Madany and Raveendran (1992)

For more details, see Table 4.1.

were reported as 508, 330, 322, 269, and 238 $\mu g/m^3$, respectively (EPC, 1991). These values are much higher than the proposed air quality limit for Saudi Arabia (24 hours mean of 340 $\mu g/m^3$, and yearly mean of 80 $\mu g/m^3$) as shown in Table 4.2. In fact, because of the desert climate, air particulates are normally higher in Kuwait and Saudi Arabia than most other areas of the world. The very strong influence of the local environment can been seen in the concentration of air particulates. For example, a very high load of particulates was observed during a sand storm (EPC, 1991). Before the Gulf war, most of measurements of particulates were based on TSP. Previously established data were therefore not available that could help to depict the contribution of the Kuwait oil fires to the PM10 concentrations in Kuwait.

Concentrations of PM10 in the atmosphere are dependent on many factors. The location of measurements and wind direction significantly influences PM10 concentrations. For example, the EPC (1991) found higher mean concentrations of particulates in Al Ahmadi (338 $\mu g/m^3$) and Fahaheel

(310 $\mu g/m^3$) as compared with those recorded at Salmiya-Kuwait City (250 $\mu g/m^3$). The Ahmadi and Fahaheel locations were nearer to the oil fire sources. Wind direction also effected the particulate levels and was different at the various sampling locations. Locations downwind of the fires had higher concentrations of particulates as compared with those located upwind. In general, relatively lower levels of air particulates were measured with westerly, northwesterly, and easterly winds at Mansoria. Northwesterly winds increased both PM10 and TSP in Kuwait as well in northeastern Saudi Arabia, Bahrain, and Qatar. Northeasterly winds pushed the smoke plume over Iran and Iraq (Basra) and, as a result, concentrations of particulates should have increased at these locations.

In addition to the wind direction at the measuring locations, the wind speed also effected air particulate load. The EPC (1991) noted that concentrations of air particulates increased with calm winds because of the limited dispersion of smoke particulates. Enhanced concentrations of particulates were observed for periods with wind speeds higher than 5 m per second, perhaps due to the inversion of the smoke layers. El Desouky and Abdulraheem (1991) reported that PM10 concentrations increased when the wind speed dropped to <1 m per second (due to a lack of dispersion) and increased when wind speed was >6 m per second (due to breaking of the inversion layer).

The USEPA was the first foreign organization to assess air quality in the region. The USEPA mainly concentrated on Kuwait, although a few observations were also taken in Saudi Arabia. The team brought several PM10 portable samplers for determining particulates. These samplers were also used to collect air particulate samples for later chemical analysis in the United States. Being portable and designed to draw small volume of air, these samplers collected small amounts of PM10 particulates. Because of the small quantities, the USEPA (1991) could only report data on elements and chemicals that were present in higher concentrations and have little health and/or environmental significance. Some of these portable PM10 samplers were donated to local organizations in Kuwait and Saudi Arabia. In Saudi Arabia, the King Fahd University of Petroleum and Minerals, MEPA, the Royal Commission of Jubail and Yanbu, and Saudi Aramco collected air quality data starting in March 1991. MEPA also sent a mobile laboratory to Kuwait to monitor air quality from April to June 1991.

It was found later that the PM10 portable samplers, provided by the USEPA, were underestimating particulate loads in the ambient air. The EPC (1991) found that portable PM10 samplers collected about 74% of those particulates collected by standard fixed PM10 samplers. However, statistical analysis of data from the portable and standard PM10 samplers revealed that both data sets were highly correlated ($r = 0.93$). Similar observations were reported by KFUPM/RI (1991, 1992; $r = 0.96$). From the foregoing, it was concluded that the USEPA (1991) underestimated PM10 concentrations by 25 percent. Therefore, the data reported by the USEPA (1991) were not considered further.

In Saudi Arabia, the air particulate measurement campaigns of KFUPM/RI, Saudi Aramco, and the Royal Commission are still continuing in the Eastern Province of Saudi Arabia. Like Kuwait, daily mean concentrations of PM10 were frequently found to be higher than the limit proposed by MEPA (1402 H), particularly during the shamal season. The mean monthly concentrations of PM10 data collected by Saudi Aramco at Dhahran are plotted in Figure 4.4. The figure indicates that PM10 concentrations increased rapidly in May 1991, and remained at about that level until July 1991, and decreased gradually after that to a minimum concentration in November 1991. The high concentration period coincides very well with the shamal season when the wind was blowing primarily from the north (Kuwait). Since PM10 sampling has recently been introduced in this region, historical data on PM10 concentrations in Saudi Arabia or in other countries in the region are not available. Ambient concentrations of PM10 in Bahrain were also measured during the Kuwait oil fire episode. Like Kuwait and Saudi Arabia, on several occasions concentrations of PM10 exceeded that of air quality limit given in Table 4.2. The smoke plume from the Kuwait oil fires and shamal winds were responsible for the excessive PM10 concentrations measured at Bahrain.

The NILU was the only organization that measured particulates inside Iraq (Sivertsen and Berg, 1991). Concentrations of air particulates were relatively

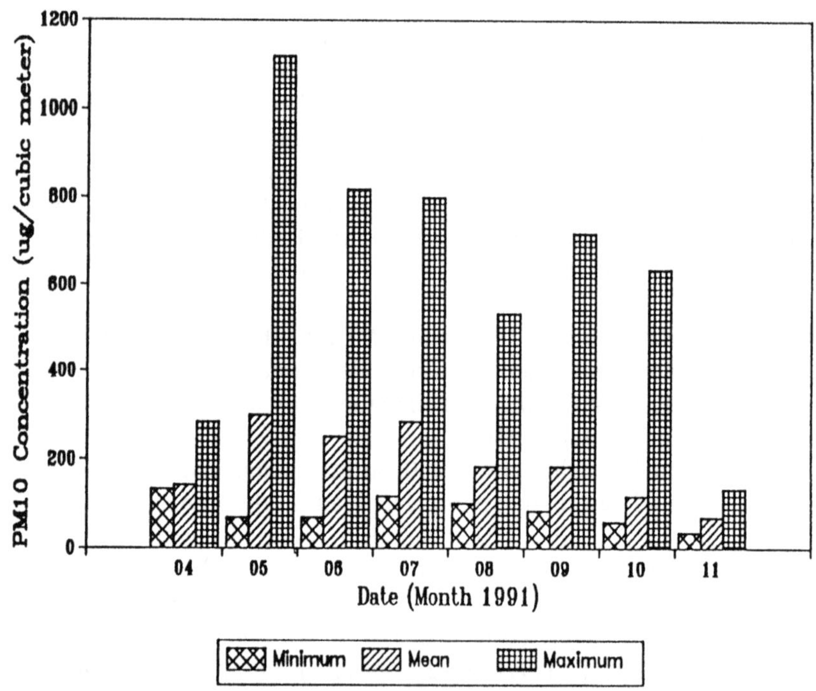

Figure 4.4. Monthly minimum, mean, and maximum concentrations of PM10 air particulates in Dhahran, Saudi Arabia.

low at this location compared to those reported in Kuwait. These observations were made for only one week in June (June 5–12) when the predominant wind direction in the area was from the north. Reviewing the data collected in Kuwait by the foreign scientific teams, the EPC (1991) also noted lower concentrations of particulates in its northern areas as compared with those located in the southern areas during April to July 1991. The predominant wind during this period was from the north. Therefore, a general trend of decreased air particulate loads was expected. More information on air particulates in Iraq was not available. The available data are too limited to assess what happened inside Iraq. Like Iraq, data on air particulates in other parts of the impacted region, except for Bahrain, are lacking. More effort needs to be invested to collect data, particularly from Iran and Iraq.

Based on the foregoing, it may be concluded that the Kuwait oil fires enhanced ambient concentrations of PM10 in the Gulf region. On many occasions, the daily mean concentration of PM10 was higher than the air quality guideline proposed by MEPA (1402 H) for Saudi Arabia. These excesses were more apparent in Kuwait, northeastern Saudi Arabia, and Bahrain. Because of the health significance of PM10 particulates and their tendency to exceed air quality guidelines, continuous further monitoring of this parameter is recommended.

4.8. Ambient concentrations of carbon dioxide, carbon monoxide, and elemental carbon

Carbon dioxide is a common constituent of ambient air. It may induce toxicity if its concentration is increased to a level where oxygen becomes a limited factor. Carbon dioxide gas, when dissolved in water, forms a weak carbonic acid. At high concentrations, CO_2 can be an irritant to respiratory and body defense systems. The environmental significance of carbon dioxide stems from its role in the greenhouse effect rather than its toxicity.

The most abundant emission product of the Kuwait oil fire was CO_2 (about 83% of all air emissions, Chapter 3). The majority of organic compounds in the oil could be converted to CO_2 during combustion. Since CO_2 is the third largest component of the atmospheric air, little interest was shown in the monitoring of CO_2 at the ground level. However, all airborne investigations monitored CO_2 concentrations in the plume (Cofer et al., 1992; Jenkins et al., 1992; Daum et al., 1992; Hobbs and Radke, 1992; Johnson et al., 1991). The environmental significance of CO_2 is largely derived from the fact that the gas contributes significantly to the greenhouse effect. Therefore attempts were made during both numerical modeling efforts and airborne investigations to calculate the total emission of CO_2 from the Kuwait oil fires. Based on the airborne measurements made by the British Meteorological Office, Jenkins et al. (1992) calculated CO_2 emission of 161 million tonnes per year, however, the uncertainty in this estimate was great, about 40 percent. The estimates of

Jenkins et al. (1991) was lower than that of Hobbs and Radke (1992). As expected, concentrations of CO_2 decreased with an increase in distance from the fire sources. It was reported that the origin of the fire influenced CO_2 emission rate. For example, comparing smoke plumes of different origin, Cofer (1991) found that oil lake fire smoke contained the lowest concentrations of CO_2 although its CO levels were the highest. Most carbon dioxide was dissipated in the upper atmosphere. At the ground surface, CO_2 was expected to induce no adverse effect on the ambient environment.

Carbon monoxide is a non-irritating gas that is readily absorbed into the blood stream from the lungs where it forms a reversible complex with the hemoglobin called carboxyhemoglobin. This complex reduces oxygen supply in the body. Because of its toxicity to humans, air quality guidelines have been proposed worldwide. Air quality guidelines proposed by the USEPA (1978) and MEPA (1402 H) are given in Table 4.2. The human health implications of this pollutant will be discussed in another chapter. We summarize below the concentration data that were collected during and after the Gulf crisis.

During the early days of the Kuwait oil fires, it was suspected that combustion of the crude oil might turn out to be incomplete because of limited oxygen supply (large quantities of CO would then be produced from the Kuwait oil fires). Concentrations of CO at ground surfaces might be enhanced to toxic levels with an inversion of the smoke plume. Because of this fear, many national and international organizations monitored CO concentrations at ground level. Some of these data are summarized in Table 4.4.

Table 4.4. Concentrations of carbon monoxide in the Gulf region during 1991.

Organization	Location	Measurement Date	CO concentration, ($\mu g/m^3$)		
			Minimum	Maximum	Mean
EPC[a] (Kuwait)	Mansoria	August–	–	55	21
	Al Ahmadi	April–	0.0	3086	514
NIST[b] (USA)	Kuwait City	May 15	–	–	350
AIRPARIF[c] (France)	4 sites in Kuwait	March 27–April 4	16	4813	1058
NILU[d] (Norway)	Kuwait	April	–	32370	565
		June	–	11900	800
KFUPM/RI[e] (Saudi Arabia)	Jubail	February–October	31	2163	–

@ Data reported by Daoud (1991)

a EPC (1991)

b Mulholland et al. (1991)

c Lameloise et al. (1991) monitored Kuwait City, Al Rawdatayn, Al Ahmadi, and Al Magwa Oil Fields.

d Sivertsen and Berg (1991)

e Data was collected by the Royal Commission of Jubail and reported by KFUPM/RI (1992). For more detail, see Table 4.1.

All data collected and reported in Table 4.4 indicate that CO concentrations were below the permissible air quality guidelines given in Table 4.2. This coincides well with the observations that inversions play a limited role in ground level dispersion of pollutants in a smoke plume. Local conditions, such as traffic load, wind speed, measurement intervals, etc., seem to have had more effect on CO concentrations at ground level than did the Kuwait oil fires. Using the available historical data on air quality in Kuwait City, El Desouky and Abdulraheem (1991) noticed a definite drop in CO concentrations during April--July of 1991 as compared with those measured during the same period in 1990. They attributed this drop to the decreased traffic load in 1991 as compared with 1990. If concentrations of CO were below the air quality limit in Kuwait, it can be argued that a similar situation might have existed in other countries of the Gulf region due to the exodus of many residents during this period. The data collected in Saudi Arabia indicate this to be the case.

A major portion of soot was elemental carbon (Chapter 3). Therefore, a knowledge of the soot produced from the Kuwait oil fires is important. High concentrations of soot are known to influence both the environment and human health (Chapter 7). Very large amounts of soot/elemental carbon lofted to the stratosphere can induce effects with global implications such as a reduction in temperature, solar radiation, upper atmosphere perturbation, etc. A number of studies have been made on workers exposed to excessive concentrations of carbon black. For example, in the rubber and printing industries exposure occurs to carbon black in conjunction with mineral oils containing PAHs. Epidemiological studies suggest an increase in stomach cancer in those workers whose jobs exposed them to various dusts containing carbon black (Blum et al., 1979; Parks et al., 1982). The toxic effects of exposure to carbon was also reviewed by IARC (1984a). Epidemiological evidence suggests diminished respiratory function, particularly in smokers. It is, therefore, important to obtain information on elemental carbon produced during the Kuwait oil fires.

As discussed in Chapter 3, the smoke plume did not reach to the stratosphere but remained largely confined to 2–5 km in altitude. Thus, elemental carbon produced during the combustion remained suspended in the atmosphere for a short time (about three weeks, Chapter 3 for smoke residence time), and was dispersed over large areas and dry deposited on the ground. In Kuwait, there are reports that suggest that desert surface was blackened during the oil well fires. During the smoky days in May to July 1991, a microlayer of soot was visible even on the seawater of the northern Gulf. A large mass of the soot was frequently seen trapped on the surface of the Arabian Gulf. Inland, all air quality monitoring stations in the impact areas yielded dark black filters (that was used for the collection of air particulates) when the sky was filled with smoke. A comparison of filter color collected at Dhahran during the period of Kuwait oil fires under different weather conditions is presented in Figure 4.3.

Unfortunately, it was difficult to separate soot particles from others collected on the particulate filters. As a result, information on the concentrations of soot or elemental carbon is scarce. However, many of the airborne studies addressed this issue. Hobbs and Radke (1992), from their airborne observations through the smoke plume, estimated a combustion rate (soot production) of 1.6 to 2.8% for lake fires and 0.3 to 0.6% for the super composite fire plume in Kuwait. They noticed that soot production in the Kuwait fires was closely associated with their appearance. They observed that little soot mass (about 4% by mass) was present in white smoke while the black smoke plume contained about 20–25% soot and a variable amount of salts. Fires associated with the oil lakes in the desert produced the blackest smoke, which contained up to 48% soot by mass. The mass of the smoke particles in the supercomposite plume was about 30% salt, 15–20% soot, 8% sulfate, and about 30% organics chemicals or elemental carbon.

From the foregoing, the following conclusions are derived:

1. More than 83% of the emission products of the Kuwait oil fires was CO_2. No measurements were made to determine the concentrations of CO_2 in the ambient atmosphere.
2. All the observations of CO concentrations in the residential areas were below the air quality guideline given in Table 4.2.
3. Local conditions, such as traffic load, seem to have influenced CO concentrations in the residential areas more than the Kuwait oil fires.
4. Information on the elemental carbon or soot concentrations at ground level during the fires is scarce. However, from the discussion in Chapter 3, it was concluded that the atmospheric residence time for elemental carbon was probably short.
5. Particulate filter papers collected under different weather conditions clearly showed enhanced concentrations of elemental carbon during smoke episodes.
6. Based on efficient burning, the short atmospheric residence time for the suspended smoke, few inversions, and low concentrations of carbon and its oxides in the ambient air, it is assumed that, except for smoke/soot and particulates, the oil fires had insignificant effect on environmental, meteorological, and health conditions in Kuwait, Saudi Arabia, and other countries in the Gulf region.

4.9. Ambient concentrations of sulfur dioxide (SO_2) and hydrogen sulfide (H_2S)

Sulfur dioxide is easily dissolved in water to form a highly acidic solution. It generally effects the respiratory tract, aggravates existing respiratory and cardiovascular problems, and in the case of a prolonged severe exposure, may result in mortality. In addition to human health hazards, high ambient concentrations of SO_2 may damage vegetation, enhance corrosion of metallic structures, and produce acid rain. Because of human health hazards and adverse environmental effects associated with it, monitoring of ambient concentrations of SO_2 is an integral part of any air quality assessment program. Air quality guidelines have been proposed worldwide. The guidelines of USEPA regarding SO_2 and H_2S are contained in Table 4.2. Considering its probable production potential and adverse environmental interactions, many national and international organizations monitored ground concentrations of SO_2 during the Kuwait oil fires (Table 4.5). This section will review SO_2 concentration data collected during this period; however, the health implications of these concentrations will be discussed separately in Chapter 7.

One of the pre-war fears shown by the international scientific community was the health consequences of the sulfur dioxide gas that would be produced if oil wells in Kuwait were ignited. Concentrations of total sulfur in Kuwait crude oils vary between 2.5 and 4.6% (Chapter 3). It was calculated that about 2.3 million tonnes of SO_2 would have been produced during the Kuwait oil fires (Chapter 3). These data indicate that the Kuwait fires had the potential to produce high concentrations of SO_2 in the plume, and during an inversion period (or smoke touch down episode), could contaminate ground level ambient air to alarmingly high levels. Considering the relatively small area of smoke production (i.e. 40 × 40 km area (Chapter 3)), and assuming that SO_2 remains in gaseous form, it appears that a very large amount of SO_2 would be emitted that could certainly result in serious environmental and human health damage in the Gulf region.

Because of the probable large amounts of SO_2 production during the Kuwait oil fire period and the hazards associated with SO_2, several local and international organizations monitored ground level SO_2 concentrations in the Gulf region. Many technical and logistic problems were encountered by the air quality monitoring teams in Kuwait. These problems are briefly discussed here.

With the invasion and occupation by Iraq, Kuwait's social and scientific infrastructures were in chaos. Their scientific staff was mostly expatriates who left the country before or during the Gulf war. According to Kuwait authorities, the majority of their scientific equipment was either damaged or taken to Iraq. This situation hampered the quality and continuity of scientific data collection in Kuwait by both national and international organizations. The data reported for Kuwait are at best fragmented, and were collected at different locations, time, under variable weather conditions, and by different organizations using different methodologies. The logistic support sought by the international organizations was not available, at least in the early days of the fires. Air quality observations were taken at many locations in Kuwait but for short durations only (for few days or weeks). Therefore, the data collected in Kuwait do not represent a continuous series of measurements but only fragmented data sets of variable accuracy. Figure 4.5 shows some meteorological and air monitoring stations in Kuwait.

Concentrations of SO_2 measured by local and foreign investigators were collected and summarized in Table 4.5. These data should be interpreted within the above mentioned limitations. Ambient concentrations of SO_2 varied in a broad range. Table 4.5 contains data collected in and around the oil fires and

Table 4.5. Concentrations of SO_2 in the Gulf region during 1991.

Organization	Location	Measurement Date	SO_2 concentration (ug/m^3).		
			Minimum	Maximum	Mean
EPC[a] (Kuwait)	Mansoria	August–October	–	261	14.7–28.7
	Al Ahmadi	April–September	0.0	4067	–
German[b] (Kuwait)	Mansoria	March 3–	1.3	3.9**	–
	Al Zoor	April 27	4.5	19.6	–
USEPA[c] (USA)	Many sites	March 13–	9.9	218.6	84.5
	in Kuwait	March 27	nd.	5214	–
AIRPARIF[d] (France)	Many sites in Kuwait	March 27–April 4	6	495	159
Japanese Team[e] (Japan)	4 sites in Kuwait	April 28–May 5	2.5	1146	–
NILU[f] (Norway)	Kuwait	April	–	38	2.3
		June	–	142	1.9
NILU[f] (Norway)	Iraq	June	5	36	11.5
MEPA[g] (Saudi Arabia)	3 sites	April–	2	22	8.5
		Kuwait	June		
Saudi Aramco[h] (Saudi Arabia)	Dhahran	January–November	8	112	–
Saudi Aramco[h] (Saudi Arabia)	Abqaiq	January–November	3	146	–
Saudi Aramco[h] (Saudi Arabia)	Rahimah	January–November	3	102	–
Saudi Aramco[h] (Saudi Arabia)	Tanajib	April–November	3	175	–
Royal Commission[i] (Saudi Arabia)	Jubail	February–October	8	112	–

[a] EPC (1991).

[b] Measurements by German laboratory as cited by EPC (1991).

[c] USEPA (1991): The real time measurements at different locations in Kuwait were below the detection limits of SO_2 analyzer. We discarded these observation. Concentrations of SO_2 were measured from March 13–24 by bubbler method at three hospitals in Kuwait and are given in the above Table. The hospitals were Adan, Mubareek Al Kabeer, Al Farwania, and Al Jahra. Saudi Arabia: Dhahran, Riyadh.

[d] Lameloise et al. (1991) monitored Kuwait City, Al Rawdatayn, Al Ahmadi, and Al Magwa Oil Fields.

[e] Okita (1991): Ahmadi, Riggae, Mansoria, Jahra.

[f] Sivertsen and Berg (1991).

[g] Measurement taken by MEPA, Saudi Arabia.

[h] Data was collected by Saudi Aramco for Dhahran, Abqaiq, Rahimah, and Tanajib and reported by KFUPM/RI (1991,1992)

[i] Data was collected by Royal Commission for Jubail and Yanbu, and reported by KFUPM/RI (1991,1992).

** USEPA (1991) air monitoring team was unable to find measurable concentrations of SO_2 at several locations in Kuwait and Saudi Arabia – even in Ahmadi and Wafra Oil Fields. It seems that the analytical technique used was not sufficiently sensitive.

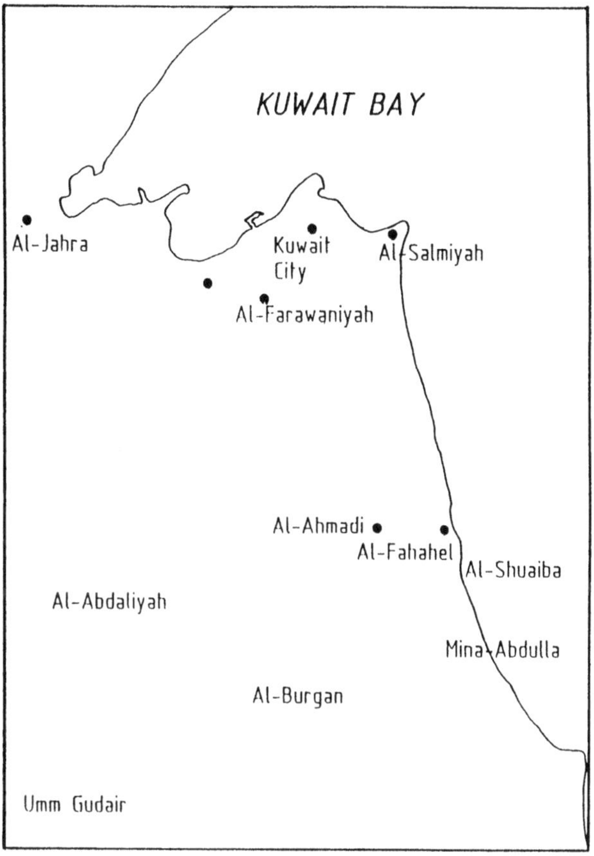

Figure 4.5. Meteorological and air quality monitoring stations in Kuwait.

several hundred kilometer away from the fire sources, therefore, significant variability in the data can be explained.

Reviewing all the collected data from Kuwait, the EPC (1991) concluded that, in the residential areas, SO_2 levels were below the air quality guidelines, especially when the wind direction was such that it carried the smoke plume away from the monitoring sites. However, there were some levels which exceeded hourly air quality guidelines (Table 4.2) when the wind was blowing from the fires. The USEPA (1991) team was the first to report that SO_2 concentrations in many parts of Kuwait were below the detection limit of the instrument used. The USEPA team detected the highest concentration of SO_2 (0.68 ppm) in the smoke plume of the Burgan Oil Field. This concentration was above the maximum 24 hour limit (Table 4.2). None of the other concentrations of SO_2 found in the populated areas exceeded that of the United States air quality limit. According to the monitoring results of the Japanese team (Okita, 1991), collected between April 30 and May 8, 1991, ambient concentrations of SO_2 were generally low in Kuwait City and the other monitored locations.

From their observations, the Japanese team speculated that SO_2 pollution might not be serious enough to induce acute health effects. However, the team found very high levels of SO_2 in one of the oil fields during an episode of smoke "touch down". The French team (Lameloise et al., 1991) monitored ground level SO_2 concentrations in different regions of Kuwait and observed relatively higher concentrations near the fire source. Elevated SO_2 concentrations were also noted when the wind direction was from the direction of the oil fires. The data in Table 4.5 suggest that there might have been incidents when SO_2 concentrations were above the air quality guidelines (Table 4.2). However, most such high concentration episodes would have been confined to near-fire areas, i.e., in or around the burning oil fields. Fortunately, the population from these areas had already moved and were not exposed.

Saudi Aramco, the Royal Commission for Jubail and Yanbu, and MEPA determined SO_2 concentrations in Saudi Arabia during the Gulf crisis. The monitoring program of MEPA was in response to the crisis. However, the air quality monitoring programs of Saudi Aramco and the Royal Commission had already started in 1982 and 1984, respectively. The Royal Commission's air monitoring program is largely confined to the industrial city of Jubail whereas Saudi Aramco monitors air quality at many locations in the Eastern Province of Saudi Arabia. Data collected on SO_2 concentrations in Saudi Arabia are listed in Table 4.5. The monthly mean concentrations of SO_2 in Saudi Arabia were found to be within the permissible limit proposed by MEPA (1402 H). A review of the Saudi Aramco data indicates that ambient concentrations of SO_2 at any of the sampling sites and for any given day did not exceed that of the air quality limits proposed by the USEPA (1978) and MEPA (1402 H) [1 hour average, 280 ppb; 24 hours mean, 140 ppb; 30 ppb]. Based on these observations, it was concluded that concentrations of SO_2 in Saudi Arabia always remained below the air quality guidelines.

The Environmental Protection Council of Kuwait started an air quality monitoring program in 1983. Thus, some historical data on SO_2 concentrations in Kuwait were available. The EPC (1991) compared SO_2 concentration data collected at the general hospitals in Kuwait during April–July in 1990 and 1991. It was noticed that the daily mean concentration of SO_2 from March–April 1991 doubled to 56.2 $\mu g/m^3$ compared to that measured during the same period and at the same locations in 1990. However, the concentrations of SO_2 were within the ambient air quality guidelines (Table 4.2) at all times at the monitoring locations in residential areas. Higher SO_2 levels were reported in oil fields and in the vicinity of fires. As expected, the wind direction also effected SO_2 concentrations. Wind blowing from the fires towards the monitoring site enhanced pollutant levels, including SO_2 levels. In general, the southern Kuwait region was subjected to relatively higher concentrations of SO_2 because of many oil fields in this area. Relatively lower SO_2 concentrations were measured in northern Kuwait along the Iraqi border. From these observations, it was assumed that Iraq was not exposed to high concentrations of SO_2 during the Kuwait oil fires. However, information on Iraq is very scarce and nothing can be said with certainty. Historical data on ambient SO_2 concentrations in Kuwait City were summarized by El Desouky and Abdulraheem (1991). They compared data collected during April–July in 1990 and 1991 and concluded that SO_2 concentrations increased in 1991 but were within the air quality limit shown in Table 4.5. The enhancement in SO_2 concentrations in 1991 was attributed to the emission from the burning oil wells. El Desouky and Abdulraheem (1991) pointed out that levels of SO_2 were not lower for the entire Kuwait region. Relative higher concentrations of SO_2 were recorded by the German mobile laboratory in the Mansoria and Al Zoor areas. El Desouky and Abdulraheem (1991) also investigated the effect of wind speed on ambient concentrations of air pollutants. They found that pollutant concentrations were significantly higher under stagnant ambient conditions (wind speed below 1 meter per second), probably due to a lack of sufficient dispersion of pollutants. Wind speeds higher than 6 meter per second also increased ambient pollution of SO_2 at ground level. This may be due to the plume breaking through the inversion layer.

As stated above, concentrations of SO_2 in Kuwait could be effected by many factors, making comparison difficult. Saudi Aramco has the most extensive air monitoring network in Saudi Arabia. Measurements of SO_2 concentrations at Dhahran, Abqaiq, and Rahimah from 1982 to 1991 were summarized by Al Sunaid (1992a, 1992b). Saudi Aramco collected data on SO_2 concentrations for several year before the Gulf war. The yearly mean concentration of SO_2 at Dhahran, Abqaiq, and Rahimah for the years 1986 to 1991 were adapted from Saudi Aramco (Al Sunaid, 1992a, 1992b) and are plotted in Figure 4.6. Sulfur dioxide concentrations before 1990 were similar in Dhahran, Abqaiq, and Rahimah. In 1990, the mean concentration of SO_2 increased in Dhahran and Abqaiq but not in Rahimah. According to Al Sunaid (1992a, 1992b), this increase can be accounted for by the increase in petroleum production to compensate for the loss of Kuwait and Iraq petroleum production. The mean concentration of SO_2 increased rapidly in 1991 at all the three locations monitored by Saudi Aramco. This increase could be attributed to the Kuwait oil fires. The Royal Commission for Jubail and Yanbu also monitored ambient concentrations of SO_2 in the industrial city of Jubail since 1984. The data showed an appreciable increase in the yearly mean concentration of SO_2 at this location for 1991 (Suthers, 1991); however, the magnitude of increase was relatively less than that observed by Al Sunaid (1992a). This may possibly be due to high background concentrations of SO_2 at Jubail because of close proximity of their monitoring stations to a petrochemical complex.

To further investigate the effects of the Kuwait oil fires, the monthly mean concentrations of SO_2 from Al Sunaid (1992a, 1992b) are plotted in Figure 4.7. Comparing monthly data for 1989, 1990, and 1991, it is obvious that levels of SO_2 were highest during the period from January to October 1991. Figure 4.7 depicts concentrations of SO_2 in 1989 which were similar throughout the year except for the months of June and July. Concentrations of SO_2 throughout 1990 remained more or less similar and were comparable to those measured

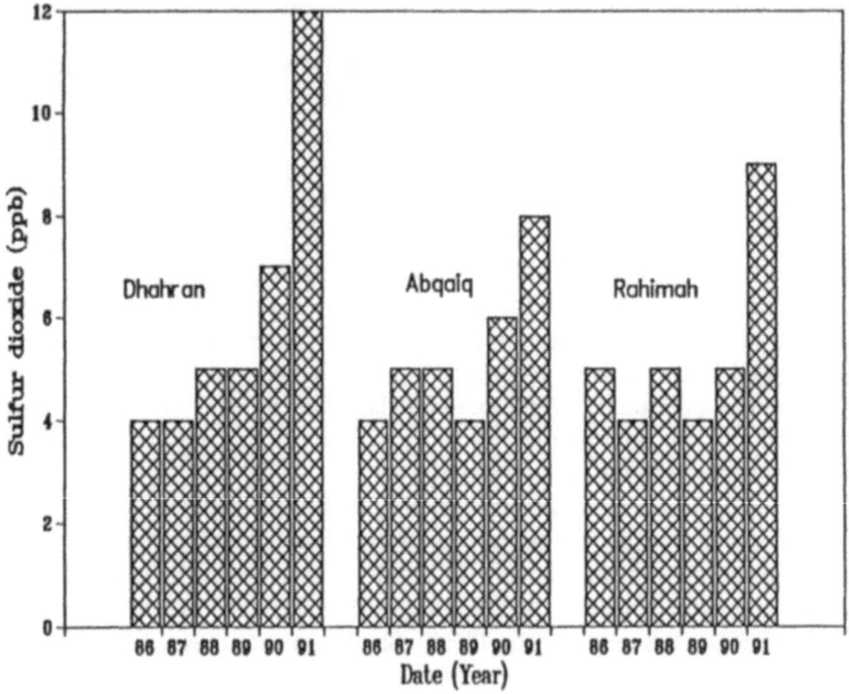

Figure 4.6. Yearly mean concentrations of sulfur dioxide at three sampling stations in Saud Arabia.

during 1989. The maximum monthly mean SO_2 concentration was recorded in July 1991. In 1991, the concentration of SO_2 increased rapidly from March through September 1991. Generally, in March the wind began to change direction and became predominantly from the north, locally known as shamal winds. These winds continued to blow till August. During the shamal period in 1991, the smoke plume from the Kuwait oil fires was pushed over to the Dhahran area. There were days in the summer when the sky at noon looked deceptively like an evening in Dhahran (see Figure 4.2 for soot concentrations). In response to smoke dispersion, concentrations of SO_2 were expected to rise. The data presented in Figures 4.6 and 4.7 illustrate this adequately. The increase in the monthly mean concentrations of SO_2 was more pronounced at Dhahran as compared with Abqaiq. This is understandable since Dhahran was located closer to the plume centerline than Abqaiq (see Figure 4.2). After the extinguishing the oil fires in Kuwait in October 1991, concentrations of SO_2 at Dhahran and Abqaiq fell to the previously established background levels. Al Sunaid (1992a, 1992b) argued that the increase in SO_2 was the result of two events. One of them was the burning oil wells in Kuwait and the second was the sudden unplanned rise in Saudi oil production to compensate for the loss of Kuwaiti and Iraqi production. The average daily crude oil production by Saudi Aramco in 1989 was 4.86 million barrels which was increased to 6.26 million barrels in 1990 (Al Sunaid 1992a, 1992b).

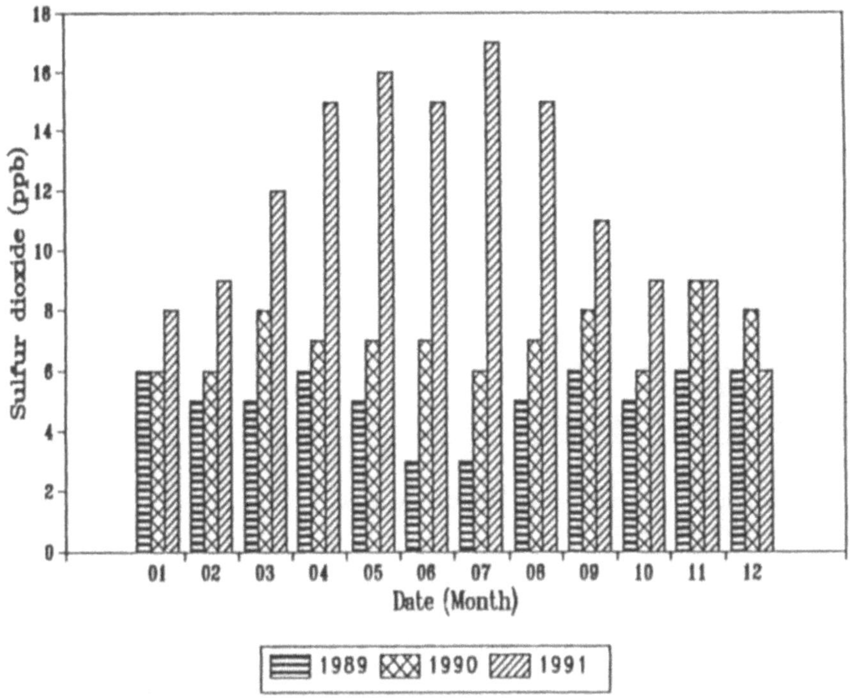

Figure 4.7. Monthly mean concentrations of sulfur dioxide at sampling stations in Saudi Arabia.

Based on the total sulfur concentration in the Kuwait crude oils and the total emission of oil from the damaged oil wells, the production of 2.3 million tonnes of SO_2 was expected. It was suspected that SO_2 concentrations at ground level could be enhanced to dangerous levels. However, except for occasional high levels in the immediate vicinity of the fire sources, concentrations of SO_2 in Kuwait, Saudi Arabia, Bahrain, and at other adjacent locations were found to be below air quality guidelines (Table 4.2). The question of what happened to the sulfur that was emitted by the Kuwait oil fires still remains unanswered. As stated in Chapter 3, airborne investigations suggested very rapid oxidation of SO_2 to SO_4 in the smoke plume (Hobbs and Radke, 1992; Jenkins et al., 1991; Johnson et al., 1991). Because of this rapid oxidation, formation of sulfuric was expected. It was suspected that ground level concentrations of acidic aerosol might be enhanced to harmful limits.

Acidic aerosol inhalation is known to adversely affect human health (Saskolne et al., 1989; Uthell, 1985). Expecting enhanced concentrations of acidic aerosols, the USEPA (1991) team determined ambient concentrations of sulfuric acid at several locations in Kuwait and two in Saudi Arabia. In Kuwait, the team found sulfuric acid concentrations ranging from below the detection limit to 31 ppb (on an air volume basis). Sulfuric acid concentrations in Dhahran and Riyadh, Saudi Arabia were appreciably lower than those found in Kuwait. The concentrations of acid found during this monitoring program were typical of urban concentrations in industrialized countries. The data reported by the USEPA (1991) are given in Table 4.6.

Table 4.6. Concentrations (ppb) of different acids reported by USEPA (1991)*.

Location	Sulfuric	Nitric	Hydrochloric
U.S. Embassy, Kuwait	nd–1	nd–2	nd–6
Camp Freedom, Kuwait	4–31	4	16
Dhahran, Saudi Arabia	5–6	2	–
U.S. Embassy, Saudi Arabia	2	3	–
Port Shuaybah, Saudi Arabia	19	–	–
Al Ahmadi Oil Field, Kuwait	27	10	9
Al Burgan Oil Field, Kuwait	30	32	15
Al Wafra Oil Field, Kuwait	nd	nd	nd

nd – Below the detection limit.
* – Measurements were made during March 13–20, 1991.
– – No information listed.

The data in Table 4.6 indicate that concentrations of all acids were higher in the oil field areas with burning oil wells. Very low concentrations were detected in residential areas. Of course, concentrations of these pollutants would have been affected by wind speed and direction. On calm days with the wind blowing from the Al Burgan and Al Ahmadi Oil Fields enhanced concentrations of sulfuric acid aerosols would occur. Concentrations of sulfuric acid were higher in Kuwait than in Saudi Arabia. Based on these data and the concentration trend, it was concluded that other areas in the Gulf region, such as, Iran, Iraq, Bahrain, Qatar, and the United Arab Emirates, were probably not exposed to excessively higher than background concentrations of acid in their ambient air. Therefore, it was concluded that the Kuwait oil fires did not increase ambient concentrations of sulfuric acid, except in the immediate vicinity of the fire sources.

It was reported that some oil fields in Kuwait also contained H_2S (HPI, 1987). In addition to its presence in the crude oil, it was possible that H_2S might be produced as a minor emission product of the Kuwait oil fires. The USEPA (1991) determined concentrations of H_2S by real–time measurements at several locations in Kuwait. Concentrations of H_2S ranged from less than the detection limit to 42 ppb in one Kuwaiti oil field. These concentrations are below the air quality guidelines suggested in Table 4.2 (one hour mean H_2S concentration of 195 μg/m3 which is about 140 ppb). Saudi Aramco monitored ambient concentrations of H_2S at Rahimah and Tanajib, Saudi Arabia. All measurements of H_2S were found to be within the air quality guidelines. The historical data collected by Saudi Aramco showed no yearly trend in H_2S concentrations at these locations. Yearly mean concentrations of H_2S remained almost constant over the monitoring period. The yearly means range was 2–4 ppb with a maximum concentration of 64 ppb recorded in 1987. These data suggest that H_2S always remained below the proposed air quality limit given in Table 4.2. The Royal Commission for Jubail and Yanbu has also monitored air quality at the Industrial City of Jubail, a very large petrochemical manufacturing complex. No yearly trend could be seen in H_2S concentrations that were collected by the Commission (Suthers, 1991). The above reports suggest that the Kuwait oil fires had no effect on ambient concentrations of H_2S in Saudi Arabia. Based on the foregoing, it was concluded that H_2S concentrations in other Gulf countries would probably have remained unaffected by the Kuwait oil fires.

From the foregoing, the following salient points were extracted.

1. If all the sulfur in the crude oil burned in the Kuwait oil fires was converted to SO_2, approximately 2.3 million tonnes of SO_2 gas would have been produced.
2. Many national and international organizations measured ground level concentrations of SO_2 in Kuwait, Saudi Arabia, Bahrain, and Iraq. All the measurements were found to be below the air quality limit proposed for this pollutant (Table 4.2).
3. Higher concentrations of SO_2 were found in and around the Kuwait oil fields with fires than other areas. Occasionally, concentrations of SO_2 in the burning oil fields were found to increase above the proposed air quality guidelines by MEPA (1402 H) and USEPA (1978).
4. Concentrations of SO_2 at a given time and location were dependent on the wind direction, wind speed, and distance from the fire sources.
5. Historical data collected in Kuwait and Saudi Arabia indicate a significant increase in yearly and monthly means of SO_2 concentrations in 1991 suggesting a strong influence of the Kuwait oil fires on ambient levels of this pollutant.
6. The monthly mean concentrations of SO_2 in the Eastern Province of Saudi Arabia was enhanced during the shamal season (predominantly north winds) and were always within the air quality guideline proposed by MEPA (1402 H).
7. Concentrations of sulfuric acid in Kuwait and Saudi Arabia were low and typical of that expected in an urban area.
8. Ambient levels of H_2S were very low. Historical data collected in Saudi Arabia since 1984 indicated no enhancement in H_2S concentrations during 1991. All the observations taken in 1991 were within the air quality guidelines proposed by MEPA (1402 H) and USEPA (1978).
9. Concentrations of SO_2, H_2S, and sulfuric acid were found to be below the air quality guidelines proposed in Table 4.2 during the Kuwait oil fires episode. Such low concentrations of these pollutants would probably have an insignificant impact on the environmental conditions in Kuwait, Saudi Arabia, Bahrain, and other Gulf countries.

4.10. Ambient concentrations of nitrogen oxides

Nitrogen oxides in ambient air are emitted primarily from combustion sources. The major emission species is NO (nitrogen oxide) which is transformed to NO_2 by photochemical oxidation. These nitrogen oxides, collectively referred as NO_x, are oxidants and primarily effect humans via the lungs. A detailed discussion of the health implications of these gases will be presented later.

Many Gulf war activities had the potential to produce large quantities of nitrogen oxides. Important among these activities are the emission products of fuel consumed during the crisis and the Kuwait oil fires. It has been estimated that about 6 billion gallons of fuel were consumed during the Gulf war and another 1.12 billion barrels of crude was burning during the oil fires at the damaged oil wells in Kuwait. Because of its potential production and probable health and environmental consequences, many national and international organizations monitored ground concentrations of NO_x in Kuwait and Saudi Arabia. The results of their investigations are given in Table 4.7.

The mean concentration of NO_x in the ambient air ranged between 17 and 51 $\mu g/m^3$ for Kuwait and Saudi Arabia. All concentrations measured are within the air quality guidelines shown in Table 4.2 for this pollutants (nitrogen Oxides - defined as NO_2: mean for one hour is 660 $\mu g/m^3$ and for one year is 100 $\mu g/m^3$). The maximum concentration of NO_x was reported by Dauod (1991) in Al Ahmadi City which is very near to the Al Ahmadi Oil Field. The value represents a day when smoke from the fires had completely engulfed the city. Few others measurements were found to be above the permissible levels of NO_x. These reports suggest that concentrations of NO_x were occasionally increased above the air quality guidelines, and that these exceedences were essentially confined to the oil fields or during smoke touch down at ground level.

As in Kuwait, all measurements of NO_x made in Saudi Arabia were also below the air quality guidelines proposed by MEPA (1402 H) and the USEPA (1978). Historical data on ambient concentrations of NO_2 collected at Dhahran and Abqaiq (Saudi Arabia) were available from Saudi Aramco (Al Sunaid, 1992a, 1992b). Figure 4.8 shows that the yearly mean concentration of NO_2 in Saudi Arabia remained unaffected by the events in the Gulf, including the Kuwait oil fires. The maximum concentrations of this pollutant recorded at the two location were below the air quality guideline. Dhahran ambient air appears to contain more of NO_2 as compared to Abqaiq. Al Sunaid (1992a, 1992b) attributed this difference to comparatively higher traffic loads in Dhahran. Similar observations were recorded by Suthers (1991) from the air quality data collected by the Royal Commission for Jubail and Yanbu in the Industrial City of Jubail in Saudi Arabia. In Kuwait, the EPC (1991) found that the yearly mean concentration of NO_x at Mansoria was higher in 1990 as compared with the 1991 level. Like Al Sunaid (1992a, 1992b), the EPC (1991) also argued that the decrease in traffic due to the Iraqi invasion was responsible for lowered yearly mean concentrations of NO_x at Mansoria.

Table 4.7. Ambient air concentrations of NO_x in the Gulf region during 1991.

Organization	Location	Measurement Date	NO_x Concentration ($\mu g/m^3$)		
			Minimum	Maximum	Mean
EPC[a] (Kuwait)	Mansoria	August	–	239	36
		September	–	511	34
		October	–	622	51
	Al Ahmadi	April–September	0.0	69771	36
NILU[b] (Norway)	Kuwait	April	–	1011	39.6
		June	–	540	29.4
AIRPARIF[c] (France)	4 sites in Kuwait	March 27–April 4	1	77	17
Japanese Team[d] (Japan)	4 sites in Kuwait	April 28–May 5	–	54	–
MEPA[e] (Saudi Arabia)	3 sites Kuwait	April–June	9	31.7	21.0
Saudi Aramco[f] (Saudi Arabia)	Dhahran	January–November	13	87	–
Saudi Aramco[f] (Saudi Arabia)	Abqaiq	January–November	2	36	–
Saudi Aramco[f] (Saudi Arabia)	Tanajib	April–November	2	30	–
Royal Commission[g] (Saudi Arabia)	Jubail	February–October	1	36	–

– Information either not available or not calculated.
[a] EPC (1991).
[b] Sivertsen and Berg (1991).
[c] Lameloise et al. (1991) monitored Kuwait City, Al Rawdatayn, Al Ahmadi, and Al Magwa Oil Fields.
[d] Okita (1991): Ahmadi, Riggae, Mansoria, Jahra.
[e] Measurement made by MEPA, Saudi Arabia.
[f] Data was collected by Saudi Aramco for Dhahran, Abqaiq, Rahimah, and Tanajib, and made reported by KFUPM/RI (1991,1992).
[g] Data was collected by Royal Commission for Jubail and Yanbu, and reported by KFUPM/RI (1991,1992)

The other nitrogen based pollutant that was be of concern is nitric acid that can be produced by solubilizing NO_x in water. Concentrations of nitric acid in ambient air were determined by the USEPA (1991) team during their visit to the area. Measurements were taken in Kuwait and Saudi Arabia. Nitric acid concentrations ranged from below the detection limit to 32 ppb found in Burgan Oil Field (Table 4.6). In general, the concentration of nitric acid was higher in the burning oil fields in Kuwait (2–32 ppb) than those reported in Saudi Arabia (2–3 ppb). The ambient levels of nitric acid found by the USEPA (1991) are typical of any urban area.

From the foregoing, it was concluded that data on NO_x and/or NO_2 collected by both national and international organizations were consistently below the proposed air quality limit shown in Table 4.2. Concentrations of

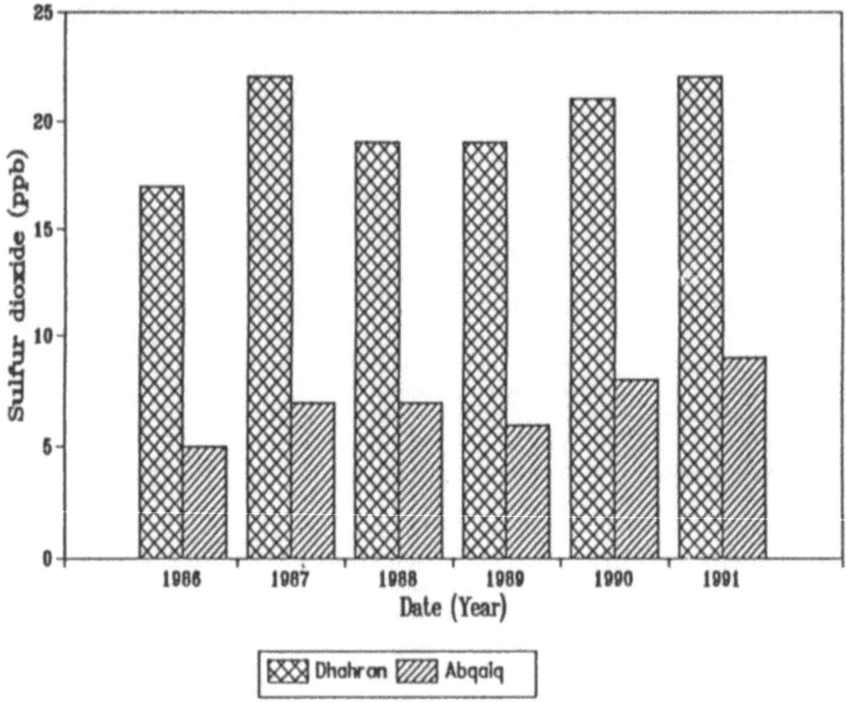

Figure 4.8. Yearly mean concentrations of nitrogen oxide at different sampling stations in Saudi Arabia.

nitric acid were also low. Previous measurements indicated no increase in NO_x that could be associated with the Kuwait oil fires. The concentrations seem to have been more affected by local conditions such as traffic load than by the Kuwait oil fires. The health significance of the data collected in Kuwait and Saudi Arabia will be discussed separately. It is anticipated that NO_x concentrations would have an insignificant effect on environmental conditions in Kuwait and other Gulf countries.

4.11. Ambient concentrations of ozone

Ozone (O_3) is a respiratory tract irritant that primarily effects the lung. Human health implications of O_3 will be discussed in Chapter 7. Depletion of ozone in the stratosphere is of greater concern than its toxicity in the atmosphere. Nitrogen oxides and some organics generated by the oil fires in Kuwait are known to react with O_3.

Ozone was not a pollutant of primary concern with respect to the Kuwait oil fires. Nitrogen oxide may react with O_3 in the atmosphere. It was suspected that the fires might produce a large amount of NO_x that would react with O_3, and thereby deplete it. It was included in many air measurement programs to determine if reaction between O_3 and NO_x occurred. Limited interaction between O_3 and NO_x was noticed in the smoke plume. A decrease in O_3 concentrations near the source and its later regeneration was reported from the data derived during airborne studies (Hobbs and Radke, 1992, Jenkins et al., 1991). Data on O_3 concentrations at ground level are summarized in Table 4.8.

Table 4.8. Concentrations of Ozone in the Gulf region during 1991.

Organization	Location	Measurement Date	O_3 Concentration (ug/m^3)		
			Minimum	Maximum	Mean
EPC[a] (Kuwait)	Mansoria	August	-	116	9
		September	-	70	23
		October	-	68	12
EPC[b] (Kuwait)	Al Ahmadi	April–September	2	52	20
NIST[c] (USA)	Kuwait City	May 15			
MEPA[c] (Saudi Arabia)	3 sites in Kuwait	April–June	1	204	7
NILU[d] (Norway)	Kuwait	April	-	161	10
		June	-	137	12
AIRPARIF[e] (France)		March 27–April 4	4	96	40
Saudi Aramco[f] (Saudi Arabia)	Dhahran	January–November	19	84	-
Royal Commission[g] (Saudi Arabia)	Jubail	February–October	8	89	-

[a] EPC (1991).
[b] Data reported by Daoud (1991).
[c] Mulholland et al. (1991).
[d] Sivertsen and Berg (1991).
[e] Lameloise et al. (1991) monitored Kuwait City, Al Rawdatayn, Al Ahmadi, and Al Magwa Oil Fields.
[f] Data was collected by Saudi Aramco for Dhahran, Abqaiq, Rahimah, and Tanajib, and reported by KFUPM/RI (1991,1992).
[g] Data was collected by Royal Commission for Jubail and Yanbu, and reported by KFUPM/RI (1991,1992).

The maximum concentration of O_3 reported in Table 4.8 is 204 $\mu g/m^3$. Even this value is within the permissible limit for this pollutant (1 hour mean as 235 $\mu g/m^3$ or 0.12 ppm). O_3 concentrations in Kuwait and Saudi Arabia were within the air quality limit listed in Table 4.2. Considering these observations, it was anticipated that normal concentrations of O_3 persisted in the Gulf region during the Kuwait oil fires. Saudi Aramco has monitored ambient O_3 concentrations at Dhahran since 1983. An analysis of these historical data yielded no general yearly trend in O_3 concentrations at Dhahran. Concentrations of O_3 were always found below the air quality limit proposed by USEPA (1978) and MEPA (1402 H).

It can be concluded from the foregoing that O_3 concentrations were within the proposed air quality guideline in the Gulf region. The impact of the Kuwait oil fires on O_3 concentrations in Kuwait or other areas could not be ascertained. Ozone might have been depleted slightly very near to the fire source in Kuwait.

4.12. Ambient concentrations of volatile organics

The Kuwait oil fires might have introduced volatile organic chemicals into the ambient air via two routes. First, the uncontrolled flow of crude from many of the wells formed large lakes of oil (Al Besharah, 1991). Volatile organics were continuously escaping to the atmosphere from these lakes. Secondly, from many of the damaged oil wells crude came out under extremely high pressure and a portion of the crude was converted to an oil mist. From the oil mist, lower molecular weight organic compounds might be volatilized (Al Besharah, 1991). These events were the major source of volatile organics at ground levels. In addition, combustion also generated volatile organics, many of which are detrimental to human health. The amount of volatile organics in the emission products might increase if combustion is incomplete. Considering the large potential for the production of volatile organic chemicals from the fire-associated activities in Kuwait and their probable toxic effect on humans inhaling them, the USEPA (1991) determined volatile organic at many locations in Kuwait and Saudi Arabia from March 12 to April 4, 1991. Results of this study are given in Table 4.9. The French air monitoring team (Lameloise et al., 1991) also determined volatile organics compounds in Kuwait. These data are also included in Table 4.9.

The maximum concentrations of volatile hydrocarbons reported by USEPA (1991) are lower than those reported by Lameloise et al. (1991). The USEPA (1991) was the first team able to visit Kuwait and then on very short notice. The team was probably not well organized and prepared. In addition, the team expected very high pollutant loading and brought with them less sensitive monitoring and sampling equipment. The mean concentrations of volatile hydrocarbons reported by Lameloise et al. (1991) were lower and comparable to what could be found in Paris. However, the maximum concentrations were generally found to be higher in the vicinity of burning oil fields (Burgan and Ahmadi Oil Fields). No historical data on the ambient distribution of volatile organic chemicals could be found which would allow a comparison with the results of French team or those of USEPA team. Fortunately, levels of volatile hydrocarbons were not excessively high in any residential area. According to the French team, long-term adverse effects from the exposure to the volatile organic chemicals in the Gulf region should be minimal.

Table 4.9. Concentrations ($\mu g/m^3$) of volatile organics in air particulates from the Gulf region.

Chemicals	AIRPARIF (1991)		USEPA (1991)
	Range	Mean	Range
Heptane	4–58	19	0.13–6.0
Octane	2–52	17	–
Nonane	2–38	17	–
Decane	3–59	21	–
Undecane	3–44	16	–
Dodecane	2–22	9	–
Tridecane	1–25	8	–
Benzene	2–12	6	0.4–9.8
Toluene	3–47	17	0.2–8.7
Ethylbenzene	1–16	6	0.1–10.0
p-Xylene	6–64	19	0.3–27.8
o-Xylene	1–36	11	0.1–24.0
Trimethylbenzene	3–27	11	–
Cyclohexane	–	–	0.1–7.7
Nephthelene	–	–	0.5–1.6

– No information available.

USEPA (1991) collected samples from Al Maqua, Al Ahmadi,, Al Burgan, Al Wafra, Al Dhuba, Mina Abdullah, Um Al Haiman, Freedom City, Kuwait City, Adnan hospital, Mubareek Al Kabeer, Al Farwaniah, and Al Jahra in Kuwait, and from Dhahran and Riyadh in Saudi Arabia.

AIRPARIF in Lameloise et al. (1991) monitored Kuwait City, Al Rawdatayn, Al Ahmadi, and Al Magwa Oil Fields.

4.13. Concentrations of metals in air particulates

Crude oil contains many metals, such as nickel (Ni) and vanadium (V), some of which may be found in high concentrations. Many of these metals are toxic to humans if present at relatively high levels or following prolonged exposure to contaminated air. Some the metals that can be found in crude oils are also known to be phytotoxic and zootoxic. Therefore, determination of metals in air particulates is necessary to gain an understanding of the consequences of toxic metal accumulation in organic tissues caused by the Kuwait oil fires.

Like other crude oil, Kuwait crude oil contain appreciable quantities of metals, especially nickel and vanadium (Sadiq et al., 1992; HPI, 1987; El Desouky and Abdulraheem, 1991). Several grades of crude oil are found in Kuwait (Chapter 3), and to assume a typical value for metal concentration in all these curde oil is therefore not logical. To overcome this limitation, several samples of oil from the oil lakes in Kuwait, Mina Al Ahmadi, and Kuwait Export crude were collected and analyzed. Concentrations of aluminum, cadmium, calcium, cobalt, chromium, copper, iron, manganese, potassium, molybdenum, lead, sodium, strontium, nickel, vanadium, and zinc were determined in these samples. Mercury and arsenic were found in trace concentrations. A mean concentration of these metals was calculated and adopted in this book as the representative metal concentration of crude oil burned in the Kuwait oil fires. As presented earlier, a total of 1.12 billion barrels of crude oil were probably emitted during the course of the Kuwait oil fires. From these data, the total input to the environment of some selected metals was calculated and these are listed in Table 4.10.

Table 4.10. Approximate amount of metals emitted by the Kuwait oil fires.

Metals	Amount (Tonnes)*	Metals	Amount (Tonnes)*
Aluminum	917	Cadmium	28
Cobalt	20	Chromium	77
Copper	57	Iron	747
Manganese	20	Molybdenum	125
Nickel	2084	Lead	114
Vanadium	7224	Zinc	432

* 1.12 billion barrels of Kuwaiti crude oil is assumed to have been emitted by the damaged oil wells (see Chapter 3 for further details). Calculations modified from Sadiq et al. (1992).

Based on the data in Table 4.10, it appeared that metal contamination of the environment in the Gulf region was a genuine threat. It was suspected that air particulates in the Gulf region could be enriched by metals. Particular concern was shown for the enhanced concentrations of toxic metals. Metals present in and/or on soot and dust particles can be absorbed through the lungs and thus may effect human health. Realizing the potential threat of metal contamination, many scientists monitored metal concentrations in ambient air particulates (Amin, 1992; Sadiq and Mian, 1993a, 1993b, 1993c, 1993d; Al Sunaid, 1992a, 1992b; Madany and Raveendran, 1992; Mitchell and Sunaid, 1992; Raveendran and Al Mahmood, 1992; EPC, 1991 and many others). Some of these findings are discussed below.

Immediately after the cessation of the hostilities in the Gulf Region, KFUPM/RI at Dhahran, Saudi Arabia, started a particulate monitoring program. Samples of total suspended particulates (TSP) and inhalable particulates (PM10) were collected at Dhahran. These monitoring activities are still continuing. Some air particulate samples from Khafji (Saudi Arabia) were also collected by MEPA and analyzed by KFUPM/RI. Metal concentrations were determined in these samples and these results are given in Table 4.12. Concentrations of nickel and vanadium were determined by Madany and Raveendran (1992) in five inhalable particulate samples (July 31 to August 4, 1991) collected in Bahrain. Saudi Aramco at Dhahran also collected several hundred air particulate samples from many locations and determined metal concentrations, particularly nickel and vanadium (Mitchell and Sunaid, 1992).

The Environmental Protection Committee of Bahrain collected many samples of air particulates that were analyzed for metals (KFUPM/RI, 1992; Raveendran and Al Mahmood, 1992). The Environmental Protection Council in Kuwait collected PM10 particulate samples from Mansoria, Reqa, and Jahra in Kuwait during May, June, and July 1991. Concentrations of several metals were determined in these samples. The United States Environmental Protection Agency collected in March 1991 many PM10 particulate samples inside Kuwait using portable samplers. Metals other than calcium, magnesium, sodium, iron, and aluminum could not be detected in these samples due to excessive dilution during sample preparation and analysis. A few particulate samples were collected during the AIRPARIF air monitoring campaign (March 27 to April 4, 1991) in Kuwait and analyzed for nickel and vanadium (Lameloise et al., 1991).

A Japanese team also visited Kuwait from April 28 to May 5, 1991 and collected some particulate samples which were analyzed for metals. These samples were collected at two locations (Al Ahmadi residential area and International Hotel in Kuwait City).

The locations where air particulate samples were collected for metal determination are shown in Figure 4.9. Some data on metal concentrations in the TSP and PM10 air particulates are summarized in Tables 4.11 and 4.12,

Table 4.11. Metal concentrations (ng/m^3) in total suspended air particulates (TSP).

Metals	Ref. 1	Ref. 2	Ref. 3	Ref. 4
Arsenic	<0.01–29.8	1–2	<0.01–2.28	–
Barium	11.6–708	36–59	9.9–59.4	–
Cadmium	<0.05–123	9–13	<0.05–5.6	1–3
Cobalt	<0.05–52.3	6–13	4.6–13.7	–
Chromium	<0.1–202	21–46	19.2–83.1	3–1155
Copper	13.0–896	14–69	15.4–139.7	3–32
Manganese	11.0–1204	77–107	17.2–176	–
Molybdenum	<0.05–56.5	6–13	5.5–20.9	–
Nickel	3.1–298.1	11–38	19.9–70.3	1–1300
Lead	58–2953	86–939	101–456	87–1062
Titanium	8.1–848	61–87	20.7–206.9	–
Vanadium	0.5–160.2	nd	7.4–38.3	3–10
Zinc	17–984	31–523	22.9–92.4	3-2021

Ref. 1 – Sadiq and Mian (1993a, 1993b, 1993c, 1993d). 230 filter samples were collected from March 1991 to July 1992 at the roof of the Research Institute Dhahran, Saudi Arabia. Air particulate monitoring is continuing.

Ref. 2 – Three filter samples collected in June 1991 were supplied by the Environmental Protection Committee of Bahrain and were analyzed by the Research Institute (KFUPM/RI, 1992)

Ref. 3 – 24 filter samples were collected (7 in May, 4 in September, 7 in October, and 6 in November 1991) by the Meteorological and Environmental Protection Administration (MEPA) of Saudi Arabia and were analyzed by the Research Institute (KFUPM/RI, 1992)

Ref. 4 – Raveendran and Al Mahmood (1992): 1 in February, 3 in March, 2 in April, 10 in May, 3 in June, 6 in July, 1 in August, 3 in September, 9 in October 1991.

nd – Concentrations below the detection limit of the analytical procedure.

– – Information not available.

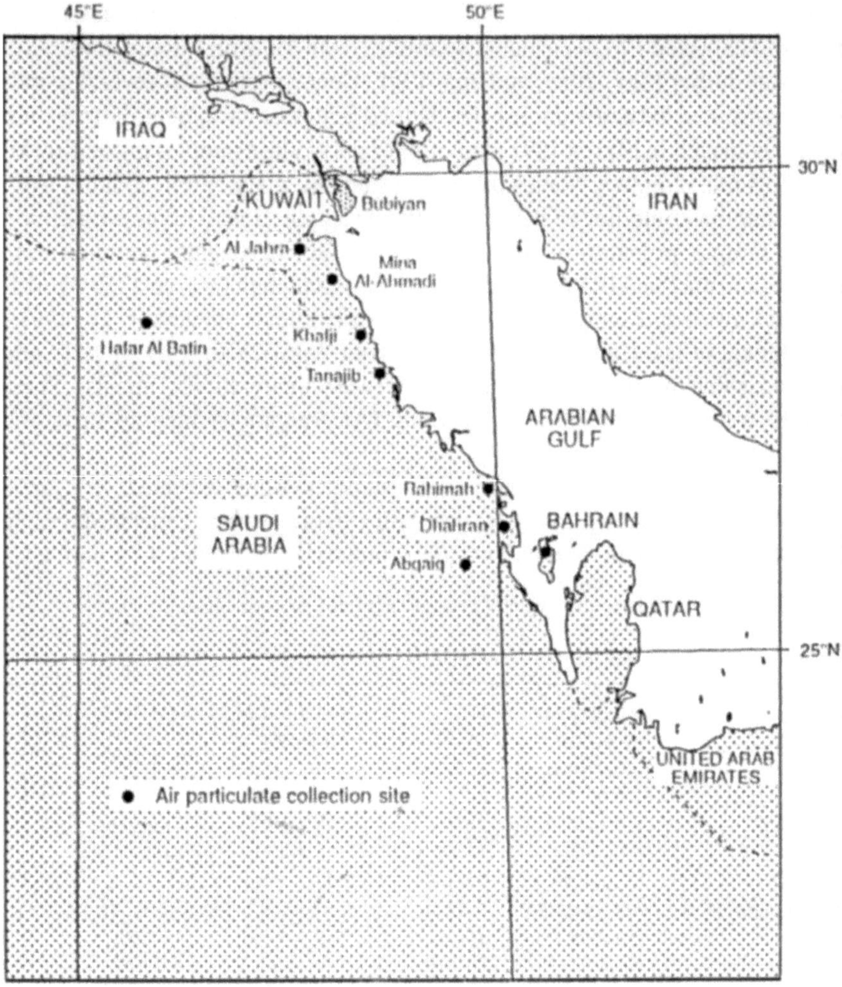

Figure 4.9. A map of air particulate sampling locations for metal analysis.

respectively. These observations suggest a wide range in metal concentrations in the air particulates. This was expected because of the variable climatic conditions. The smoke plume and dust storms would tend to increase the amount of particulates in air and thus enhance metal concentrations.

Many of the above metal determinations were taken over short periods (for a few weeks only). A clear picture of the contributions of the Kuwait oil fires to metals in air particulates cannot be obtained from these short-term observations. Tables 4.11 and 4.12 suggest that the air particulate monitoring program of KFUPM/RI (Sadiq and Mian, 1993a, 1993b, 1993c, 1993d) was more comprehensive. Therefore, these data will be used to evaluate the impact of the Kuwait oil fires on metal concentrations in air particulates. Monthly

Table 4.12. Metal concentrations (ng/m^3 air) in inhalable air particulates (PM10).

Metals	Ref. 1	Ref. 2	Ref. 3	Ref. 4	Ref. 5	Ref. 6
Arsenic	0.1–8.9	2–4	0.2–15.4	–	–	–
Barium	8.2–249	14–37	54.3–598	–	–	–
Cadmium	2.3–27.3	1–3	<0.05–167	1–3	–	–
Cobalt	1.7–30.1	7–9	nd	–	–	–
Chromium	11.0–114	8–42	12.4–387.3	3–1155	80	30–152
Copper	2.7–594	4–20	16.3–390.2	3–32	–	19–138
Manganese	7.6–613	20–112	16.3–630.5	–	198	59–242
Molybdenum	2.1–32.4	7–10	nd	–	–	–
Nickel	9.5–1170	17–33	18.1–352.1	1–1300	83	5–113
Lead	13.9–1222	129–451	80.4–1118	87–1062	258	136–771
Titanium	7.0–1159	18–96	nd	–	–	–
Vanadium	1.8–1165	14–23	37.2–187.1	3–10	28	2–37
Zinc	11.1–1183	31–133	25.3–546.1	3–2021	–	–

Ref. 1 – Sadiq and Mian (1993a, 1993b, 1993c, 1993d). 184 filter samples were collected from April 1991 to July 1992 at the roof of the Research Institute Dhahran, Saudi Arabia. The air particulate monitoring is continuing.
Ref. 2 – Six filter samples collected in June 1991 were supplied by the Environmental Protection Committee of Bahrain and were analyzed by the Research Institute (KFUPM/RI, 1992)
Ref. 3 – 14 filter samples were collected (6 in May, and 8 in November 1991) by MEPA of Saudi Arabia and were analyzed by the Research Institute (KFUPM/RI, 1992)
Ref. 4 – Raveendran and Al Mahmood (1992): 1 in February, 3 in March, 2 in April, 10 in May, 3 in June, 6 in July, 1 in August, 3 in September, 9 in October 1991.
Ref. 5 – EPC (1991) (Environmental Protection Council): Collected PM10 samples during May, June, and July 1991 from three locations in Kuwait. The total number of samples or range of concentrations were not available.
Ref. 6 – El Desouky and Abdulraheem (1991). Six PM10 samples were collected in June 1991. The samples were collected from Mansoria and Reqa using a fixed PM10 sampler.
nd – Concentrations below the detection limit of the analytical procedure.
– – Information not available.

mean concentrations of metals (on a volume basis) were calculated for the PM10 and TSPs collected at Dhahran. In general, metal concentrations both in PM10 and TSP air particulates increased rapidly from April to July 1991, and decreased sharply in August–September 1991 to December 1991–January 1992. In February–March 1992, metal concentrations increased slightly again. To illustrate this general behavior of metal distribution, concentrations of nickel and vanadium in PM10 and TSP samples are plotted in Figure 4.10. The period from April to July is the shamal season when winds are predominantly from the north, relatively strong, and dust storms are more frequent in the Gulf region. As a response to the climatic condition, the particulate load in the air increases, which in turn enhances metals concentrations in the air. This is very clearly demonstrated by Mitchell and Sunaid (1992). They found a very close relationship between vanadium concentrations in the TSP samples and the total TSP amounts at Dhahran in April 1991. After making a correction for background levels, they also showed a decrease in nickel and vanadium concentrations in air particulates from north (Tanajib) to south (Dhahran). The health implications of the metal concentrations reported in Tables 4.11 and 4.12 will be discussed in Chapter 7.

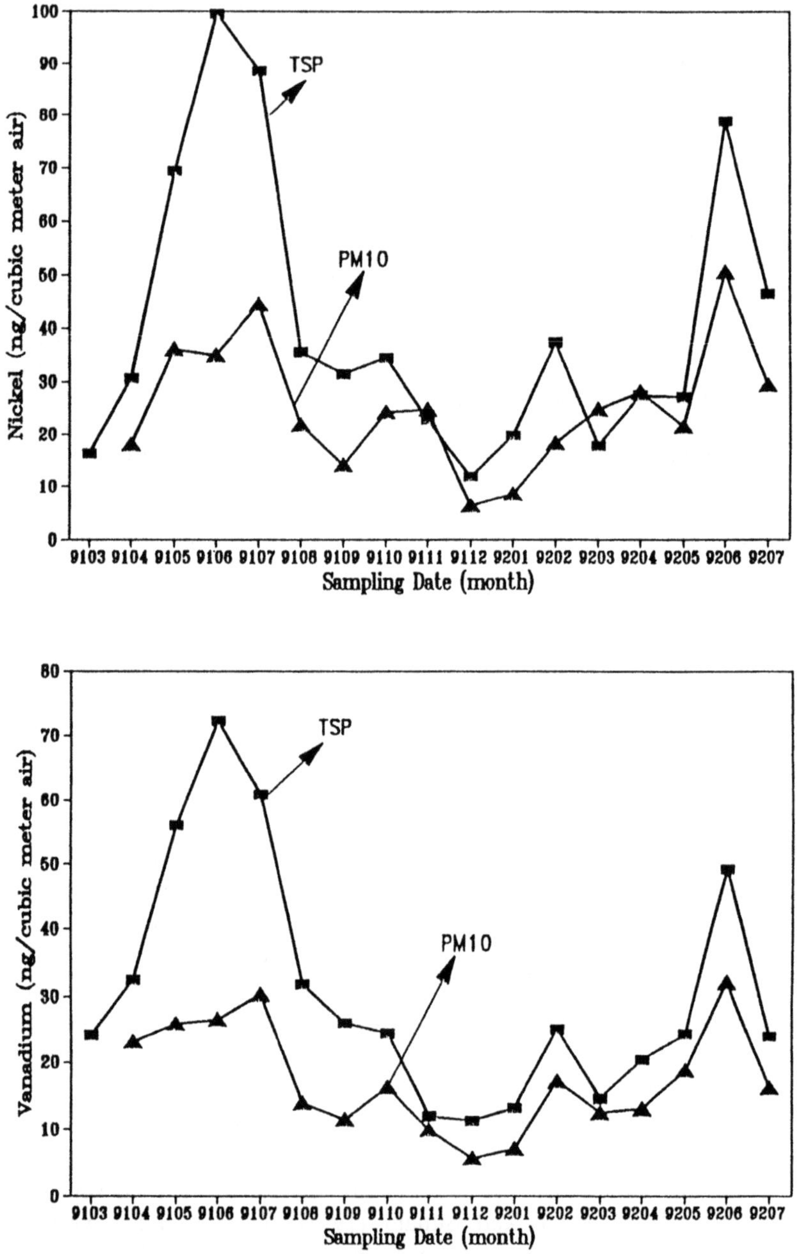

Figure 4.10. Monthly mean concentrations (on volume basis) of nickel and vanadium in the TSP and PM10 samples collected at Dhahran, Saudi Arabia.

Expressing metal concentrations on an air volume basis (ng/m^3 air) is useful to assess human exposure to toxic metals. However, this approach is insufficient to evaluate metal contributions from the Kuwait oil fires to the Gulf region where dust storms are frequent and where high background levels of particulates would in turn enhance metal concentrations when expressed on an air volume basis. However, if expressed on a weight basis, a source-related enhancement in metal concentrations in the air particulates can be seen. Therefore, monthly mean concentrations of metals in air particulates on a weight basis were calculated. To illustrate general trends in metal distribution on a weight basis, the mean concentrations data for nickel and vanadium are plotted in Figure 4.11. Both metals followed a similar trend in the PM10 and TSP samples. Unlike Figure 4.10, the highest concentrations of nickel and vanadium were found in particulates collected during March 1991, and, in general terms, these decreased gradually from April to September 1991, increased suddenly in October–November 1991, and continued their gradual decrease through March 1991. Concentrations of nickel and vanadium increased slightly from April to June 1991. The surge in nickel and vanadium concentrations air particulates in March 1991, may be attributable to the Gulf war. Unfortunately, no particulate samples were collected before March to substantiate this theory. The increase in nickel and vanadium concentrations in October–November 1991 cannot be explained at present. It can be concluded that the Kuwait oil fires contributed to enhanced concentrations of nickel and vanadium in air particulates at Dhahran based upon the concentrations in air particulates collected during 1991 and 1992. A similar behavior was observed for arsenic, cadmium, cobalt, copper, chromium, and molybdenum but not for lead, titanium, and barium. To further strengthen the above conclusions, concentrations of metals in TSP samples collected in 1982/83 and 1991 at the same locations are compared in Figure 4.12. Concentrations of cadmium, molybdenum, chromium, copper, nickel, vanadium, and titanium were higher in TSP samples collected in 1991 as compared with those collected in 1982. Contrary to this, concentrations of lead, barium, and cobalt were found to be higher in air particulate samples collected in 1982 than in samples from 1991.

One of the prime concerns as a possible aftermath of the Gulf war was the health effects of human exposure to war-related environmental contamination. The health implications of metals in air particulates will be discussed separately. Data on metal concentrations in other parts of the world were collected and are compared with metal concentrations found in the Gulf region in Table 4.13.

As expected, relatively higher concentrations of nickel, vanadium, and chromium were found in the inhalable particulates from this region. Considering the level of industrialization in the Gulf region, the concentrations of metals measured in air particulate samples seem high. This comparison suggests contamination of ambient air particulates in the Gulf region. Desert soils had been contaminated, especially with metals, from the activities related to the Gulf war (Chapter 6). It is feared therefore, that resuspension of contaminated soils during high winds may maintain exposure of the local population to metals in the future. Continuous monitoring of particulate metals is recommended in the future for the Gulf region.

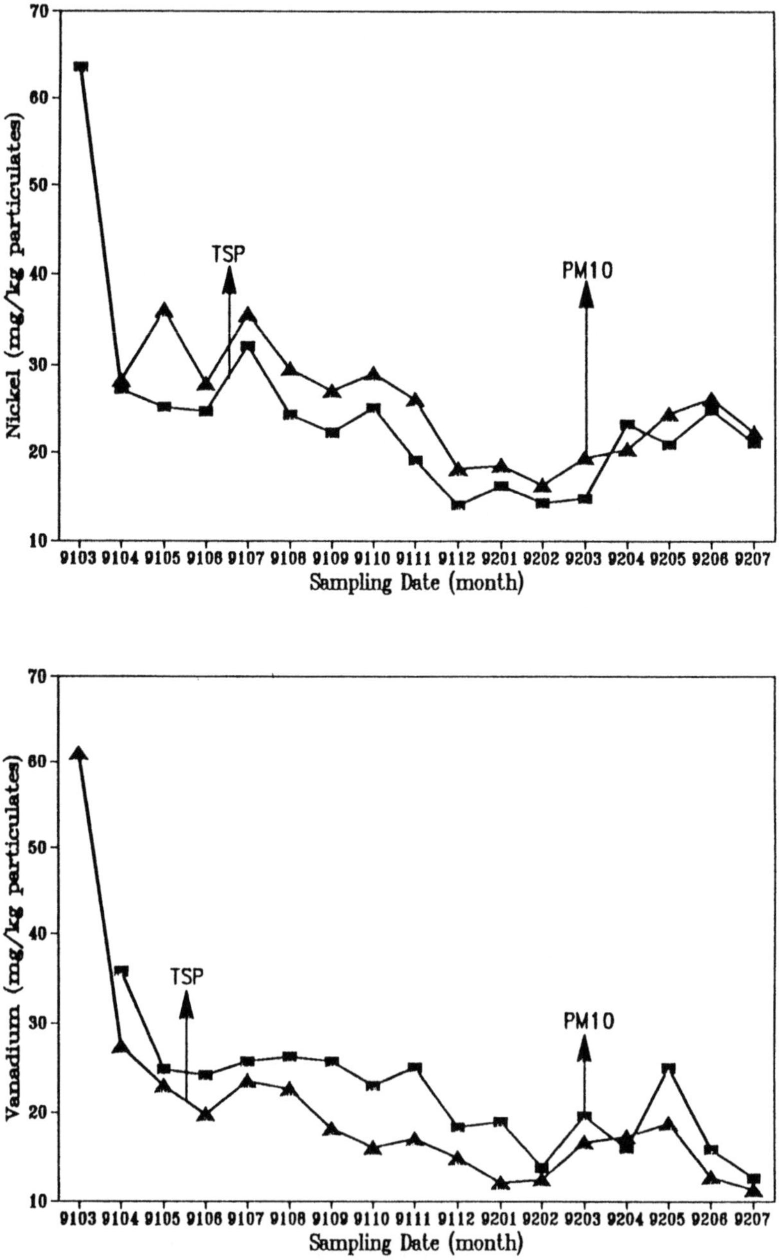

Figure 4.11. Monthly mean concentrations (on weight basis) of nickel and vanadium in TSP and PM10 samples collected at Dhahran, Saudi Arabia.

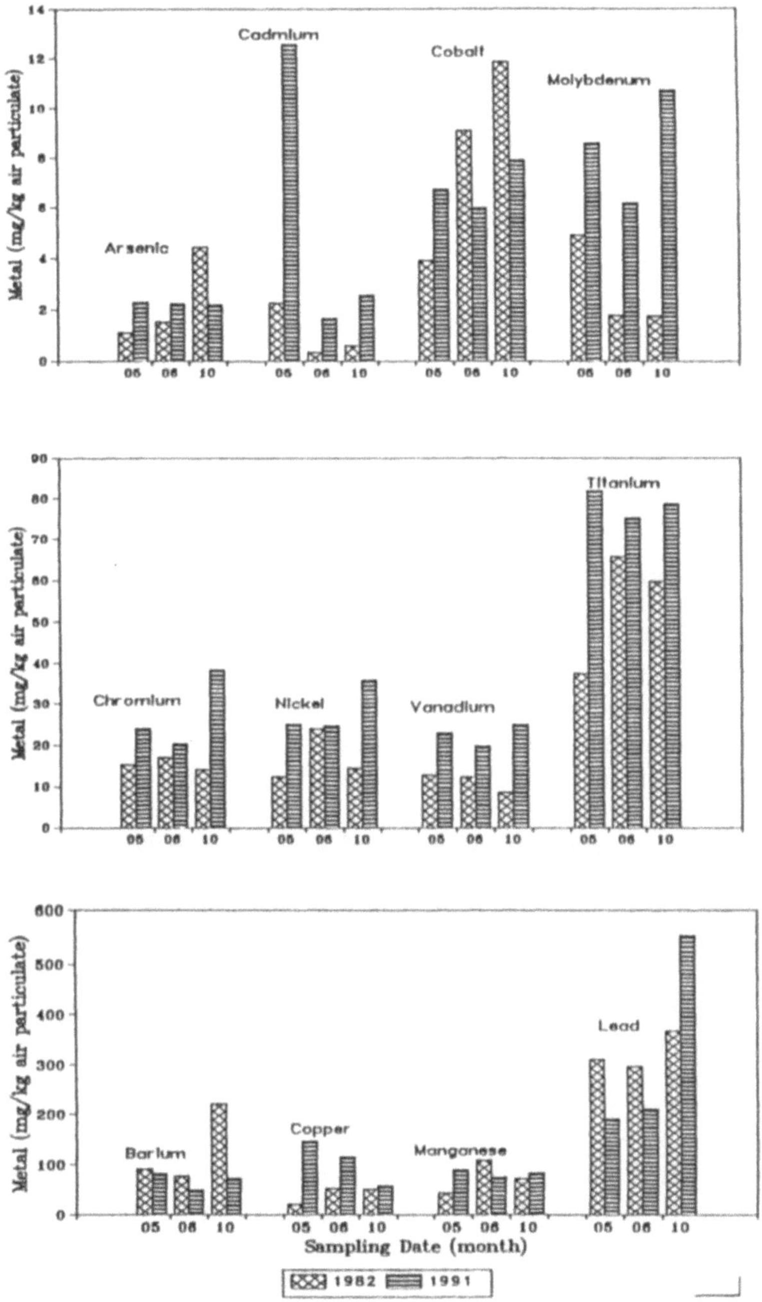

Figure 4.12. A comparison of metal concentrations in TSP collected at Dhahran, Saudi Arabia during 1982 and 1991.

Table 4.13. Metal concentrations (ng/m^3) in air particulates from different areas of the world.

Metals	Canada[1]	USA[1]	Europe[1]	Others[1]	Gulf Region[2]
Arsenic	7.7–626	2–2320	5–330	20–85	0.1–15.4
Cadmium	10–51	0.2–7000	0.4–260	0.6–177	2–167
Nickel	4–371	1–328	0.3–1400	2–158	1–1300
Lead	353–3416	30–96270	10–9000	1–11,020	14–1222
Vanadium	10–130	0.4–1460	11–73	2–180	2–1165
Zinc	55–1390	15–8328	160–8340	110–2700	3–2021
Cobalt	1–8	0.2–83	0.4–18	0.3–10	2–30
Chromium	4–26	2–124	4–227	0–227	8–1155
Copper	17–500	3–5140	13–2760	2–6810	3–594
Manganese	20–70	4–448	23–850	2–590	7–630

[1] Data were taken from Schroeder et al. (1987). Schroeder et al. (1987) reviewed a large body of data collected worldwide. These data include those collected from contaminated locations.
[2] Data extracted from the literature cited in this book.

4.14. Concentrations of organics in air particulates

Over 80% of crude oil is organic in origin and its organic fraction consists of several thousand organic chemicals. Therefore, emissions from the combustion of fossil fuel are expected to contain hundreds of organic chemicals. Identification of these chemicals is not an easy task. The chemical composition of the emission product from the burning of crude largely depends on the type of crude and combustion conditions. With reference to the Kuwait oil fires, several types of crude oil were burned and, since it was free burning, combustion conditions were also variable. Because of the complexity of combustion conditions, hundreds of organic chemicals were expected in the emission from the Kuwait oil fires. Broadly speaking, the organics expected in the fire emission included polycyclic aromatic hydrocarbons (PAH), aliphatic hydrocarbons, and volatile organics. We have already discussed volatile compounds in the air. The remaining two groups of organic compounds will be discussed here.

PAHs enter the environment as a result of the incomplete combustion of fossil fuels, petroleum distillates, automobile exhausts, and tobacco smoke. Several PAH compounds have both mutagenic and carcinogenic properties. The health significance of PAH is discussed separately (Chapter 7). Here the results of monitoring in the Gulf region are discussed. Recently, many studies on PAH concentration in the ambient air particulate samples have been carried out at several places in the world (Ang, 1986). These studies have shown that the concentrations of PAH in air particulates vary widely, particularly with location and season. In rural areas, the concentration of PAH varies from about 0.001 to 2.00 μg/1000 m^3, and from 0.1 to 60 μg/1000 m^3 in urban areas (Sawicki et al., 1960). Benz(a)pyrene concentrations ranging from 20 to 400 μg/1000 m^3 have been reported in some polluted atmospheres (Egan and

Sawyer, 1975). The proposed allowable limit for benzo(a)pyrene in air was set by OSHA at 0.2 $\mu g/m^3$. Recently the proposed total PAHs criteria for human health protection was set at a value of 0.0109 $\mu g/m^3$ in air. However, these criteria are still controversial (Eisler, 1987). Estimates of the annual intake by humans of PAH are between 1 and 10 mg with benz(a)pyrene contributing 0.1 to 1.0 mg and 0.35 to 1.5 mg, respectively, per person in western countries (Borneff, 1975; Fritz, 1975).

Because of the health hazards associated with PAHs and the high potential for atmospheric contamination by these compounds from the burning oil wells in Kuwait, the levels of oil hydrocarbons in the atmosphere, especially the polycyclic aromatic compounds (PAHs), were measured by many researchers. Some of the data gathered during these investigations are summarized in Table 4.14.

Table 4.14. Concentrations of PAHs (ng/m^3) in ambient air particulates in the Gulf region.

Compound	Lameloise et al. (1991)	NILU	Harwell	Madany	USEPA (1991)	EPC (1991)
Naphthalene	–	94–228	–	–		
Acenaphthylene	–	1.4	1.6–3.6	–		
Fluorene	–	7–35	–	–		
Phenanthrene	–	22–49	0.1–0.6	–		
Anthracene	–	6–11	–	–	3–7	
Fluoranthene	19–110	7–17	4–20	0.1–0.11	5–17	0.4–71.4
Pyrene	–	6–16	11–101	0.11–0.11	5–13	6.1–40.1
Benzo(a)anthracene	1.3–5.6	0.4–1.3	0.9–2.5	–		0.1–35.7
Chrysene	–	–	0.9–4.3	0.08–0.29		
Benzo(b)fluoranthene	3–20	–	0.0–7.2	0.45–0.77	1–10	0.7–44.3
Benzo(k)fluoranthene	2–12	–	–	0.08–0.17		
Benzo(e)pyrene	4–26	0.6–1.3	–	0.22–0.74	1–6	
Benzo(a)pyrene	5–30	0.3–0.8	0.3–5.1	0.42–1.12		0.1–30.6
Dibenzo(a,h)anthracene	0.1–9	–	–	0.07		0.1–50.9
Benzo(g,h,i)perylene	4–26	1–15	–	0.05–0.1	1–6	
Indeno(1,2,3-cd)pyrene	6–41	1–13	0.1–2.9	0.4–1.05	3–40	
Perylene	–	–	–	0.06–0.19		
Total PAHs	44–280	146–387	19–147	2–5	19–99	7–237

Lameloise et al. (1991).	Air particulate samples were collected on 3/28, 3/31, 4/1, 4/3, 4/4, 1991 from Kuwait City.
NILU	Sivertsen and Berg (1991). Air particulate samples were collected on May 14/15, 21, 27, and June 15, 1991.
Harwell	Harwell Laboratory as cited by El Desouky and Abdulraheem (1991). Air particulate samples were collected on March 25, 26, April 1, 3, 6, 7, 16, 18, and June 26, 1991.
Madany	Madany and Raveendran (1992) analyzed air particulates collected in Bahrain.
USEPA (1991)	Air particulate samples were collected from different places from Kuwait using portable PM10 samplers and were analyzed for PAH in the United Sates.
EPC (1991)	Samples were collected from Mubareek Hospital (3), Mansoria (4), and Shuwaikh industrial area (11) during April–July 1991.

Many factors can effect concentrations of individual PAH compounds. In addition to wind direction, particulate load, and distance from the source, the analytical procedure also effects concentrations of PAHs. Significantly different results can be obtained using different analytical procedures. Therefore, the data in Table 4.14 should be interpreted carefully as different methodologies were used to collect the air particulates, for sample preparation, and for determination of PAHs. In this book, the above data will be used to determine general trends in PAHs distribution.

As shown in Table 4.14, many investigators determined concentrations of PAHs in PM10 samples collected from Kuwait, Bahrain, and Saudi Arabia. An evaluation of the data in Table 4.14 shows that concentrations of individual PAH compounds vary widely even within the same set of data. For example, Lameloise et al. (1991) found that concentrations of fluoranthene (9–110 ng/m^3) was the highest followed by Indeno(1,2,3-cd)pyrene (6–41 ng/m^3). The highest concentrations of naphthalene (94–228 ng/m^3) were reported by Sivertsen and Berg (1991). No other investigator reported this compound. The second highest compound found by Sivertsen and Berg (1991) was phenanthrene (22–49 ng/m^3). The EPC (1991), citing the data from Harwell laboratories, reported that pyrene concentrations (1–101 ng/m^3) were the highest followed by fluoranthene (4–20 ng/m^3). Concentrations of PAH reported by Madany and Raveendran (1992) were the lowest of all the data reported in Table 4.14. Though the data are limited, it may be concluded that concentrations of PAH were lower in PM10 particulates collected in Bahrain than those collected in Kuwait. The EPC (1991) reported data on ambient concentrations of PAHs in particulates collected from different locations in Kuwait. In the EPC's review, it was found that concentrations of fluorenthene (0.7–44.3 ng/m^3) and pyrene (6.1–40 ng/m^3) were higher than any other organic compounds. The wide variations in the concentrations of PAHs in air particulates was attributed to the episodic nature of the smoke at the sampling locations (EPC, 1991). Mulholland et al. (1991) collected soot samples from two locations in the Al Magwa Oil Field and determined concentrations of PAHs. Their work was of preliminary nature and sampling was not quantitative. Therefore, these results are not included here. Particulate samples were also collected from Kuwait and Doha during August 4–9, 1991 by the NOAA/Arabian Gulf Program and were analyzed for PAHs. Most of the PAHs detected in these Kuwait samples were petrogenic and not emission products (Henry, 1991).

Immediately after the Gulf war, an air particulate monitoring program was started by KFUPM/RI in Dhahran, Saudi Arabia. In addition to PM10, samples of TSP were obtained. Samples of PM10 were also obtained from Khafji, the northern most Saudi town on the Arabian Gulf coast, and from the State of Bahrain. The collected samples were analyzed for aliphatic hydrocarbons and PAHs. The presence of 22 PAHs was observed in the PM10 samples (Fayad et al., 1993). The results of the study showed that these samples were contaminated with aliphatic hydrocarbons and PAHs. Concentrations of

PAHs found in the inhalable (PM10) air particulate samples collected from Dhahran, Khafji, and Bahrain ranged from 0.0 to 2.4, 0.0 to 7.2, and 0.42 to 6.67 ng/m^3, respectively. The concentrations of the n-alkanes in the Dhahran and Khafji air particulate samples ranged from 0.0 to 66.5 ng/m^3 and 0.0 to 242.6 ng/m^3, respectively. Concentrations of aliphatic compounds in the PM10 samples were higher than those of PAHs. It was also found that concentrations of oil hydrocarbons were higher in PM10 samples than those found in TSP samples collected simultaneously at the same location (Fayad et al., 1993).

Concentrations of total PAHs in PM10 air particulate samples ranged from 7 to 387 ng/m^3. The proposed allowable limit for benzo(a)pyrene in air was set by OSHA at 0.2 $\mu g/m^3$. Concentrations of this compound ranged between 0.1 to 31 ng/m^3. This range is much lower than the proposed guideline. Recently, the proposed total PAH criterion for human health protection was set at 10.9 ng/m^3 in air. However, this criterion is still controversial (Eisler, 1987). Concentrations of total PAHs found in many of the air particulate samples, except those reported by Madany and Raveendran (1992) were higher than the proposed controversial limit. The health implications of the PAHs will be discussed in a separate chapter. To further elaborate on the PAHs situation in the Gulf region, concentrations of PAHs in several cities around the world are compared in Figure 4.13. From this figure it is evident that the concentrations of PAHs in Dhahran and Berlin are the lowest. Comparable high concentrations of PAHs were found in Kuwait and Paris. All other locations monitored for PAHs contained lower concentrations in the air particulate samples when compared with those found in Kuwait and Paris.

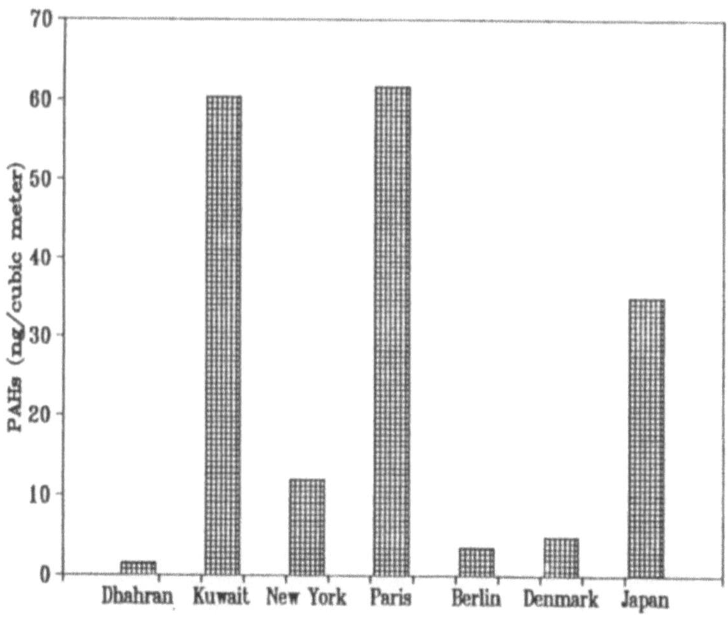

Figure 4.13. A comparison of PAHs concentrations found in the Gulf region with the similar data from other locations in the world.

4.15. Future response and research needs

The following is suggested:

1. The Kuwait oil fires was a unique episode that has shown several weaknesses in the national and international coordination. Local organizations were in such a disarray that their overall contributions to data collection was smaller than those of foreign organizations. Foreign scientific teams either came on their own initiative (like Greenpeace, Harvard University, etc.) or on instructions of their government (for example USEPA, NOAA, the British Meteorological Office, etc.). The UN interest was mainly limited to meteorological investigations, especially modeling activities (WMO, 1991, 1992). The UN failed to coordinate international activities in the Gulf region. We strongly feel that, in environmental disasters such as the Kuwait oil fires and the Arabian Gulf oil spill, the United Nations Environmental Program (UNEP) should have funds and expertise available to respond quickly and effectively.
2. Concentrations of inhalable air particulates were enhanced as a result of the Gulf crisis. A long-term monitoring program is recommended for the Gulf region. The contribution of soil erosion to this problem should be investigated and mitigative measures should be adopted by local governments.
3. Concentrations of many metals and organic compounds were found to be higher in the Gulf region than in many parts of the world. These chemicals should be identified and monitored on a long-term basis. Inhalation and retention of these chemicals by the human body should also be investigated during this program.
4. A regional organization, such as the Regional Organization for the Protection of Marine Environment (ROPME), should be established to collect all environmental data produced in the region, synthesize these data, and recommend mitigative measures to governments. The organization should develop an environmental database for the Gulf region.

References

Al Besharah, J. 1991. The Kuwaiti oil fires and oil lakes - facts and numbers. In Environmental and Health Impact of the Kuwaiti Oil Fires. Proc. of an International Symp. held at the University of Birmingham, 17th October, pp. 12–15.

Al Sunaid, A.A. 1992a. Saudi Aramco AMMNET historical air quality and Kuwait fire SO_2 removal mechanism. Paper presented at the Second WMO Meeting of Experts to Assess the Response to and Atmospheric Effects of the Kuwait Oil Field Fires, Geneva, 25–29 May.

Al Sunaid, A.A. 1992b. Saudi Aramco air quality monitoring program (AMMNET). Proc. Ist Bahrain International Conference on Environment, 24–26 February, pp. 82–93.

Amin, M.B. 1992. Environmental and health impact assessment study in the Arabian Gulf region due to oil spill and the Kuwait oil fires. Volume 1, pp. 121–148. Proceedings of the 1st Bahrain International Conference on Environment. Bahrain 24–26 February.

Ang, K.P., Gunasingham, H., and Tay, B.T. 1986. The distribution of polynuclear aromatic hydrocarbons in ambient air particulate in Singapore. Environ. Monit. Assess. 6:171–180.

Bakan, S., Chlond, A., Cubasch, U., Feichter, J., Graf, H., Grassel, H., Hasselman, K., Kirchner, I., Latif, M. Roeckner, E., Sausen, R., Schlese, U. Schriever, D. Schult, I., Schumann, U., Sielmann, F., and Welke, W. 1991. Climate response to smoke from the burning of oil wells in Kuwait. Nature 351:367–371.

Blum, S., Arp, E.W., Smith, A.H., and Tyroler, , H.A. 1979. Stomach cancer among rubber workers: an epidemiological investigation. In Dement, J.M. and Lemen, R. (ed.), Dusts and Diseases. Pathotox Publishers, Park Forest South, USA.

Borneff, J. 1975. Drinking water production from service water. Example of the Lake of Zurich with regard to polycyclic, metals, and chlorinated hydrocarbons. Gas. Wasser. Abwasser, 55:467.

Browning, K.A., Allam, R.J., Ballard, S.P., Barnes, R.T.H., Bennetts, D.A., Maryon, R.H., Mason, P.J., McKenna, D., Mitchell, J.F.B., Senior, C.A., Slingo, A., and Smith, F.B. 1991. Environmental effects from the burning oil wells in Kuwait. Nature 351:363–367.

Cahill, T.A. 1992. Aerosol measurements in the Kuwait oil fire plume. pp. A14–A15. Paper presented at the WMO Meeting of Experts on the Atmospheric Part of the Joint UN Response to the Kuwait Oil field Fires, Geneva, April May 25–29, Final Report on the WMO meeting of experts on the atmospheric part of the Joint UN. Response to the Kuwait Oil field Fires. Geneva, April. 27–30.

Cofer, W.R. III, Stevens, R., Winstead, K., Pinto, J., Sebaacher, D., Abdulraheem, M., Al Shafi, M., Mazurek, M., Rasmussen, R., Cahoon, D., and Levine, J.S. 1992. the Kuwait oil fires: composition of source smoke. Cited by Daum et al. (1992).

Daoud, M.W. 1991. Ahmadi City under the smoke. pp. 27–31, In Environmental and Health Impact of the Kuwaiti Oil Fires. Proc. of an International Symp. held at the University of Birmingham, 17th October.

Daum, P.H., Al Sunaid, A., Busness, K.M., Hales, J.M., and Mazurek, M. 1992. Studies of the Kuwait oil fires during mid-summer 1991. Brookhaven National Laboratory - USA, Battelle Pacific Northwest Laboratory - USA, and Saudi Arabian Oil Company - Saudi Arabia, Final Report.

Eisler, R. 1987. *Polycyclic Aromatic Hydrocarbons, Hazards to Fish, Wildlife and Invertebrates.* United States Fish and Wildlife Service, LEAD 87-189825.

El Desouky, M and Abdulraheem, M.Y. 1991. The impact of oil well fires on the air quality in Kuwait. pp. 16–26, In Environmental and Health Impact of the Kuwaiti Oil Fires. Proc. of an International Symp. held at the University of Birmingham, 17th October.

EPC, 1991. State of the Environment Report. A Case Study of Crimes Against the Environment. Environmental Protection Council, Kuwait, pp. 1–49.

Fayad, N.B., El-Mubarek, A.H., Edora, R.L., Ghamdi, A., and Polancos, A.B. 1993. Organic pollutants in ambient air of the Eastern Province of Saudi Arabia after the 1991 Kuwait oil well fires. Paper presented at the Industrial Conference in Riyadh, February 1993.

Fritz, W. 1972. Extent and source of contamination of our foods with carcinogenic hydrocarbons. Ernahrungsforschung, 16:547.

Henry, C.B. 1991. Draft report on Kuwait PAHs analyses. Letter to Robinson, J. Director NOAA/Arabian Gulf Program, Institute for Environmental Studies, Louisiana State University.

Hobbs, P.V. and Radke, L.F. 1992. Airborne studies of the smoke from the Kuwait oil fires. Sci. 256:987–991.

HPI Consultants Inc. 1987. Crude Oil Assay Handbook. First Edition, Austin, Texas, USA.

IARC, (1984). Monograph on the evaluation of the carcinogenic risk of chemicals to humans, 33:35–86.

Jenkins, G.J., Johnson, D.W., McKenna, D.S., and Saunders, R.W. 1992. Aircraft measurements of the Gulf smoke plume. Weather 47(6):212–220.

Johnson, D.W., Kilsby, C.G., McKenna, D.S., Saunders, R.W., Jenkins, G.J., Smith, F.B., and Foot, J.S. 1991. Airborne observations of the physical and chemical characteristics of the Kuwait oil the smoke plume. Nature 353:617–621.

KFUPM/RI, 1992. Gulf Oil Spill 1991. Arabian Gulf Atmospheric Pollution Research Program. 2nd Interim Report. The Research Institute, King Fahd University of Petroleum and Minerals, Dhahran, Saudi Arabia.

KFUPM/RI, 1991a. Gulf Oil Spill 1991. Arabian Gulf Atmospheric Pollution Research Program. 1st Interim Report. The Research Institute, King Fahd University of Petroleum and Minerals, Dhahran, Saudi Arabia.

Lameloise, P., Thibaut, G., Masse, R., Lafuma, J., Person, A., and Pasquereau, M. 1991. Measurement Campaign of the Regional Mobile Laboratory for Measurements of Air Quality in Kuwait. Final Report, March 27 to April 4, Paris, France.

Linden, E. 1991. Getting blacker everyday. Time, May 27.

Madany, I.M. and Raveendran, E. 1992. Polycyclic aromatic hydrocarbons, nickel and vanadium in air particulate matter in Bahrain during the burning of oil fields in Kuwait. Sci. Total Environ. 116:281–289.

MEPA (1402 H). Environmental Protection Standards Document No. 1401–01, Meteorology and Environmental Protection Administration, Ministry of Defense and Aviation, Kingdom of Saudi Arabia.

Mitchell, D.P. and Al Sunaid, A.A. 1992. AMMNET air particulate chemical characterization and metal Analysis. Proc. the International Conference on Chemistry in Industry. November 14–16, Bahrain.

Mulholland, G.W., Benner, B.A., Fletcher, R.A., Steel, E., Wise, S.A., May, W.E., Madrzykowski, D., and Evans, D. 1991. Report of Test, FR 3985 - Analysis of smoke samples from oil well fires in Kuwait. The United States Department of Commerce, National Institute of Standards and Technology, Gaithersburg, USA.

Okita, T. 1991. Review of the survey by Japanese team. Final Report to Sponsor of the Kuwait Oil Fires Conference. !5 October, Harvard School of Public Health, Boston, USA. pp. 35–36.

Parks, H.G., Veys, C.A., Waterhouse, J.A.H., Cooke, D., and Peters, A.T. 1982. Cancer mortality in British rubber industry. Br. J. Indust. Medicine, 39:209–220.

Penner, J.E. 1991. Kuwaiti oil fires. World Meteorological Organization. Global Atmosphere Watch. No 72, Draft Report of the WMO Meeting of Experts on the Atmospheric Part of the Joint UN Response to the Kuwait Oilfield Fires, Geneva, April 27–30, pp. 40–43.

Raveendran, E. and Al Mahmood, A.M. 1992. Trace metals in the atmosphere of Bahrain during the Kuwait oil fires. Environmental Technol. 13:1–4.

Sadiq, M. and Mian, A.A. 1993a. Concentrations of chromium and copper in air particulates of Dhahran, Saudi Arabia. Arabian J. Sci. Eng. (accepted)

Sadiq, M. and Mian, A.A. 1993b. Concentrations of nickel and vanadium in air particulates collected at Dhahran during and after the Kuwait oil fires. Submitted to Atmospheric Res.

Sadiq, M. and Mian, A.A. 1993c. Concentrations of lead, cadmium, cobalt and titanium in air particulates collected at Dhahran during and after the Kuwait oil fires. Submitted to Atmospheric Res.

Sadiq, M. and Mian, A.A. 1993d. An investigation of the Kuwait oil fires on metal concentrations in air particulates. Water Air Soil Poll. (submitted)

Sadiq, M., AlThagafi, K.M. and Mian, A.A. 1992. A Preliminary Evaluation of Metal Contamination of Soils From the Gulf war Activities. Bull. Environ. Contam. Toxicol. 49:663–639.

Saskolne, C.L., Pagano, G., Cipollaro, M., Beaumont, J.J., and Giordano, G. 1989. Epidemiological and toxicological evidence of chronic health effects and the underlying biological mechanisms involved in sub–lethal exposures to acidic pollutants. Archiv. Environ. Health 44:180–191.

Sawicki, E., Houser, T.R., Fox, F.T., and Stanley, T.W. 1960. Benzo(a)pyrene content of the air of American communities. Amer. Ind. Hyg. Assoc. 21:443.

Schroeder, W.H., Dobson, M., Kane, D.M., and Johnson, N.D. 1987. Toxic trace elements associated with airborne particulate matter: a review. JAPCA 37:1267–1285.

Shaw, W. 1992. Smoke at Bahrain during the Kuwait oil fires. Special Issue Gulf War Meteorology. Weather 47:220–227.

Sivertsen, B. and Berg, T.C. 1991. Air quality measurements in Kuwait. Norwegian Institute for Air Research (NILU), July 1991.

Suthers, J.R. 1991. Air and water pollution in Jubail. pp. 32–39, In: Environmental and Health Impact of the Kuwaiti Oil Fires. Proc. of an International Symp. held at the University of Birmingham, 17th October.

Tawfiq, N.I. 1991. Response by Saudi Arabia to the Environmental crisis caused by the Gulf War. pp. 40–43, In Environmental and Health Impact of the Kuwaiti Oil Fires. Proc. of an International Symp. held at the University of Birmingham, 17th October.

USEPA, 1991. the Kuwait oil fires. Interagency Interim Report. Air sampling and monitoring of the Kuwait oil fires. April 3, pp. A1–A15.

USEPA 1987. Summary Review of the Health Effects Associated with Copper. Health Issue Assessment. United States Environmental Protection Agency, EPA/600/8-87/001.

USEPA, 1978. National Primary and Secondary Ambient Air Quality Standards. Fed. Register 43:46246–46277.

Utell, M.J. 1985. Effect of inhaled acid aerosols on lung mechanics: An analysis of human exposure studies. Environ. Health Pers. 63:39–44.

Warner, S.F. 1991. The Environmental consequences of the Gulf war. Environment 33(5):7–9, 25–26.

WMO, 1991. World Meteorological Organization. Global Atmosphere Watch. No 72, Draft Report of the WMO Meeting of Experts on the Atmospheric Part of the Joint UN Response to the Kuwait Oil Field Fires, Geneva, April 27–30. Report on the WMO meeting of experts on the atmospheric part of the Joint UN. Response to the Kuwait Oilfield Fires. Geneva, April. 27–30.

WMO, 1992. World Meteorological Organization. Global Atmosphere Watch. No 82, Draft Report of the WMO Meeting of Experts on the Atmospheric Part of the Joint UN Response to the Kuwait Oilfield Fires, Geneva, April May 25–29, Report on the WMO meeting of experts on the atmospheric part of the Joint UN. Response to the Kuwait Oilfield Fires. Geneva, April. 27–30.

CHAPTER 5

Air pollution from military operations

5.1. Summary

The following are highlights of this chapter.

- In response to the occupation of Kuwait by Iraq, the United States and its allies deployed military forces in the Gulf region from August 1990 through January 1991. By mid January 1991, the strength of the Allied troops was over 700,000 personnel. The Iraqi troops were estimated to range between 300,000 to 500,000 in Kuwait territory.
- Maneuvering of both the Allied and Iraqi troops, erection of defensive structures, and preparation of military facilities before the War disturbed the ground surface in areas of operations, thereby enhancing wind erosion of the desert surface. This resulted in an increase in airborne particulates in the area of conflict and adjacent areas.
- The Gulf war left about 40,000 km^2 of Kuwait, Iraq, and Saudi Arabia littered with live and spent ammunition, explosives, mines, and other military-related wastes.
- The ground operation (including preparatory exercises, etc.) of the Gulf war was estimated to have consumed about 7 billion gallons of fuel during the Gulf crisis. Air operations used more than 600 million gallons of jet fuel during the Gulf crisis. In total, the fuel consumption for the Gulf crisis is estimated at about 8 billion gallons.
- The major pollutants emitted as emission products during ground and air operations were carbon dioxide about 71 million tonnes, total particulate matter (smoke/soot) about 6,959 tonnes (of which elemental carbon was about 4,911 tonnes), carbon monoxide 42,160 tonnes, nitrogen oxides 107,086 tonnes, sulfur dioxide (as sulfuric acid) 270 tonnes, toluene 44 tonnes, anthracene 43 tonnes, fluorathene 9 tonnes, pyrene 11 tonnes, and a large but unknown quantity of toxic metals and other organics. The majority of these pollutants were emitted during February 1991 – within a short period of about 2–3 weeks. This suggests that concentrations of some of these pollutants might have been enhanced several times greater than background levels during this period.

- It is estimated that a total of about 120,000 tonnes of munitions was used in the Gulf war, including over 84,000 tonnes of bombs that were dropped on Iraq installations in Kuwait and Iraq. This estimate does not include Iraqi munitions destroyed by the Allied bombardment. Destruction of Iraqi munitions stockpiles were a top priority of the Allied air forces and some air strikes were reported to have been very successful. Therefore the total amount of munitions used in the Gulf war may be much higher than the value quoted above. Contrary to media reports during the Gulf war, only about 10% of the bombs used in the Gulf war were smart bombs (that accurately hit their targets), the remaining were conventional bombs.
- The amount of carbon dioxide, carbon monoxide, and nitrogen oxide produced during military operations were 53, 5, and 133% of that emitted by the Kuwait oil fires, respectively. The majority of the pollutants associated with the war were emitted in just 43 days, whereas emissions from the Kuwait fires were spread over eight months. The short period of the war and large quantity of pollutants may have had more adverse effect on air quality than the Kuwait oil fires.
- Understandably, no information was collected during the Gulf war as people were more concerned with their lives rather than with air quality. Later investigations suggested that there may have been high concentrations of toxic metals in the atmosphere during the Gulf war. This may be true for other pollutants. From the information discussed in this chapter, it can be concluded that air quality was adversely affected due to war-related activities. The health implications of this pollution episode will be discussed in Chapter 7.
- Several Iraqi nuclear facilities were bombed by the allied air force. It is not known if radiation was released from these facilities. Therefore, the environmental impact caused by damage to these facilities is unknown. The Allied Forces also bombed chemical weapon plants and stockpiles. The exact damage to these facilities may never be known.
- Many chemicals that were hazardous to health and/or the environment were used in the Gulf war as a military hardware and supplies, for example, inserts for the 30 mm cannon shell for A-10 airplane were manufactured from spent uranium. Military personnel and civilian populations may have been exposed to these toxic substances. Such information is regarded as a military secret and is classified.
- Information on the chemical composition of munitions, fuel, and its additives is classified. This secrecy has complicated our efforts to reasonably assess the environmental impact of the Gulf war, especially on humans.

It is concluded that many more questions remain to be answered with respect to the environmental impact of military related activities. Without the cooperation of military authorities of both sides, the true extent of the environmental damage from the Gulf war may never be realistically assessed.

5.2. Introduction

Soon after the occupation of Kuwait by Iraqi army on August 2, 1990, the United States expressed its determination to evict Iraq from Kuwait using force if peaceful means failed. To emphasize its commitment, the United States began to deploy its forces in the Gulf area. The deployment operation was started in August 1990 and continued through January 1991. Twenty seven other countries also joined the United States. By January 1991, the total strength of the Allied Forces increased to over 700,000. Table 1 shows the deployment of forces by different countries. In addition, between 350,000 to 600,000 Iraqi soldiers were committed to Kuwaiti territory.

Table 5.1. Deployment of allied forces during the Gulf war.

Country	Troops Deployed
The United States	570,000
Saudi Arabia	40,000
Egypt	38,500
The United Kingdom	35,000
Syria	19,000
Kuwait	11,500
Pakistan	11,000
France	7600
Bangladesh	2000
Morocco	1700
Afghan Mujahedeen	310
Total Allied troops	*736,610*
Iraqi troops	300,000–500,000

With the repeated failure of diplomatic efforts to resolve the Kuwait conflict peacefully, both Iraq and Allied Forces stepped up their defensive and offensive preparations. These activities damaged the already severely stressed desert ecosystem. Finally, the Gulf war began on January 16, 1991. The Allies' strategy was to humble Iraq by using air power before launching a ground attack. As a result of this strategy, the Allied air forces flew 110,000 sorties in 43 days, dropping about 84,000 tonnes of bombs on Iraqi forces and military installations. Emission products from the ammunition and fuel combustion further added to environmental deterioration in this area. Air quality and land resources were affected substantially in the war areas by the military actions (Lathrop, 1983). The effects of the Gulf war on the land resources will be discussed in Chapter 6.

Air quality is a crucial component to the terrestrial ecosystem in the Gulf region. The Gulf crisis had an adverse effect on air quality, not only in the impacted region, but far from its boundaries. The emphasis in this section will be to identify actions that could have adversely effected the air quality. For the evaluation of air pollution, military activities such as troop movements, fallout from the use of explosives, and emissions from the transport and combat missions, etc. will be discussed.

5.3. Impact of troop deployment on air quality

The major Allied troop deployment was in Saudi Arabia, especially in the Eastern Province and Riyadh region. The deployment started in August 1990 and continued through January 1991. Each new group of military personnel brought with them equipment, ammunition, and other support gear. Transportation of military assets, both personnel and equipment, were undertaken by sea and air. Troop transportation significantly intensified air activities, especially at the Dhahran and Riyadh airports. After landing in Saudi Arabia, most of these troops were moved inland to the desert, primarily in the Eastern Province of Saudi Arabia. Troop deployment activities effected air quality by enhancing soil erosion and emitting air pollutants from the military. The impact of the combustion products on air quality in the Gulf region will be discussed in another section.

Military activities during August 1990 through January 1991, were most damaging to the desert crust. The troops, once stationed in the desert, dug fox holes, erected defensive barricades, and dug trenches to slow down enemy advances (see Figure 6.6). The army also conducted intensive training exercises during this period. These activities disturbed the vegetative ground cover and broke hard desert layers. Two weeks before the ground attack, the Allied troops were moved several hundred kilometers to their respective attack positions. This massive redeployment of the Allied ground forces further damaged desert cover along the Saudi/Kuwait border area.

In Kuwait and southern Iraq, military activities were parallel to that of the Allied forces in the northeastern Saudi Arabia. As a matter of fact, the Iraqi military in Kuwait, including their armor, was well "dug in" (Cutter, 1991). In addition, the Iraqis planted more than one million antipersonnel and antitank mines in Kuwait, especially along the Gulf coast and Saudi/Kuwait border. Iraqi military activities were even more aggressive and abused the desert crust more severely than those of the Allied forces.

The areas of military deployment is shown in Figure 5.1 and includes northeastern Saudi Arabia, Kuwait, and southern Iraq. As a result of military build-up, the land cover (both the desert crust and vegetation) was severely damaged in this area resulting in enhanced soil erosion. Naturally, an increase in soil erosion increases airborne particulates, including inhalable particles. The data discussed in Chapter 4 indicated excessive concentrations of the particulates in the ambient air of Kuwait, Saudi Arabia, and Bahrain during March to September 1991. Comparing air particulate data collected in Dhahran during March to September 1991 and 1992 with similar previously established data for 1980/83 period, it is concluded that the Gulf crisis had enhanced air particulates during the monitoring period (Chapter 6). However, the exact contribution of troop deployment to the enhancement of air particulates in the Gulf atmosphere could not be ascertained.

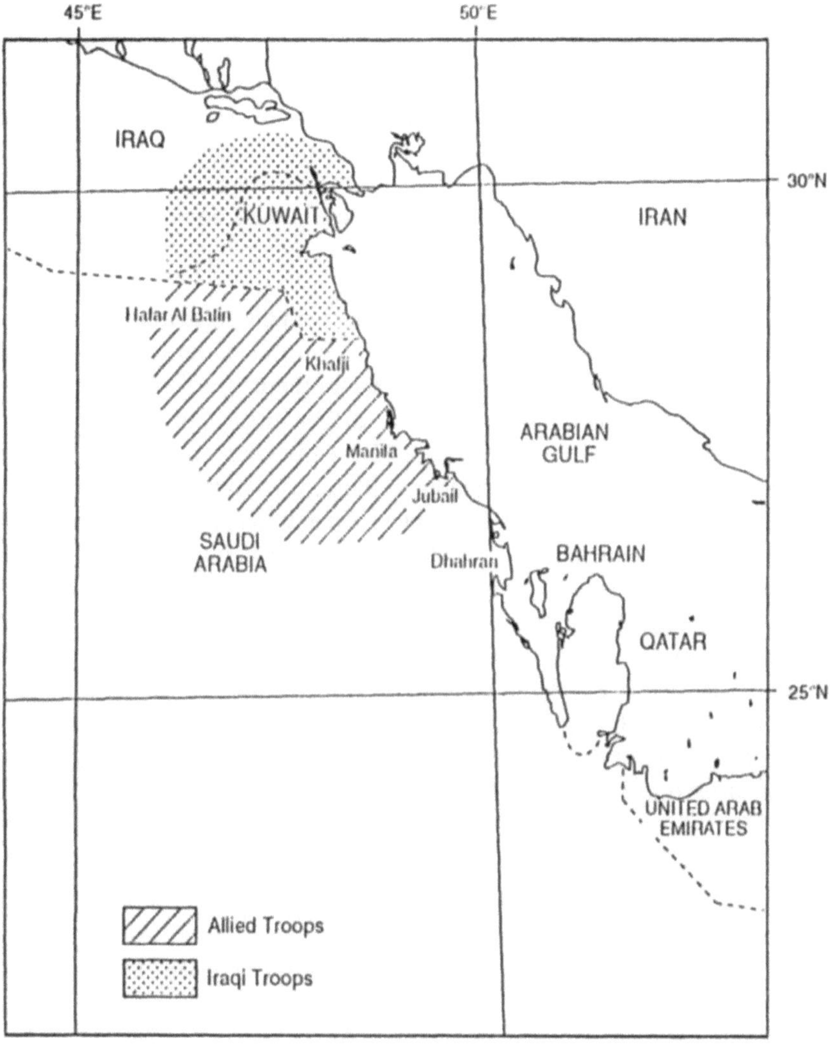

Figure 5.1. A map showing general deployment of troops during the Gulf crisis.

Qualitative data on the increase in soil erosion could not be obtained. During the military build-up period (August 1990 to January 1991), Many people in the Eastern Province of Saudi Arabia were trying to leave. Because of this chaotic situation, little scientific data could be collected. When the region returned to normalcy in March 1991, much attention was focused on the massive oil spill in the Arabian Gulf and the Kuwait oil fires. The edaphological damages from the military operations were completely overlooked.

After the cessation of hostile activities in the Gulf region, when most of the Allied forces had already left Saudi Arabia, the authors surveyed a part of Saudi desert. We found numerous vehicle tracks, camp bases, garbage burial sites, and signs of many other military-related activities. At certain places, large volumes of earth were moved to cover solid wastes. Because of the political limitations, we could not travel to Kuwait or southern Iraq to survey impacts on the ground cover. However, we anticipate that edaphological conditions in those areas would be similar to what was found in northeastern Saudi Arabia.

5.4. Effects of military-generated combustion products on air quality in the Gulf region

The deployment of more than a million troops (both the allied and Iraq military personnel), along with their military equipment and munition stockpiles in northeastern Saudi Arabia, Kuwait and southern Iraq consumed large quantities of fuel. Fuel consumption was further intensified during the Gulf war, especially during ground operations. The emission products from vehicles certainly must have affected air quality in the conflict area. The magnitude of air pollution from the combustion products depends mainly on the amount and type of pollutants, which in turn, is determined by the amount of fuel consumed by the military vehicles and planes.

The assessment of total fuel consumption during the Gulf crisis is a difficult task. Many types of military vehicles were used in the crisis, some of them were tested for the first time in desert conditions. Information regarding the number of each type of vehicle, fuel consumption rate and requirements, and the magnitude of each type used were not available. Some information on the equipment used during the Gulf crisis were collected and are summarized in Table 5.2. Information listed in this table illustrates the complexity of the problem. Estimation of the total amount of fuel consumed during the Gulf crisis is complex.

Table 5.2. A brief summary of military hardware that were used during the Gulf war (extracted from Abdullah, 1992 and Amin and El Kadi, 1991; Black, 1992).

Military Hardware	Manufacturer	Country of Acquisition	Number used during the Gulf war
Strike Planes			
B-52	USA	USA	26
A-4	USA	Kuwait	20
F-111	USA	USA	50
Tornado/IDS	UK	UK/KSA	92
S.2 B Buccaneer	UK	UK	12
F-15E	USA	USA	48

Table 5.2. Continued

Military Hardware	Manufacturer	Country of Acquisition	Number used during the Gulf war
F-117 Stealth	USA	USA	44
B-2	USA	USA	20?
A-12 Avenger	USA	USA	20?
Jaguar	UK	UK/France/GCC	59
F-16	USA	USA/Bahrain	136
F-18	USA	USA	192
Mirage F-1	France	France/GCC	40
Mirage 2000	France	France/GCC	45
F-18	USA	Canada	24
F-4 Phantom	USA	USA	24
A-6	USA	USA	160
A-7	USA	USA	48
TU-16	USSR	Iraq	
TU-22	USSR	Iraq	17
H-6D	USSR	Iraq	
SU-24	USSR	Iraq	16
Mirage F-1	France	Iraq	120
MIG-27	USSR	Iraq	90
SU-20, SU22	USSR	Iraq	80
SU-7	USSR	Iraq	30
L-39	USSR	Iraq	24
SU-25	USSR	Iraq	24
Xian-F-6	China	Iraq	40
Interceptor Planes			
F-15E (Interceptor)	USA	USA/KSA	189
F-14	USA	USA	120
Tornado/IDS	UK	UK/KSA	42
Tornado/IDS	UK	Italy	8
A-10	USA	USA	120
A-8 Harrier	USA	USA	60
Hawk	USA	GCC	120
F-5	USA	GCC	111
MIG-25	USSR	Iraq	30
MIG-21	USSR	Iraq	150
Xian-F-7	China	Iraq	60
MIG-23	USSR	Iraq	90
MIG-29	USSR	Iraq	40
Electronic & Reconnaissance Planes/Helicopters			
F-4G Wild Weasel	USA	USA	24
EF-111 Raven	USA	USA	24
A-6 Prowler	USA	USA	36
Lynx	UK	UK	4
Seaking	UK/USA	USA	4
P3 Orion	USA	USA	
C-3 Viking	USA	USA	
E-2 Hawkeye	USA	USA	
E-3 Sentry (AWACS)	USA	USA	115
RC-135	USA	USA	
EC-135	USA	USA	

Table 5.2. Continued

Military Hardware	Manufacturer	Country of Acquisition	Number used during the Gulf war
Nimrod MR	UK	UK	4
MIG-25R	USSR	Iraq	8
AWACS	USSR	Iraq	4
Helicopters			
AH-1S Cobra	USA	USA/KSA	1200 (AH-1S Cobra through UH-60 Black Hawk)
AH-64 Apache	USA	USA	
WG.13 Lynx	UK	UK	
S-61R Seaking	UK/USA	USA	
SA 321 Super Frelon	France	France	
SA 342 Gazelle	France	France	
CH-46A Sea Knight	USA	USA	
CH-47C Chinook	USA	USA/KSA	
CH-53E Super Stallion	USA	USA/KSA	
OH-6 Defender	USA	USA	
OH-58 Kiowa	USA	USA	
UH-60 Black Hawk	USA	USA	
MBD BO 105	Germany	Iraq	550 (MBD BO 105 through Others)
MI-8 HIP	USSR	Iraq	
MI-24 Hind	USSR	Iraq	
Others	Many countries	Iraq	
Transport/Refueling Planes			
C-5A Galaxy	USA	USA	600 (C-5A Galaxy through KC-135 Stratotanker)
C-130 Hercules	USA	USA/GCC	
C-141 Starlifter	USA	USA	
Victor K-2	UK	UK	
KC-135 Stratotanker	USA	USA	
AN-12/IL-76	USSR	Iraq	4
AN-12 (transport)	USSR	Iraq	40 (AN-12 through IL-76)
AN-24 (transport)	USSR	Iraq	
AN-26 (transport)	USSR	Iraq	
IL-76 (transport)	USSR	Iraq	
Tanks			
Challenger	UK	UK	160
Leaopard	Germany	USA	40
M-60	USA	USA	360
M-1 Abrams	USA	USA	1850
M-60	USA	Egypt	300
T-62	USSR	Syria	80
Light tanks	USA/UK	USA/UK	48
T-80	Yugoslavia	Kuwait	?
AMX-30/M-60	France/USA	KSA/GCC	900
T-54/55	USSR	Iraq	1000
T-62	USSR	Iraq	1200
T-72	USSR	Iraq	500
Chieftain	?	Iraq	50
PT-76	USSR	Iraq	100
Type-59/69	USSR	Iraq	1500
M-77	?	Iraq	200

Table 5.2. Continued

Military Hardware	Manufacturer	Country of Acquisition	Number used during the Gulf war
Tam	?	Iraq	60
Self-propelled Guns			
AVF-1	France	France	24
ZSU-23-4	USSR	Iraq	
M-109 A3/M-110	USA	USA	600
M-109 A3/M-110	UK	UK	100
M-109	USA	Egypt	54
AVF-1	France	Iraq	24
ZSU-23-4	USSR	Iraq	} 4000
M-1973	USSR	Iraq	
M-1974	USSR	Iraq	
Armored Vehicles			
AMX-10RC	France	France	96
AMX-10P	France	France	36
FAB	France	France	36
Fahd-30/M-113	Egypt	Egypt	220
M-3 Bradly	USA	USA	1300
Fox	USA	Germany	8
M-113	USA	USA	1200
LFTAB	USA	USA	600
LAV	USA	USA	400
M-113/AMX-10P	USA/France	GCC	3000
Warrior	UK	UK	135
Semiter	UK	UK	60
Striker	UK	UK	24
BMP-1	USSR	Iraq	1000
PTR-50/60	USSR	Iraq	152
Benhard-M3	France	Iraq	200
Casel	Brazil	Iraq	600
Jaravca	Brazil	Iraq	300
Others	?	Iraq	7700
Miscellaneous			
Multi rocket launcher	USA	USA/KSA	100
		Iraq	500-600
Anti-tank/anti-plane Missile Launchers*	Allied	Allied	Many thousands
Anti-tank/anti-plane Missile Launchers@	?	Iraq	10000
Balance between Land Forces			
Country	*Helicopters*	*Tanks*	*Armored Vehicles*
USA	1700	1200	2200
UK	12	163	
France	72	40	260
KSA	82	550	1900
Egypt		300	
Syria		300	
UAE	46	220	235
Oman	35	53	66

Table 5.2. Continued

Military Hardware	Manufacturer	Country of Acquisition	Number used during the Gulf war
Qatar	11	24	65
TOTAL-Allied	*1958*	*2930*	*4726+*
Iraq	380	4000	2500

? Not sure or information missing.
* SAM-7, Milan, Crotal, Shaheen, Mistral, and many others.
@ SAM 2, SAM 3, SAM 6, SAM 8, SAM 9, Roland, Scud, Frog, Seakboard.

From August 1990 through January 1991, there was rigorous training of military personnel in the desert. Armored divisions were involved in many of such exercises. The exact nature of these preparatory military exercises are not known. However, we may assume that fuel consumption from these training exercises was high, perhaps 25% of all fuel consumed during the ground operation. In addition to the Allied forces, the Iraqi army also conducted some fairly large size exercises before the war. We assume that pre-war preparatory activities by the army consumed about one billion gallons of fuel.

Fuel consumption by the military services is relatively high even in peace time. For example, in peace time, the military uses about 2–5% of the national oil consumption in the United States. During a war situation, petroleum uses generally increase significantly to about 15–30 percent. According to a February 22, 1991 CBS (Columbia Broadcasting Service of the United Sates) news report, the US military used about 3 billion gallons of fuel in one week, just to move the front lines nearer to the Iraqi border in preparation for the ground assault. Redeployment immediately before the ground operation was completed in about two weeks. This would double the amount of fuel consumed only in the redeployment of the troops. We conservatively estimate a figure of about 5 billion gallons for the total redeployment operation. About 100 million gallons of fuel should be added to the above estimates for other Allied forces (Cutter, 1991). The above estimates seems reasonable when one considers that one tank consumes about 1,700 gallons of fuel per day. Armor is so petroleum intensive that fuel consumption is measured in gallons per mile rather than miles per gallon (Cutter, 1991). One armored division (about 348 tanks) uses about 600,000 gallons of fuel per day. The six divisions that took part in the ground campaign, plus support vehicles, are estimated to have consumed about 10 million gallons of oil during the ground assault. We estimate that about one billion gallons of fuel were consumed by the Allied forces during the ground operation.

The United States and allied countries transported more than 700,000 troops to the Gulf region (Table 1) from September 1990 to January 1991. As a result, air transport activities in Saudi Arabia were very intense during this period. The authors live about 500 meters from Dhahran airport/airbase, one of the most frequently used facilities. Galaxy and Hercules transport planes were used to relocate most of these troops. An almost constant stream of transport planes flew over our residences and offices at the King Fahd University of Petroleum and Minerals. As the ground war approached, these flights intensified. From August 1990 to January 1991, there was a flight departing the airport about every 10 minutes throughout the day and night. Based on this experience, we assumed that an average of 100 flights per day occurred during this period. A similar level of aerial activity could safely be assumed for the remaining operation in Saudi Arabia. Each flight flew for at least 15 minutes over the Gulf region after takeoff. Information on fuel consumption by Galaxy and Hercules transport planes were not available. However, based on our discussion with some civilian pilots and as proposed by Cutter (1991), we have assumed a consumption of 2600 gallons of jet fuel per hour for these planes. If transportation activities continued for 120 days at the assumed level, the total consumption of jet fuel would be about eight million gallons. The combustion products of this fuel would have been deposited over the Gulf region during the transportation of military personnel and equipment.

Like army, air force personnel were being trained from August 1990 to January 1991. At Dhahran, we noticed three jets, mostly of the F-15 type, in the air for most of the day time. The Riyadh airbase also conducted a similar air force training activities. In addition to Riyadh and Dhahran, several airports in the Kingdom of Saudi Arabia, Bahrain, Qatar, and the United Arab Emirates were used by the Allied air force. It was assumed that the air training activities for all other locations combined were about double that of Dhahran. For Dhahran, we have assumed a total flying time of 12 hours for all three planes. Training and surveillance activities continued from September 1990 to January 15 1991, a period of about 140 days. It was further assumed that fuel consumption by the planes used in the training and surveillance activities were equivalent to the F-4. Cutter (1991) reported a fuel consumption rate of about 1700 gallons per hour for the F-4. From the foregoing, a total consumption of about 9 million gallons of fuel was estimated.

The second important air activity with respect to fuel consumption were combat missions (bombardment) over Kuwait and Iraq. The air campaign started on January 16, 1991 and continued for 43 days. From the beginning to the end of the operation, Iraqi troop concentrations and strategic installations were bombarded very intensively. On the average, there were about 2,500 sorties each day. In total, 110,000 sorties were flown by the Allied air force. Iraqi resistance or response to the Allied bombardment was minimal and need not be considered. As listed in Table 5.2, more than 25 different types of striker and interceptor planes, in addition to many transport and refueling planes, were involved in the air campaign. Many of these planes were designed for a special task and hence their fuel type and requirements vary. In addition, these planes had widely variable fuel consumption rates and requirements depending upon the level of maintenance and operating conditions. Actual information on the total amounts and type of fuel consumed during Gulf war air operation was not available because of military censorship, thus we must make educated assumptions.

As shown in Table 5.2, the majority of military equipment used by the Allied air forces were manufactured in the United Sates. It was, therefore, assumed that planes from other countries did not vary substantially in fuel consumption and emission products from those of the United States. It is well known that different fighters and bombers consume fuel at different rates and perhaps require a variety of fuel types. However, for the sake of simplicity and the lack of availability of data, it was assumed that all fuel used was diesel.

Table 5.2 shows 25 different types of planes that were used in the Gulf crisis. The nature of the use of each type of plane was classified. Therefore, to determine fuel consumption, we made some educated assumptions. Cutter (1991) reported that an F-4 fighter consumes about 1700 gallons while B-52 bomber needs about 3600 gallons of fuel per hour. It appears reasonable to assume an average fuel consumption of F-4 and B-52 planes (2600 gallons per hour) as representative for each sortie. A similar assumption was used by Cutter (1991) who further assumed a mean time of three hours per sortie. Combat planes are generally flown for a shorter time and therefore, we have adopted a time of 2 hours per sortie for use in our calculations. The Allied air forces flew about 110,000 sorties during the 43 days of air campaign (Lane, 1991). From the above information, we calculated a total consumption of about 600 million gallons fuel for the air campaign. Diesel fuel consumed by military services during the Gulf crisis are summarized in Table 5.3.

Table 5.3. A summary of diesel consumption during the Gulf crisis (from August 1990 to February 1991).

War Activity	Fuel Consumed (in million gallons)
Pre-war preparatory army activities	1000
Redeployment of troops to assault position	5000
Ground war (100 hours)	1000
Troop and equipment transportation via air	8
Pre-war air force training and field survellance	9
Air campaign (43 days and 110,000 sorties)	600
Total fuel consumption during the Gulf crisis	7617 or approx. 8000

Types of engine differ in their combustion products. Table 5.2 shows the wide diversity of the military equipment used during the Gulf crisis. These equipment were powered by a variety of engines, a review of the number and types of engines involved is beyond the scope of this book. In general, diesel and gasoline engines are the major power source used in military vehicles. In diesel engines, the fuel is self-ignited whereas in gasoline engines this function is performed by spark plugs. The fuel used in the diesel and gasoline engines differs, with diesel fuel consisting of higher boiling point petroleum fractions. Primarily because of its density, diesel contains approximately 13% more energy than gasoline (IARC, 1989a). Because of the nature of work to be performed and high demand for power from military machines, diesel engines are used extensively in military applications. It is assumed that most of the 8 billion gallons of fuel was used in diesel engines.

Two types of diesel engines are in common use; direct and indirect injection engines. Direct injection diesel engines are preferred for heavy duty applications whereas indirect injection diesel engines are commonly used for light duty applications (NRC, 1982). Since the majority of the military machinery are heavy duty and extreme power is required, we can safely assume that most military fuel usage was in direct injection diesel engines.

The major products of the complete combustion of petroleum based fuel in an internal combustion engine are carbon dioxide (13%) and water (13%) with nitrogen from the air comprising most of the remaining exhaust (73%). A very small portion of the nitrogen is converted to nitrogen oxides (NO_x) and some nitrated hydrocarbons are also produced (IARC, 1989b). For all practical purposes, 100% combustion efficiency is not attainable and combustion is always incomplete. Incomplete combustion results in the emission of carbon monoxide (CO), unburned fuel, lubricating oil, and oxidation and nitrification products of the fuel (Yamaki et al., 1986). These incomplete combustion products may consist of thousands of compounds in gaseous and particulate phases (Zaebst et al., 1988). The types and concentrations of incomplete combustion products is a function of many factors including engine type, engine operating conditions, fuel and lubricating oil composition, and the emission control system used in the vehicle (Johnson, 1988). An exact determination of exhaust or combustion products emitted during the Gulf war was not feasible. To simplify the matter, emission products for different types of diesel engines are listed in Table 5.4.

Table 5.4 indicates that heavy-duty diesel engines emit larger quantities of particulate matter (smoke/soot), carbon monoxide (CO), nitrogen oxide (NO_x), and polycyclic aromatic hydrocarbons (PAH). In addition to the above mentioned emission products, the particle phase largely consists of elemental carbon (60–80% – Ball, 1987; Zaebst et al., 1988), sulfuric acid (2–7% – IARC, 1989b) and some metallic species, especially iron from the exhaust system (IARC, 1989b) barium from fuel (Hampton et al., 1983), and zinc from lubricating oil (Hare and Baines, 1979).

The data given in Table 5.4 refer to per unit distance and are probably biased in favor of light-duty diesel engines. Heavy-duty diesel engines may not release more pollutants but they consume more fuel and, as a result, emit larger quantities of exhaust per unit distance as shown in Table 5.4. The information given in Table 5.4 is not directly applicable to the Gulf war situation. To calculate emissions, one needs to convert data in Table 5.4 from unit distance to unit volume of fuel. Lilly (1984) reported a diesel consumption rate of 2.5

Table 5.4. Emissions from various diesel engines (data extracted from IARC, 1989b)

Exhaust Parameter	Heavy-duty Diesel Engine	Light-duty Diesel Engine
Gas phase (mg/km)		
Total particulate matter	1036	246
Total extractable matter	188	124
Carbon monoxide	6250	794
Formaldehyde	-	13
Benzene	-	15
Nitrogen oxides	17500	794
Toluene	11	-
Gas phase Polycyclic Aromatic Hydrocarbons (PAH), (μg/km)		
Anthracene	5600	1313
Fluorathene	775	188
2-Nitrofluorene	-	56
Pyrene	988	472
Particulate Phase Polycyclic Aromatic Hydrocarbons (PAH), (μg/km)		
Anthracene	274	66
Fluorathene	538	140
2-Nitrofluorene	-	61
Pyrene	739	178
Total PAH	-	122–625

liters per 10 km for a bus carrying about 50 passengers under calm conditions. A passenger bus is no match for heavy-duty diesel engines such as a tank or plane engine. Therefore, we propose a consumption rate of about 3 liter per kilometer as representative of heavy-duty diesel engines that were used in the Gulf conflict. From the foregoing, the total emission products produced during the Gulf crisis were calculated and are given in Table 5.5.

Major air pollutants generated due to military activities were carbon dioxide, carbon monoxide, nitrogen oxide, and many organic and inorganic compounds. Carbon dioxide and carbon monoxide produced during the ground operation was about 48 and 1 percent, respectively, of that emitted by the Kuwait oil fires. It is estimated that about 80% of the fuel was consumed during February 1991. Air pollution resulting from war activities would have been at its maximum during this month. The data presented in Table 5.6 suggest significant air pollution in February 1991. The concentrations PAHs, metals, CO and NO_x that might have been discharged to the atmosphere during a short time and in a small area are remarkable (Table 5.6).

No air quality data could be collected in the war zone. However, the authors were staying in Dhahran during those days. There was an unusual smell in the air, probably from the elevated levels of hydrocarbons and other pollutants. The atmosphere was hazy even before the smoke from the Kuwait oil fires was blown over Dhahran. One could very often smell gun powder in Dhahran as a result of the intensive bombing activities to the north. If the population in Dhahran could smell powder, the exposure of Kuwaitis, Iraqis, Saudis, and army personnel

Table 5.5. Emission estimates@ of heavy-duty diesel engines

Parameter	Estimates
Carbon dioxide (g/gallon of fuel)	9753
Total particulate matter (mg/gallon of fuel)	980
Elemental Carbon (70% of total particulate matter, mg/gallon of fuel)	686
Total extractable matter (mg/gallon of fuel)	178
Carbon monoxide (mg/gallon of fuel)	5938
Nitrogen oxides (mg/gallon of fuel)	16625
Sulfur dioxide as sulfuric acid (4% of total particulate matter, μg/gallon of fuel)	39000
Toluene (μg/gallon of fuel)	6000
Anthracene (μg/gallon of fuel)	5580
Fluorathene (μg/gallon of fuel)	1247
Pyrene (μg/gallon of fuel)	1641
Metallic residue	*

@ The estimates are extracted from information listed in Table 5.2.
* No information was available but a preliminary study (Sadiq et al., 1992a) suggested that military operations may be a major source of metal contamination in the Gulf region.

Table 5.6. Total emissions from fuel consumption during the Gulf war

Parameter	Estimate
Carbon dioxide	65 (million tonnes)
Total particulate matter	6370 (tonnes)
Elemental carbon	4500 (tonnes)
Total extractable matter	1157 (tonnes)
Carbon monoxide	38597 (tonnes)
Nitrogen oxides	98036 (tonnes)
Sulfur dioxide as sulfuric acid	254 (tonnes)
Toluene	39 (tonnes)
Anthracene	36 (tonnes)
Fluorathene	8 (tonnes)
Pyrene	10 (tonnes)

to air pollutants within the war zone must have been high. Unfortunately, this suspicion cannot be verified because no information could be collected on air quality in the Kuwait territory of operation.

5.5. Impact of the explosives on air quality

Collection of data on any military-related activities during the Gulf war were not easy as most information is classified. Moreover, the scientists in Kuwait and in the Eastern Province of Saudi Arabia were more worried about evacuation from the area than indulging in any air quality monitoring program during the Gulf crisis. No conventional air monitoring laboratory is equipped to analyze the unknown chemicals produced due to the use of high explosives. Without cooperation from the military, nobody could be sure of what to look for. This situation was further complicated since high explosive chemicals are unstable and rapidly transform into stable by-products. From the complexity described above, it is understandable that actual data on ambient concentrations of explosive chemicals could not be collected. The objective of this section is not to present actual data but make our readers aware of pollution potential due to explosive use.

The air campaign started on January 15, 1991 and lasted for 43 days. During this period, a total of 110,000 sorties were flown by the Allied air forces. Bombardment of Iraqi troop concentrations and military installations was very intensive – rather relentless and unprecedented. The Allied forces had an array of military hardware from which to choose (Table 5.2). The types of ammunition and explosives used in the Gulf war were numerous and at least as many as the available delivery systems could be expected. The amounts and types of ammunition used in the Gulf war is classified. Any information extracted from the literature will, therefore, be incomplete, and in many cases speculative. The topic of explosives is very sensitive to military personnel, therefore it will be treated as in-depth as the circumstances allow.

The Allied forces dropped about 84,200 tonnes of bombs (The Middle East, 1992). This estimate seems very conservative in view of the extent of carpet bombing. Additionally, large quantities of munitions were used during the ground war. In all about 120,000 tonnes of ammunition were estimated to have been used during the Gulf war (Barnaby, 1991). During the occupation and in anticipation of an attack from the multinational Allied Forces that had been gathering in Saudi Arabia, the Iraqis transported and stored a large quantity of munitions in Kuwait. Iraqi munitions stockpiles in Iraq and Kuwait were considered priority targets for the Allied air force. As a result, a major portion of the Iraqi munitions stockpiles was destroyed. The magnitude of this destruction remains unknown. The amount of Iraqi munitions destroyed during the bombardment may be even higher than the amount of explosive used by the Allies during the Gulf war. Additionally, the Kuwaiti Defense Ministry reported that, after the liberation, a total of 2549 tonnes of munitions were diffused and 7751 tonnes destroyed before March 1992 (Arab News, 1992).

The explosives were largely used in Kuwait and Iraq and a small amount in Saudi Arabia (such as Khafji). The total area of the State of Kuwait is 17,818 km^2 (6960 sq. miles – Al-Shatti, 1991). Iraqi territory that was intensively bombed, was probable twice the size of Kuwait. Over 40,000 km^2 area of Kuwait, Iraq, and the Kingdom of Saudi Arabia were estimated to be impacted by explosives and ammunition use.

To assess the environmental impact of explosives used during the Gulf war, one must consider its composition, type, and the area and period of use. As stated above, the Allied Forces had access to several hundred types of munitions from which to choose from. The exact quantities and type of munitions used in the Gulf war is a military secret and so specific data could not be obtained.

Because of our inability to gather classified information, the discussion on the environmental impact of the munitions will be on a general basis instead of specific type-by-type basis. In very general terms, the ammunition used consisted of two parts: the metallic and/or plastic casing(s) and the explosive train. The environmental impact of both these parts will be addressed below.

On impact, the casing is broken into large pieces and fine particles, littering the ground and contaminating the air around the impact area. The fine particles that become airborne might remain suspended for some time. Considering the amount of ammunition used in the Gulf war, it is suspected that all Kuwait, and parts of Iraq and Saudi Arabia was littered with bomb shells, spent casing, cartridges, etc. (Al Hassan, 1991, 1992). Particles that were generated during explosion would have been transported over large areas. As a result, metal concentrations in the air and air particulates would have been enhanced. Metal pollution from the use of explosives could contaminate large areas downwind. No air particulate samples could be collected in Saudi Arabia, Kuwait, or Iraq during the Gulf war period, and therefore, no data can be presented on ambient concentrations of metals. After returning to normalcy in Saudi Arabia, Sadiq et al. (1992a) collected surface layer soil samples from northeastern Saudi Arabia and analyzed them for many metals. They reported that concentrations of many metals in soil surface decreased with an increase in the distance from the Saudi/Kuwait border. Sadiq et al. (1992a) noticed that metal concentrations in the soil samples collected parallel to the Saudi/Kuwait border were statistically similar. The surface soil samples were collected shortly after the cessation of hostilities in Kuwait (end of March 1991). They suspected that the enhancement in metal concentrations in the border area probably represents metal input from Gulf war activities. However, the data collected by Sadiq et al. (1992a) were too limited to draw any definitive conclusion.

Once in the air, these metallic particulates can be inhaled both by humans and animals, scavenged to the ground by dry and wet deposition, or scattered to other areas. Sadiq et al. (1992b) reported that hair samples from resident in the war impacted area (Dhahran, Hofuf, Hafr Al-Batin, Riyadh) contained higher concentrations of metals as compared with those collected from a city more distant from Kuwait (Jeddah). The result of this investigation is discussed in detail in a Chapter 7. The impact of metal contamination on vegetation and agricultural crops is given in Chapter 6.

The explosive train part of ammunition is generally complex, highly unstable, and variable in chemical composition. Data on the exact chemical composition of military explosives are classified, and therefore, information on the compounds that were present in the explosives used in the Gulf war will probably never be known. In an attempt to focus on the potential of environmental impacts of these explosives, the available literature was reviewed. Some of the information is summarized below.

Explosive material can be classified into two very broad categories, namely deflagrating or burning materials (which undergo combustion) and high explosives (that undergo detonation or a high order explosion). In general, all explosive materials are unstable by the virtue of their chemical nature. As a general rule, more powerful explosives are more difficult to initiate (ignite). This is why there are at least two parts to an explosive device. The material that is used to initiate the main high explosive is referred as priming material or first fire. The typical commonly used priming explosive materials include (Brauer, 1974):

Lead azide
Lead styphnate (lead trinitroresorcinate)
Lead mononitroresorcinate (LMNR)
Potassium dinitrobenzofurozan (KDNBF)
Barium styphnate
Zirconium potassium perchlorate

Other priming agents that are not commonly used include:

Mercury azide
Potassium chlorate
Mercury fulminate

Deflagrating materials can be loose power, pressed grain, or solid propellant billets. These materials are generally good gas or heat producers. Deflagrating materials are generally more stable than the priming materials and are produced by mixing different chemicals. Some of such chemicals are:

Nitroglycerin
Nitrocellulose
Dinitrotoluene
Graphite
Potassium nitrate
Trimethylol ethane trinitrate
Triethylene glycol dinitrate
Diphenylamine
Zirconium potassium perchlorate
Barium nitrate

High explosives are chemical mixtures which upon initiation dissociate almost instantaneously into other more stable compounds. Dissociation or detonation generally produces a small amount of gas and high temperatures. Detonation is generally so fast that a shock wave is created that acts on its surroundings with great shattering effects. Some of the commonly used high explosives are:

PETN or Penthrite, pentaerythrite tetranitrate
RDX, Cyclotrinethylenethrinitramine
TNT, Trinitrotoluene
Tetryl, trinitrophenylmethylnitriamine
Hexanirodiphenylaminoethylnitrate
Hexanitrodiphenyloxide
Tetranotromethane
Manganese dioxide

The above high explosives are commonly used in military ammunition as filling for shells, bombs, and the warheads of rockets (Fordham, 1980). They are used in varying combinations to best suit specific needs and requirements. There are several other components of the explosives which contain chemicals, both organics and inorganics. Information on the chemical composition of these formulations is a military secret. Moreover, it is beyond the scope of this book to thoroughly review these formulations.

Considering the short period of the war (only 43 days), the small area (about 40,000 km^2) to which it was confined, and the quantities of ammunition used (the Allied usage of munitions was estimated at 120,000, and a similar amount for Iraqi losses was suspected from Allied bombardment), it is suspected that during the peak war period, the population in the area along with the army personnel would have been exposed to many unknown chemicals. The airborne concentrations of some of these chemical might have been elevated to levels where adverse effects on human health become more likely. The issue of human health is dealt separately in Chapter 7. Air quality during the war may have been affected by the emission of organic and inorganic chemicals in large but unknown quantities.

The use of explosives considerably disturbed desert surface cover. The desert crust, underlying soil, and vegetation were thrown into the air. Explosions from the bombs created craters several meters in width. Considering the carpet bombing by B-52s and other bombers, damage to the ground, especially in those areas occupied by the Iraqi Republican Guards, would have been extensive. The ferocity of bombardment was reflected in an Allied briefing with the announcement that "the Republican Guards have been dessimated". The Allied Forces liberally used "fuel-air explosives" during the Gulf war. These were largely used to open up paths in Iraqi minefields along the Kuwait/Saudi border. The pressure generated by these devices loosened upper soil surface in the impact areas.

After an explosion, a large cloud of dust mushrooms hundreds of meters in the air and then fans out over large areas. Explosions can thereby enhance concentrations of air particulates. The increase in the particulates, especially the inhalable fraction, has been known to cause detrimental health effects (Chapter 7). Along with the dust, there will be numerous chemicals. In summary, all military operations involving the use of explosives adversely effect the land surface, increase air particulates, and introduce new and unknown pollutants to the ambient atmosphere.

5.6. Miscellaneous military activities and air quality

A discussion of other military activities that probably adversely effected air quality in the war area is described below.

5.6.1. Radiation from the Iraqi nuclear facilities

On January 23, 1991, it was announced that United States bombers had destroyed two of Iraq's nuclear reactors at Al-Tuwaitha, some 20 km north of Baghdad (an 800 kilowatts thermal unit sold by France in 1980 and another 5000 kilowatts supplied by Soviet Union in 1968). These Iraqi reactors were fueled with highly-enriched uranium sold to Iraq by France (12.3 kilograms of 93% enriched uranium) and the USSR (33 kilograms of 80% enriched uranium) (Nuclear Material in Iraq, 1991). Iraq claimed that the two reactors were small and used to produce radioactive isotopes for medical purposes. On the other hand, the United States insisted that these facilities were being used to develop nuclear weapons. In addition to these reactors, other Iraqi nuclear facilities were also bombed. The location and type of activities at these facilities are not known; however, the uranium mines in the Gara mountain region and an establishment probably in Mousal, are other publicly known nuclear facilities in Iraq.

The amount of radioactivity present at the time when the reactors were bombed, depends largely on how long the fuel had been used. Information on the actual use of these reactors are not available. As a comparison, if the fuel was used for a year, the amount of radioactivity produced in the reactor would have been about 0.00003% of that released during the Chernobyl accident. As stated above, the Iraqi nuclear facilities that were bombed were commissioned in 1968 and 1980. Because of their long period of use, an appreciable amount of radioactivity might have been consumed in these reactors. In addition to the fission products, radioisotopes would have also been produced in the reactor as a result of neutron bombardment of the fuel in the reactor core. Two radioisotopes, that could be produced in the core of Iraqi reactors which pose the greatest danger to humans are cesium-137 and iodine-131. If ingested, these isotopes can induce cancer and impair thyroids. Children are particularly at risk as their thyroids are usually small.

The amount of radioactivity produced in the cores depends on the age of the reactor. These reactors have probably been in use for 10 and 20 years, respectively, for the smaller and larger nuclear reactor. The amount of radioactivity induced in the cores should have been considerable at the time of allied bombardment. There was no report suggesting that the radioactivity in the bombed reactors was released. If released, the population in Al-Taiwatha and other nearby cities may have been exposed to radiation. These people might have also ingested radioisotopes from the contaminated air, soil and food. Such exposure may increase the incidence of cancer in the exposed population and cause genetic mutation in their offspring. Because of the censorship by Iraqi authorities and state-of-war at the time of bombardment, the real situation was not revealed.

5.6.2. Use of ozone depletors in the Gulf war

Additives may be added to jet fuel to satisfy specific needs. For example, CFC's (chlorofluorohydrocarbons) are added to fuel used by the Stealth bomber, primarily to reduce particulate emissions so that the plane might remain undetected by a radar. The particular applications of these chemicals are classified, and therefore, where and how much CFC's were used in the Gulf war remains a military secret (Cutter, 1991). It is extremely difficult to collect information from the military on additives and chemicals that were used during the Gulf war. Fuel additives might have been more damaging to the environment than diesel emission products. Without military cooperation, an accurate assessment of environmental damage from the military activities due to the Gulf war will remain speculative.

For several months after deployment, Allied Forces were stationed in the Saudi desert, especially in the northeastern part. There were problem of flies and mosquitoes. Aerial sprays to kill flies and mosquitoes were used. Pesticide sprays were frequently used to control them. Many of the these sprays may have contained ozone damaging chemicals. No data are available to draw any conclusion regarding the impact of their use on the Gulf environment.

5.6.3. Uranium hazards from the Gulf war

The United States Army relied heavily on M-1 tanks (about 1850 in number – Table 5.2) and A-10 attack airplanes (about 120 in number) during the Gulf war. These war machines fired high velocity 120 mm and 30 mm canon shells to pierce Iraqi armor (Dietz, 1991). Both shells contained spent uranium metal. Upon impact, a high velocity projectile of uranium metal partially burns and generate a huge number of micrometer size particles of uranium oxide. These uranium oxide particles along with dust, might have been scattered over a large area, even beyond the war boundaries. The depleted uranium oxide particles might have been breathed by the population that stayed in the area during the Gulf war. Uranium oxide particles, if retained by the lungs, could increase the probability of lung cancer in the exposed population.

Information on radiation levels in the war-zone is scarce. Immediately after ceasefire, a British team measured radiation in Kuwait and, in a seminar at the Research Institute of King Fahd University of Petroleum and Minerals in Dhahran, the team claimed that the radiation was not significantly higher than background levels. No further details were given and no data were released later.

5.7. Future response and research needs

Effects of military operations on air quality during the Gulf crisis is discussed in this chapter. Relevant information on many aspects of air quality evaluation was either classified or not collected. One of the important factors that effected air quality was increased concentrations of air particulates. The enhancement in air particulates is dependent on the extent of damage to the desert crust. Military authorities should cooperate to delineate severely affected areas. A research program to monitor soil erosion and recovery of the disturbed desert crust in such areas should be initiated. The effect of the Gulf war on desertification in Saudi Arabia and Kuwait should be monitored on long-term basis.

A large amount of munitions was used during the Gulf war. Information on the type, quantity, chemical composition, and location of use is classified by the military authorities. It appears that, without the cooperation from the military authorities of both sides (Allied and Iraqi), deterioration of air quality in the war zone can never be truly assessed. Preliminary data reported in this chapter suggest that military operations were probably responsible for the enhancement of metal concentrations in surface soil samples collected several hundred kilometers from Kuwait. A research program should be initiated to investigate metal contamination from the Gulf war and delineate the most contaminated areas.

The Gulf crisis had consumed many billion gallons of fuel and produced large amounts of emission products. Many additives were added to fuel to suit specific tasks. Information on such additives was classified by the military authorities. These additives can be more damaging to environment than emission products. An international organization should be created to establish cooperation between the military and scientific communities in the world.

References

Abdallah, H. 1992. Gulf war – Weapons and Tactics (in Arabic), Arab Establishment for Studies and Publications, Beirut, Lebanon.

Al Hassan, J.M. 1992. The Iraqi Invasion of Kuwait, An Environmental Catastrophe. Fahad Al Marzouk Printing Press, Kuwait.

Al Shatti, A.K.S. 1992. Introduction to Kuwait. In Environmental and Health Impact of the Kuwaiti Oil Fires. Proc. of an International Symp. held at the University of Birmingham, 17th October, pp. 1–5.

Al-Hassan, J.M. 1991. The Iraqi Invasion – Environmental Catastrophe. In Environmental and Health Impact of the Kuwaiti Oil Fires. Proc. of an International Symp. held at the University of Birmingham, 17th October, pp. 6–11.

Amin, M. and El Kadi, A. 1991. Secrets of the Weapons of Destruction (in Arabic). Published in Cario/Nicosia.

Arab News, 1992. Kuwait continues battle to clear Gulf war mines. P. 3, April 30, 1992, Arab News, Saudi Arabia.

Ball, D.J. 1987. Particulate carbon emissions and diesel vehicles. Proc. Vehicle Emissions and their Impact in European Air Quality, Institute of Mechanical Engineer, Edmond, Surrey, U.K.

Barnaby, F. 1991. The environmental impact of the Gulf war. Ecologist 21:166–172.

Black, I. 1992. Desert Air Force. Osprey Publishing Limited, London.

Brauer, K.O. 1974. Handbook of Pyrotechnics. Chemical Publishing Co. Inc., New York.

Cutter, S.L. 1991. Ecocide in Babylon. Summer 41:26–31.

Dietz, L. 1991. Uranium health hazard. Chem. Engineer. News, February 4, pp. 2.

Fordham, S. 1980. High Explosives and Propellants. 2nd Edition, Pergmon Press, Oxford.

Hampton, C.V., Pierson, W.R., Schuetzle, D., and Harvey, T.M. 1983. Hydrocarbon gases emitted from vehicles on the road. 2. Determination of emission rates from diesel and spark-ignition vehicles. Environ. Sci. Technol. 17:699–708.
Hare, C.T. and Baines, T.M. 1979. Characterization of particulates and gaseous emission from two diesel automobiles as function of fuel and driving cycle. Tech. Paper Ser. 790424, Society of Automotive Engineers, Warrendale, Pennsylvania, USA.
IARC, (1989a). Monograph on the Evaluation of the Carcinogenic Risk of Chemicals to Humans. Occupation Exposures in Petroleum Refining; Crude Oil and Major Petroleum Fuels. 45:159–201, 219–237.
IARC, (1989b). Monograph on the Evaluation of the Carcinogenic Risk of Chemicals to Humans. Diesel and Gasoline Engine Exhaust. 46:41–188.
Johnson, J. 1988. Automotive emissions. In Watson, A.Y., Bates, R.R. and Kennedy, D. (eds), Air pollution, the Automobile, and Public Health. National Academy Press, Washington, DC.
Lane, C. 1991. What just lies ahead. pp. 28–29. March 4, Newsweek.
Lathrop, E.W. 1983. Recovery of perennial vegetation in military maneuver areas. pp. 265–277. In Webb, R.H. and Wilshire, H.G. (eds.), Environmental Impacts of Off-road Vehicles. Impact and Management in Air Regions. Springer-Verlag, New York.
Lilly, L.C.R. 1984. Diesel Engine Reference Book. 22.3 Diesel engine fuel consumption. Butterworths, London.
NRC, 1982. Diesel Technology - Impacts of diesel powered light duty vehicles. National Research Council, National Academy of Sciences, Washington, DC.
Nuclear Material in Iraq. Nuclear Fuel, 13, May 1991, p.5. (as cited by Barnaby, F. 1991.)
Pierson, W.R. and Brachaczek, W.W. 1983. Particulate matter associated with vehicles on
Sadiq, M., AlThagafi, K.M. and Mian, A.A. 1992a. A Preliminary Evaluation of Metal Contamination of Soils From the Gulf war Activities. Bull. Environ. Contam. Toxicol. 49:633–639.
Sadiq, M., Mian, A.A., and Al Thagafi, K.M. 1992b. Inter-city comparison of metals in scalp hair collected after the Gulf war 1991. J. Environ. Sci. Health 27(6):1415–1431.
The Middle East, 1992. The Gulf war. Blind man's bluff. The Middle East, January, P. 23.
Yamaki, N., Kohno, T., Ishiwata, S., Matsushita, H., Yoshihara, K., Iida, Y., Mizoguchi, T., Okuzawa, S., Sakamoto, K., Kachi, H., Goto, S., Sakamoto, T., and Daishma, S. 1986. The state of the art of the chemical characterization of diesel particulates in Japan. In: Ishinishi, N., Koizumi, A., McClellan, R.O., and Stober, W. (eds) Carcinogenic and Mutagenic Effects of Diesel Engine Exhaust. Elesevier, Amsterdam, pp. 17–40.
Zaebst, D.D., Blade, L.M., Morris, J.A., Schuetzle, D. and Butler, J. 1988. Elemental carbon as a surrogate index of diesel exhaust exposure. Proc. of the Am. Indust. Hyg. Conf. 15–20 May, San Francisco, California, USA.

CHAPTER 6

Gulf war impact on land resources

6.1. Summary

- The impact of the Gulf war on the land resources of Kuwait, Saudi Arabia, and Iraq was expected to be disastrous. The approach adopted to evaluate damages to the land resources in these countries is to consider all the collected data and to summarize war-related activities that could effect land resources, identify vulnerable components of land resources, and then assess damage. The impact of the Gulf crisis on the land resources are discussed with reference to soil, vegetation, wildlife, agriculture, and water.
- About 40,000 km^2 areas of Kuwait, northeastern Saudi Arabia, and southern Iraq were littered with solid wastes from the Gulf crisis. Solid wastes were generated mainly from destroyed military hardware (over 5000 Iraqi tanks and armored vehicles, over one million mines in Kuwait, etc.), residue of explosives and ammunition (over 80,000 tonnes of bombs were dropped and about 120,000 tonnes of ammunition was used), sanitary residues (over 4 million tonnes of wastes from humans, etc.), etc. Solid wastes generated during the Gulf crisis pose a serious problem to land resources in the war zone. Collection and disposal of these wastes is going to be a daunting task.
- Desert soil was badly damaged in the whole of Kuwait, southern Iraq, and northeastern Saudi Arabia. It is suspected that breakdown of the crust would enhance wind erosion in the Gulf area. Comparison between the distribution of air particulate during 1980/83, 1991, and 1992 suggest that soil erosion partially contributed to the enhancement of air particulates in the ambient air of Dhahran during 1991 and 1992.
- Oil lakes in Kuwait contain about 20 to 50 million barrels of oil and more than 50% of which can be recovered, leaving a large volume of oil in the bottom of the lakes. Soil contamination from the oil lakes was limited as penetration of oil was reported to be limited to less than 0.5 meter of top soil. It was estimated that about 35 million cubic meter of soil were contaminated by oil. Much of the contaminated soils are feared to be buried under dunes. Biodegradation of petroleum in the soil is expected to be slower due to anoxic conditions, high temperature, and the chemical nature of the oil (highly

weathered to a very dense cake). Therefore, decontamination will be slow and the complete recovery of oil contaminated soil may require many years or even decades.

- Military wastes are spread over an estimated area of 40,000 km^2 of Kuwait, northeastern Saudi Arabia, and southern Iraq. Corrosion of these wastes can contaminate surrounding soil. Soil contamination from the sanitary waste can be even more serious. It was estimated that the United States troops alone produced over 4 million tonnes of human wastes from December 1990 to February 1991. The troops collected these wastes, burned or buried the wastes underground, thus making it likely to contaminate surrounding soil. In addition to chemical contamination, biological contamination of the desert soil from the human waste is likely.
- Vegetation in southern Iraq, Kuwait, along the Saudi/Kuwait border was seriously damaged. Activities such as ground troop deployment and movement, erection of defensive structures, relentless Allied bombing over Iraqi positions, and the formation of oil lakes were a few of many activities that severely damaged desert vegetation. Information suggests that the recovery of vegetation in heavily impacted areas in the war zone can take several decades because of the harsh climatic conditions in the Gulf region.
- Available data indicate that vegetation in the war impacted area was contaminated by dry deposition of dust and soot from the Kuwait oil fires, especially with metals and organic chemicals. Oil contamination of desert vegetation was severe within 50 km of the oil fire sources, predominantly in the downwind direction. However, no direct data were collected to quantify oil contamination of natural vegetation in Kuwait.
- Gaseous pollutants generated by the Kuwait oil fires and the emissions from the military vehicles and planes can cause injury to plants. Ground measurements showed relatively high concentrations of sulfur dioxide in Kuwait, mainly in the vicinity of the oil fire sources. These concentrations were sufficiently high to induce toxic effects in the natural vegetation. However, in other areas, gaseous air pollutants were largely below their plant injury thresholds. Therefore, no adverse effect is expected on desert vegetation in Kuwait, Saudi Arabia, and other countries in the Gulf region.
- The Gulf crisis has adversely effected agriculture in the region. Iraqi troop deployment completely destroyed agricultural areas in Kuwait. In Saudi Arabia, it was estimated that, as a result of labor scarcity and other war-associated factors, crop productivity in the Eastern Province of Saudi Arabia dropped significantly. Crops which were grown in winter, such as wheat, were seriously affected. The magnitude of the adverse effect of troop deployment in Saudi Arabia could not be ascertained. It is believed that the impact of troop deployment on Iraqi agriculture was minimal. Troops were not deployed in large numbers in other Gulf countries, and therefore, agricultural activities in these areas should not be effected.
- Like crops, the range resources in Kuwait and the Eastern Province of Saudi Arabia were adversely effected by the Gulf crisis. There is substantial

evidence suggesting that animals and their products could be contaminated from the toxic chemicals produced by the Gulf war and the Kuwait oil fires. However, data to substantiate these fears are not available.

- It is anticipated that the UN embargo on Iraq has adversely effected agricultural activities in that country. Lack of fertilizers, shortages of power and irrigation water, spares for machinery, and limited labor supply would adversely effect the agricultural capacity of Iraq. The combined effects of all the above factors may result in a significant reduction in Iraqi crop productivity. There are no reports of mass starvation in Iraq. However, many of those who visited Iraq reported that malnutrition has reached alarming proportions, especially among children.
- Soot/smoke from the Kuwait oil fires decreased solar radiation, ambient temperatures, and visibility during February through September 1991. The decrease in air temperature was in the range of 0.8 to 4°C. It is anticipated that alteration in the meteorological conditions would not effect agricultural activities in the Gulf region. From the available data, it was concluded that the Gulf crisis did not impact the distribution of monsoon in the Indian Subcontinent. It was found that precipitation samples collected in Dhahran, Saudi Arabia during March 1991 contained toxic metals. However, concentrations of major cations (calcium, sodium, magnesium, and potassium) were within the historical range.
- Wildlife in the Gulf region was affected due to damage to food resources, military vehicles running over animals and their burrows, noise and air pollution, oil lakes and trenches, and heat dissipated from the Kuwait oil fires. Many of the animals (gazelle, camel, the Arabian hare, wild ass, porcupines, the Syrian hyena, Asiatic jackal, Persian and Arabian wolves, and wildcats) that were resident before the troop deployment probably migrated to other less disturbed areas. Rodents such as the jerboa, mice, hedgehogs, snakes, and lizards were not able to leave the area of deployment. Many of them were probably crushed by the vehicles or killed by trigger-happy soldiers. Principal rodents in the Gulf areas (jerboa – two species, gerbil – 6 species, mice – 3 species, rats – 4 species, and hedgehog – 1 species) and other invertebrates, lizards, Turkish and fan-footed gecko, the desert monitor, dice, horned, sand and Montpelier snakes, and the Levant viper would be most vulnerable to adverse effects due to military vehicle movement. A few camels were killed daily by the vehicles on desert roads (personal communication). No data could be found on accidental killings of wildlife in the war region.
- Data from other studies suggest that noise from the troop deployment, sonic booms from the bombardment, troop redeployment, and ground operations may have seriously damaged the hearing capacity of many animals, thus making them more vulnerable to predation. This damage may not be reversible. In addition to the physiological effect, pollution from the Gulf crisis could adversely effect wildlife. For example, concentrations of sulfur dioxide were found to be very high in the vicinity of the burning oil fields,

particularly in downwind areas. Wildlife in such highly contaminated areas could be adversely affected by the synergistically acting particulate matter and sulfur dioxide. A Kuwaiti scientist reported that many sheep in Kuwait City died due to inhalation smoke and other pollutants. It was shown that other pollutants that can cause acute poisoning of wildlife were below the permissible limits for humans. It is anticipated that such concentrations will not alter wildlife dynamics. Chronic effects could be manifested by ingesting toxic chemicals which might be deposited on food. Many of the organic and inorganic chemicals that were produced during the Gulf crisis are known to be biotoxic.

- Many of the wild animals might have been affected by oil mist inhalation. Reports from Kuwait indicate that many rodents, lizards, snakes, and birds fell into the oil lakes and trenches and died. A significant number of wild animals could have been trapped and killed when the wells were suddenly sabotaged by Iraq. The flames from the Kuwait fires could have easily burnt any animal in their reach. More animals would have been killed by the intense heat if they tried to leave their burrows or were forced to remain underground and eventually starve to death.
- It was concluded that contamination of deep groundwater aquifers due to the Gulf crisis is a possibility; however, the probability of contamination of shallow groundwater, especially in the northeastern Saudi Arabia and Kuwait, remains high. The contamination of groundwater aquifers could come from the wastes buried underground, seepage of contaminants deposited on the ground, and seepage of oil from the lakes and trenches. Luckily, the shallow groundwater aquifers in the area are of limited utility. Freshwater resources of Iraq were badly contaminated due to the direct and indirect impact of the Allied bombardment. During the War, clean drinking water supply was interrupted in Baghdad. It was suspected that unhealthy drinking water would cause epidemics of digestive tract disorders. Additional information on the surface water contamination in Iraq was not available.
- In conclusion, land resources of the War region were adversely affected. Quantitative documentation of the damages to the land resources were limited. From the information collected, it is anticipated that recovery of the land resources will probably require several decades or may never be restored.

6.2. Introduction

The Gulf war is regarded as the major ecological disaster in the Gulf region. The terrestrial environment was under continuous assault during the Gulf war. Environmental consequences of the Gulf war were probably more devastating than expected because the predominant desert ecosystem is considered very fragile. Many anthropogenic activities associated with military operations in the area could potentially damage the land resources. The objective of this chapter is to summarize the impact of the 1991 Gulf war on the land resources of the Gulf region. Our discussion will be confined to the following:

1. Solid wastes generated during the Gulf crisis.
2. Destruction of desert surface crust and contamination of soils from war related activities.
3. Damages and contamination of natural vegetation in the war zone.
4. Impact of the Gulf war on range resources of the Gulf region.
5. Effect of the war on desert wildlife.
6. Miscellaneous indirect effects (such as temperature, solar radiation, precipitation, etc.) of the Gulf crisis.

6.3. Solid waste pollution from the Gulf war

Solid wastes generated by the Gulf war can directly effect natural resources in the war zone, and therefore information was collected on the following:

- Residue from munitions, anti-personnel mines, explosives, etc.;
- Sanitary wastes and their management; and
- Damaged and discarded military hardware.

6.3.1. Munition wastes

Figure 6.1 shows the general deployment of troops during the Gulf crisis. In the deployment area, many locations were used for military exercises during the pre-war period. In these locations, live and spent cartridges, mortar shells, etc., could still be found in July–August 1992. Figure 6.2 shows two such target practicing areas in northeastern Saudi Arabia. Many kinds of munition wastes can be seen in this figure. There may be similar military practice areas in Kuwait and Iraq that were used by the Iraqi army. It was estimated that over 120,000 tonnes of munitions were used in the Gulf war (Barnaby, 1991), a good portion of that tonnage would have become solid waste. An assessment of the magnitude of munition waste is difficult because military exercises and the War were spread over a large territory, and most of the wastage was probably quickly buried by drifting sands in the war zone.

6.3.2. Wastes from anti-personnel and anti-tank mines

Iraqi deployed over one million anti-personnel and anti-tank mines in seven major fields in Kuwait covering about 60% of Kuwait (Khuraibet, 1992). As a result, mines and unexploded shells can be found scattered all over Kuwait's desert. Recovery of these mines is a daunting task. The Kuwaiti Defense Ministry reported that a total of 700,000 mines were dismantled by March 1992. The actual estimates of mines deployed by Iraq could be several million as opposed to the one million mines estimated by the Kuwaiti Defense Ministry (Khuraibet, 1992).

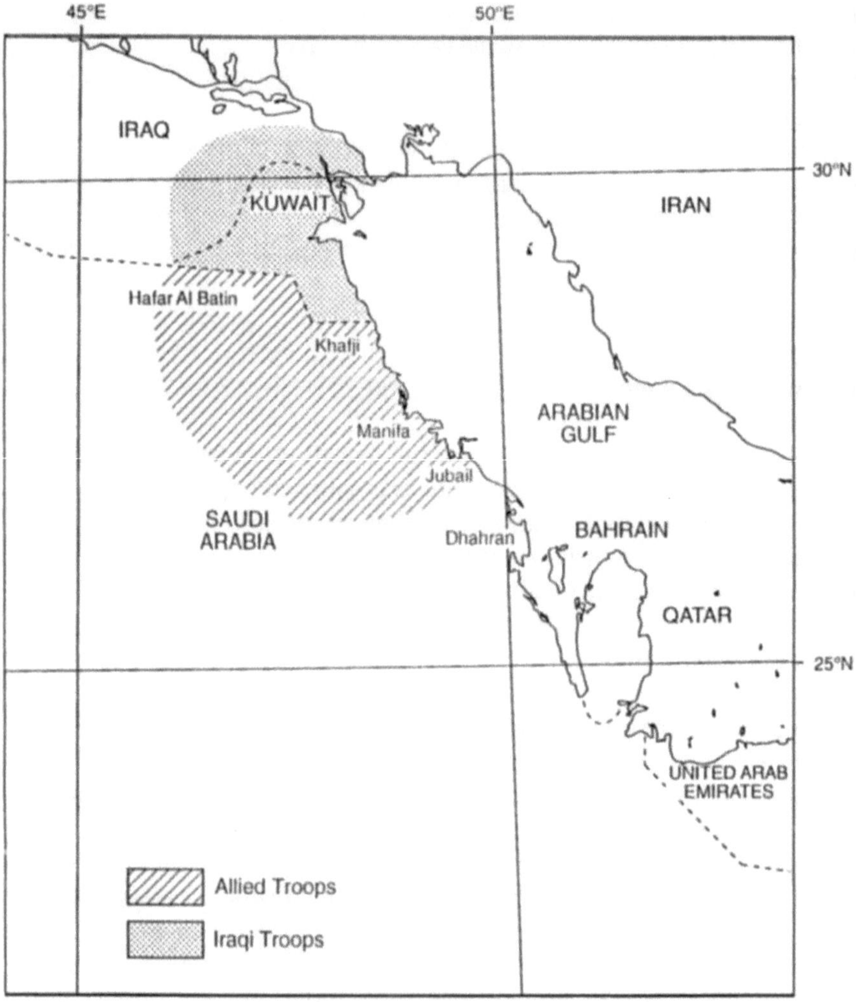

Figure 6.1. Map showing general areas of troop deployment during the Gulf crisis.

6.3.3. Residue from bombardment

It was reported that the Allied Forces dropped a total of about 84,200 tonnes of bombs (The Middle East, 1992). Fragments of exploded bombs and many live bombs can be found littering much of Kuwait and selected areas of Iraq such as southern border areas where Republican Guards were deployed around the strategic Iraqi installations, etc. The exact magnitude of solid waste from this

Figure 6.2. Munition waste scattered at a target practice area in northeastern Saudi Arabia.

source will probably never be known. It seems reasonable to assume that about half of the total tonnage of bombs were scattered as solid waste.

6.3.4. Waste from damaged tanks, armored and other vehicles

The Allied Forces predominantly bombarded Iraqi military strategic locations, such as armored units, guns, and other military transport facilities in the theater of war. It was reported that the Allied Forces destroyed about 70% of the tank force (more than 3000) and artillery power (more than 2100 pieces) of what Iraq had committed to the Gulf war theater (MacLeod, 1991; Church, 1991). At one of the roads, now commonly referred to as "death road", thousands of cars, tanks, and other military vehicles were destroyed by relentless Allied bombardment and ultimately abandoned by Iraqi military personnel. The road was packed with damaged vehicles – literally thousands of them. The amount of waste from damaged military hardware was overwhelming on this road. Several reports suggest that half of Kuwait was littered with damaged vehicles, both military and civilian, during the Gulf war (Church, 1991; Al Hassan, 1991, 1992). Solid waste generated from the damaged military hardware was mainly confined to Kuwait and southern Iraq.

6.3.5. Sanitary wastes and others

Sanitary wastes mainly consisted of the food left overs and human excretment. More than half a million troops were deployed in the northeastern Saudi desert. United States troops daily consumed two "Meals Ready to Eat" (MRE), about 6 million bags per week. Cutter (1991) estimated that the MREs alone would generate solid wastes of over 250,000 tones per year. In addition, he estimated about 10 million gallons of human wastes per day. It was calculated that about 4 million tonnes of human wastes would have been produced during December 1990 through February 1991 alone. The actual amount of human wastes would have been much higher than this estimate, perhaps as much as 10 million tonnes.

During a field reconnaissance trip in July 1991, it was noticed that at certain military camps, large quantities of barbed wire were left. Figure 6.3 depicts a location in one of such camps. This type of waste was largely confined to northeastern Saudi Arabia. Little attempt has been made to collect barbed wire from the Saudi desert. At certain locations, wood and other wastes were also found. This problem is illustrated in Figures 6.4 and 6.5 and is expected to be found largely in Saudi Arabia. In general, various military commands have done an acceptable job of burying or burning its garbage.

Vehicles require extra maintenance in the desert. There were many servicing units scattered in the desert to maintain vehicles of the military and support services. Spent lubricating oil was collected and burned. Its residue was buried. This could be a source of contamination for the surrounding soil. Quantitative estimates of this type waste were not available.

Figure 6.3. Barbed wire found at one of the military camps in northeastern Saudi desert.

Figure 6.4. Wood and other wastes scattered in one of military camp in Saudi Arabia.

Figure 6.5. Miscellaneous waste scattered in one of the military camp in Saudi Arabia.

6.4. Damage to desert crust

In natural desert systems, a dynamic equilibrium is maintained among different features. The role of surface stabilization is critical in this balance. In the Gulf war zone, the natural vegetation was sparse and the surface stabilized by biological, chemical, and even mechanical factors. Biological crusting is highly vulnerable to physical abuses. Chemically precipitated crusts are mainly salts of sodium and calcium and are widely spread in the Gulf region, especially in the sabkhas and their peripheral areas. In the California desert chloride salts dominate towards the center of the chemically crusted areas, followed by sulfates and finally carbonates (Dregen, 1968; Hunt, 1966). Similar conditions could be present in the Gulf region. Mechanical crusts in the desert soils are largely provided by the silty-clay fraction (Kodova et al., 1979; Cooke and Warren, 1973).

Because of the arid climate in the Gulf region, most soils in the region are desert soils with limited alluvial planes along the Euphrates and Tigris rivers in Iraq (Tandy, 1978). There are many active dunes in the area, especially in Saudi Arabia. The majority of the desert areas is relatively loose sandy soil with sparse vegetation, barren salt pans (sabkhas), and pebble beds. In general, soil profile development is minimal in the desert ecosystem. Some general characteristics of world deserts are compared with the Arabian desert in Table 6.1. This table shows that the major portion of the Arabian desert is covered by dunes and desert flats that are relatively susceptible to wind erosion. Many war related activities would have disturbed the desert crust. Attention was focused on the following:
- Troop deployment and military installations,
- Bombardment of strategic installations,
- Redeployment of army and ground war, and
- Activities associated with extinguishing the Kuwait oil fires.

Table 6.1. Some characteristics of World deserts (extracted from Webb and Wilshire, 1981)

Characteristics	Southwestern US	Sahara Desert	Libyan Desert	Arabian Desert
Surface type (%)				
Desert flat	21	10	18	16
Fans and bajadas	31	1	1	4
Dunes	1	28	22	26
Desert mountains	38	43	39	47
Wind audibility by surface types (%)				
Fans, flats, and bajadas	84	25	35	40
Dunes	1	64	41	52
Playas	2	2	2	2

In addition to the effects of the Gulf war on the desert crust, recovery of the surface soil and soil erosion will also be discussed in this section.

6.4.1. Troop deployment and military installations

To occupy Kuwait, the Iraqis moved several armored divisions and at least 300,000 troops into Kuwaiti territory. This military deployment broke the upper hard layer of desert soil in most of Kuwait. In response to the Iraq occupation of Kuwait, Allied Forces, about 700,000 military personnel, converged on the Gulf region. The majority of the military personnel were stationed in the northeastern Saudi Arabia. The area of high troop concentration is shown in Figure 6.1. Both Iraqi and the Allied troops destroyed surface soils by digging trenches of several thousand km, erecting barricades to a height of about 10 to 20 feet (Figure 6.6), clearing of soil surface for camping, building desert roads for transportation, and constructing necessary utilities and communication systems. Military facilities were erected for stockpiling ammunitions and helicopter bases were constructed. During the pre-war period, we regularly visited the deployment areas in Saudi Arabia. It was simply unbelievable. As far as one could see, there were military installations, armored vehicles, tanks, helicopters, etc. Several months after the troops were gone, there were still many visible tracks in July–August 1992. It was estimated that about 25% of the desert surface in the deployment area was disturbed. The situation in Kuwait and Iraq was not better than in northeastern Saudi Arabia. Barnaby (1991) estimated that as much as 25% of the land masses in Kuwait and the northeastern part of Saudi Arabia were disturbed by military maneuvers.

6.4.2. Bombardment of strategic installations

The Gulf war started on January 16, 1991 with a massive air strike on Iraqi strategic installations. Later, the air campaign concentrated on Iraqi troop concentrations and ammunition stockpiles, both inside Iraq and Kuwait. Deployment areas of Republican Guards were subjected to carpet bombing. The bomb explosions created craters many meters wide and deep. Iraqi troop concentrations in Kuwait were also heavily bombarded. During the pre-war period, the Iraqi military fringed Kuwaiti border areas with anti-personnel and anti-tank mines. The Allied air force concentrated on many of the mine fields and subjected them to carpet bombing with devices that created about 200 to 500 PSI (pound per square inch) pressure on the ground surface. An accurate estimate of the magnitude of damage to the desert crust from bombardment was not possible. Qualitatively, it can be concluded that desert soil crust was badly damaged in the areas that were bombarded.

Figure 6.6. An earthen baricade in Saudi Arabia. In Kuwait, these baricades run for miles.

6.4.3. Redeployment and the ground war

During the two weeks before the ground operation, about 200,000 Allied troops, mainly from the United States. Several armored divisions were moved to the Saudi-Iraqi border. The redeployment was so swift and extensive that it consumed about 5 billion gallons of fuel (Cutter, 1991). The redeployment exercise completely damaged the soil surface along the Saudi/Kuwaiti border.

At the start of the ground operation, six armored divisions and many infantry divisions stormed through the Iraqi positions in Iraq and Kuwait. A part of the Allied armored dashed towards the Euphrates river above Basra. The ground surface of Kuwait and border areas of Iraq was completely pulverized under armored vehicles, personnel carries, troops, etc. In summary, during the Gulf war, the ground was severely disturbed. The remnants of this abuse are still very visible two years after the Gulf war.

6.4.4. Kuwait oil fires

The Kuwait oil fires impacted the ground surface in two ways. First, there was a rain of oil mist all over Kuwait. The oil formed a compact layer by cementing soil grains together. Second, fire fighters had to dig several hundred kilometers of water channels and ponds, construct access roads, and prepare for the final assault on the oil well fire. These activities disturbed the surface soil appreciably.

6.4.5. Recovery of the desert crust

From the foregoing, it can be concluded that surface soil in the war zone was severely disturbed. The area where the surface soil was badly damaged by military related activities is shown on Figure 6.7. This includes all of Kuwait, the northern border area of Saudi Arabia and the southern border areas of Iraq. No information was available on the recovery of disturbed desert soils in the region. However, data collected in the California desert after military exercises suggest that soil crust recovery may require several years (Lathrop, 1983).

6.4.6. Effect of Gulf war on soil erosion

Soil erosion is expected to be enhanced in areas where the surface is physically damaged and the extent of damage is wide. In the Gulf war zone, ground surfaces were badly damaged over large areas as shown in Figure 6.7. The vulnerability of soil to wind erosion enhances if coupled with high wind speed, such as shamals in the Gulf region (McCualey et al., 1980; Wilshire et al., 1980). All circumstantial evidence suggest that soil erosion may be a serious environmental problem in the Gulf region. Unfortunately, no information was available that directly depicts the effect of the Gulf war on soil erosion in the Gulf region. However, it was assumed that the air particulate load can be taken

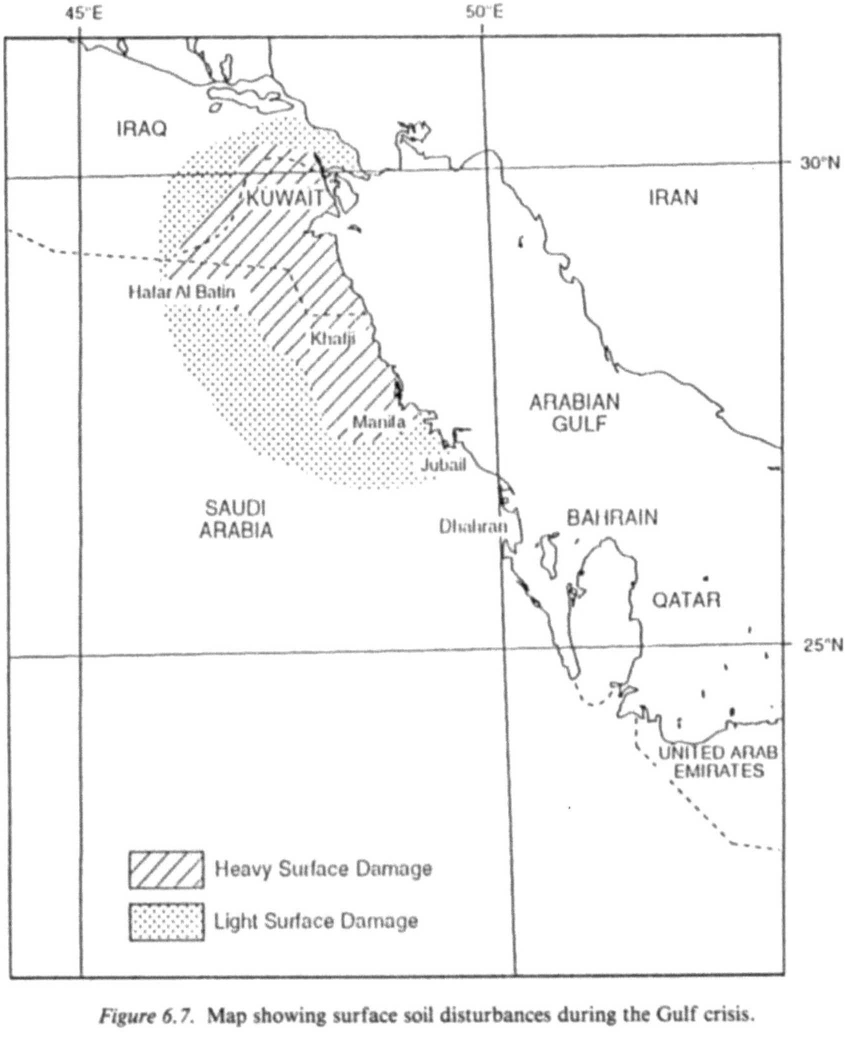

Figure 6.7. Map showing surface soil disturbances during the Gulf crisis.

as an indication of soil erosion – high soil erosion in the area will increase concentrations of ambient air particles. Total suspended particulate (TSP) were collected in Dhahran in 1980/83 (KFUPM/RI, 1985). These data were compared with similar observations taken during 1991 and 1992. The monthly mean concentrations of TSP in the ambient air of Dhahran are plotted in Figure 6.8. The highest concentration of TSP were found during June for all the years plotted in Figure 6.8. In general, concentrations of particulates were higher during March–August period. The predominant northerly winds, locally known as shamal winds, was considered responsible for the enhancement in particulate concentrations. A comparison clearly indicates that TSP concentrations were higher in 1991, especially during March to September

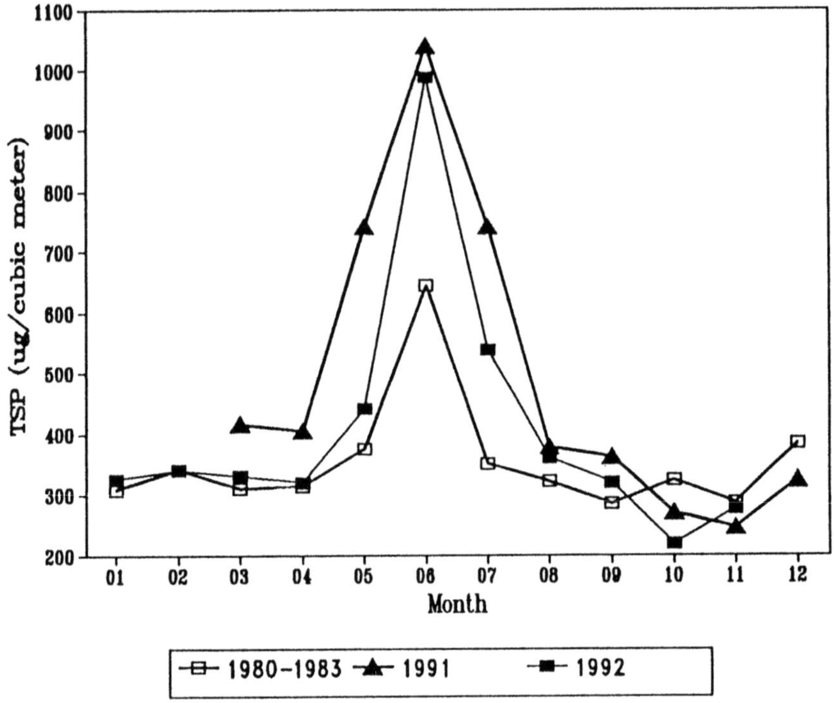

Figure 6.8. A comparison of total suspended particulate in the ambient air at Dhahran (Saudi Arabia) during 1980–83 and 1991-1992.

1991, followed by 1992, as compared with other years. Along with the fallout from the Kuwait oil fires, soil erosion could be partially responsible for the increase in TSP during March to September 1991. In 1992, the increase in TSP could again be related to soil erosion. This explanation seems plausible but the time gap between 1980-83 and 1991-1992 is considerable.

6.5. Soil contamination from the Gulf crisis

Soil is an important component of the terrestrial environment. Many activities associated with the Gulf war had the potential to contaminate desert soils in the impacted areas. Soil contamination will be discussed as:

1. Contamination of soils from the Kuwait oil fires, and
2. Contamination of soil from the military operations.

6.5.1. Soil contamination from the Kuwait oil fires

Iraq sabotaged more than 800 oil wells in different Kuwait oil fields. As a result, 613 oil wells caught fire and 76 were gushing crude oil. From a soil contamination point of view, this sabotage created twin problems; contamination from the emission products as atmospheric fallout, and formation of oil lakes on the surface of the desert.

6.5.1.1. Emissions from the Kuwait oil fires

Dispersion of smoke and associated pollutants were discussed in Chapters 3 and 4. It was concluded that Kuwait, Bahrain, and the northeastern part of Saudi Arabia along the Arabian Gulf were covered more frequently with smoke. Soils of these areas were expected to be contaminated with pollutants from the Kuwait oil fires. From a soil contamination point of view, quantities of metals, organic chemicals, soot, and oil mist emitted by the Kuwait oil fires are important and these will be discussed here.

Crude oil was emitted under high pressure which resulted in the escape of a small portion of unburned oil. This portion formed an oil mist that was observed throughout Kuwait. The magnitude of the oil spray, in terms of area and amount, was dependent on wind speed and wind direction. There were no data available that indicate the magnitude of soil contamination from oil mist.

The Kuwait oil fires produced huge amounts of soot particles (about 8 million tonnes). It was reported that about 50% of the soot particulates were scavenged within 100 km of the fires (Hobbs and Radke, 1992). Significant contamination of desert soils from soot particles would have occurred within 100 km of the fires, i.e. mainly in Kuwait and northern Saudi Arabia near the Kuwait border. The Kuwait Environmental Protection Department investigated extractable organic matter as a function of distance from the fires and depth of soil samples (EPC, 1991). Extractable organic matter in the upper soil samples was appreciably higher as compared with that extracted from the deep soil samples. Extractable organic matter was also related to distance from the source. The preliminary study concluded that the emission products of Kuwait oil fires, including oil mist and soot, contributed to the enhancement of extractable organic matter in the upper soil samples.

The Kuwait oil fires enhanced concentrations of polycyclic aromatic hydrocarbons, aliphatic hydrocarbons, and many other organic chemicals in the ambient air particulates (Chapter 4). These contaminated air particulates were later dry deposited over large areas of Kuwait and Saudi Arabia (Chapter 4). Concentrations of organic chemicals in soils in these areas would have increased as a result of dry deposition of contaminated air particulates. It was reported that concentrations of polycyclic aromatic hydrocarbons were higher in the surface soil samples and could be related to the distance from the source (EPC, 1991). A few dust samples were collected from wheat leaves in the Eastern Province of Saudi Arabia. The presence of polycyclic aromatic hydrocarbons was confirmed (personal communication). Because of the limited samples from one location only, the effect of Kuwait oil fires on the concentrations of organic compounds could not be established in Saudi Arabia.

Large quantities of brine water mixed and was emitted along with crude oil. Concentrations of organic salts were so high that the color of many of the Kuwait oil fires changed from dense black to gray white (Hobbs and Radke, 1992; Jenkins et al., 1991; Johnson et al., 1991; Daum et al., 1992). The emitted inorganic salts were adsorbed on soot particles to such an extent that the majority of soot particles were transformed from being hydrophobic to hydrophilic. Hobbs and Radke (1992) reported that the smoke particles collected from the super- composite plume (smoke plume where fires from the individual wells combine together to form one plume) consisted of about 30% salt, 15–20% soot, 8% sulfate crystals, and about 30% organic chemicals. From the foregoing, inorganic contamination of desert soils near the fire sources is to be expected.

Sodium and chloride were the major constituents of inorganic salts emitted by the Kuwait oil fires (Hobbs and Radke, 1992). High sodium concentrations adversely effect soils, especially its structure. Exact information on the sodium contamination of soils due to the Kuwait oil fires was not collected. It is, however, expected that serious sodium contamination will be found near the fire sources to a distance of about 200 km downwind direction. It is suggested that sodium contamination of the desert soil should be investigated in the Gulf region, particular in Kuwait and the northeastern part of Saudi Arabia. Sulfate produced by the fires would have reacted with either the sulfate or carbon dioxide and water to form naturally occurring calcium carbonate or calcium sulfate (Mitchell and Al Sunaid, 1992; Daum et al., 1992; Al Sunaid, 1992). Therefore, adverse edaphological effects due to the addition of these compounds were not expected. Another inorganic pollutant emitted by the Kuwait oil fires was sulfate that would have formed either calcium sulfate or carbonate – both are naturally occurring minerals in the Gulf region (Mitchell and Al Sunaid, 1992; Daum et al., 1992; Al Sunaid, 1992). Therefore, adverse edaphological effects due to the addition of these compounds were not expected.

Crude oils contain metals (HPI, 1987) that would have been emitted along with the well fires. The Kuwait fires were believed to emit about 7,224, 2,084, 917, 747, and 432 tonnes of vanadium, nickel, aluminum, iron, and zinc, respectively. Most of the soil in the region contained appreciable concentrations of either calcium sulfate or calcium carbonate or both. The emission quantities of many other metals, including arsenic and mercury, were below 150 tonnes (Sadiq et al., 1992). Concentrations of metals, especially nickel and vanadium, would have increased in soil where smoke was frequently visible (Chapter 4). Al Sunaid (1992) reported that concentrations of nickel and vanadium increased from Tanajib (from the direction of the Kuwait oil fired) to Dhahran, Saudi Arabia. Systematic investigations on the effects of the Kuwait oil fires on metal contamination of soil in the Gulf region was not conducted.

6.5.1.2. Contamination of soils from oil lakes and trenches

Over 76 oil wells in Kuwait were gushing crude oil and forming oil lakes. Oil lakes were largely formed in Burgan, Sabriya, Rawdatain, Magawa, and Ahmadi Oil Fields (Al Besharah and Saad, 1992). The oil lakes vary widely in their size, shape, depth, characteristics of the oil, and the oil well head involved in forming them. Most of the oil lakes were irregular in shape. The depth of oil lakes varied from a few centimeters to over 1.5 meters. Because of the continuous evaporation of oil and oil seepage through the soil, the exact depth of some of the oil lakes may never be known.

The characteristics of oil in the lakes varied with time. The oil on the periphery of the lakes was heavy as compared with the center. Similarly, with the passage of time the oil increased in density. The characteristics of oil in the lakes can only be generalized. Like other aspects of oil lakes, an estimate of their size is difficult. Heat and smoke from the burning oil wells, presence of mines and live explosives in and around the lakes, lack of equipment, and inadequate human resources are some of the many limitations in collecting the data on the size of lakes. In addition, lake size changed due to evaporation of crude and covering of the lakes by dunes.

The first systematic attempt was made to estimate the extent of this problem and a topographic map of oil lakes was developed in August–November 1991 (Al Besharah and Saad, 1992). Later, another attempt was made by the Kuwait Oil Company. The Company conducted an extensive ground survey to collect data on the extent and geography of the problem. The Kuwaiti Oil Minister, Homoud Abdulla Ragba estimated the total volume of oil in the surface lakes at 25 to 50 million barrels (Parmelee, 1991). By October 1992, the Kuwait Oil Company had recovered about 4.5 million barrels of oil. It further estimated that about 14 to 20 million barrels more can be recovered. About 5 to 30 million barrels of oil probably remains at the bottom of the lakes. The above estimates exclude the oil volume that has soaked in the soil (Al Besharah and Saad, 1992). Information on the area of these lakes is not available at this time. However, there is an indication that some of the lakes were quite large and contained more than 250,000 barrels of oil (Al Bashara, 1991). Recently, it was reported that 6.5 million barrels of crude was recovered from 250 oil lakes and 15–20 million barrels still remain in the lakes (Arab News, 1992b). Kramer (1991) reported that about 35 million cubic meters of soil have been contaminated in the oil lakes.

Iraqi dug several thousand kilometers of trenches along the Saudi-Kuwaiti border and coast of the Arabian Gulf. During the air war period, these trenches were filled with a large but unknown quantity of oil and ignited. The oil was partially burned leaving an unknown portion in the trenches. Since then, most of these trenches have been filled with sand and their embankments have since caved in. These trenches may not show on the surface, but the contaminated soil buried in these trenches will stay for a long time to come.

It was suspected that oil in the lakes and trenches might seep slowly to the lower soil profiles. Kuwaiti scientists examined several locations in the Al Ahmadi, Magwa, and Burgan Oil Fields for oil penetration. They reported that at many of these locations, oil had penetrated only to 30 to 35 cm (Al-Besharah, 1991; Al Besharah and Saad, 1992).

Petroleum hydrocarbons in the desert soil will eventually be biodegraded to carbon dioxide and water that will escape to the atmosphere. However, concentrations of oil-associated metallic pollutants, especially nickel and vanadium, will remain in the upper soil layers. The increase in soil metal concentrations will be proportional to the degree of oil contamination. In conclusion, oil lakes have badly contaminated soils underneath. Although oil soaking was found to be within 0.5 meter of the suface, the soils underneath the oil lake will remain unproductive for several years to come. The recovery of oil contaminated soil may take years.

6.5.2. Contamination of soil from the Gulf war operations

In general, soil contamination from the Gulf war could result from ammunition waste, sanitary waste, chemicals from the explosive use, and combustion products of fuel consumed during the Gulf war. The impact of these activities will be discussed here.

6.5.2.1. Soil contamination from military waste

The magnitude of military waste produced during the Gulf crisis has been discussed above. There are numerous media reports stating that the whole of Kuwait was littered with military waste (Al Hassan, 1991; Al Hassan, 1992; Khuraibet, 1992). Military waste can also be found scattered over large areas of Saudi Arabia and probably Iraq. Soil contamination will occur when corrosion products of the military waste are dispersed and mixed with the surrounding soil. Appreciable quantities of such wastes have already been buried by drifting sands and corrosion has already started. The magnitude of soil contamination from military waste is difficult to assess at present.

6.5.2.2. Sanitary and other wastes

Soil contamination from sanitary waste could result from solid residue and liquid waste. As stated above, approximately 10 million tonnes of sanitary waste were produced during the period of the Gulf crisis (see Section 6.3). The Allied armies collected the solid portion of human wastes and burned it along with other wastes, such as wood, spent MREs, used oil from the vehicles, etc. The unburned residue and MREs bags were buried. Ultimately, these residues will be broken down by chemicals and biological processes. The byproducts may disperse and contaminate surrounding soils.

The liquid portion of human waste was directly discharged into pits underneath the lavatories from where it seeped into the soil without any treatment. The liquid portion contained microbes and some of them can be infectious. The army had buried all human waste and it was difficult to determine locations of the lavatories during a field reconnaissance survey in July–August 1992. Information of the extent of this contamination could not be collected during the war period and it seems very difficult to collect such data now.

6.5.2.3. Soil contamination from the use of explosives

Soil contamination from munitions and explosives can result from the solid wastes and chemicals produced on detonation. Soil contamination from the military solid wastes have already discussed. The effects of chemicals on soil will be discussed here.

Explosives are actually highly unstable chemicals and would produce many organic and inorganic chemicals on detonation. A discussion of the chemistry of these chemicals is beyond the scope of this book. During the War period, data on chemical contamination of soils in the area were not collected. About two months after the war, a preliminary study was undertaken to demonstrate the probable metal contamination of soils in Saudi Arabia. The results of this study are not necessarily restricted to the explosives; they rather represent the overall impact of the Gulf war. In this study, Sadiq et al. (1992) collected several upper soil samples along the Kuwait border. Concentrations of nickel, vanadium, cadmium, cobalt, copper, cobalt, iron, lead, titanium, and zinc were found to be statistically similar but higher when compared with soils from the Al Hassa Oasis (140 kilometers west of Dhahran) and from an agricultural farm in Fadly (120 kilometers north of Dhahran), Saudi Arabia. These results

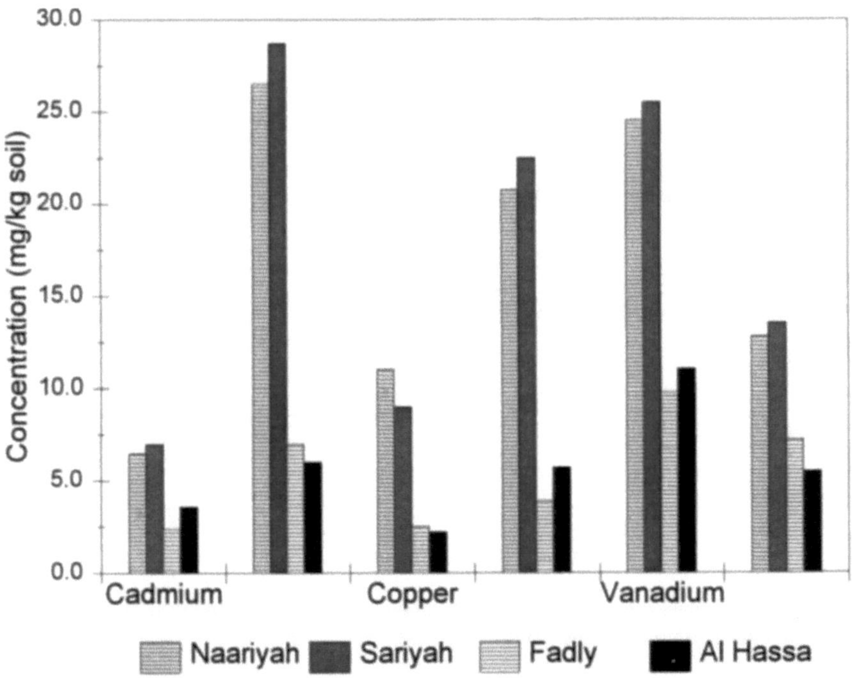

Figure 6.9. Distribution of selected metals in surface soil samples collected from the northeastern Saudi Arabia.

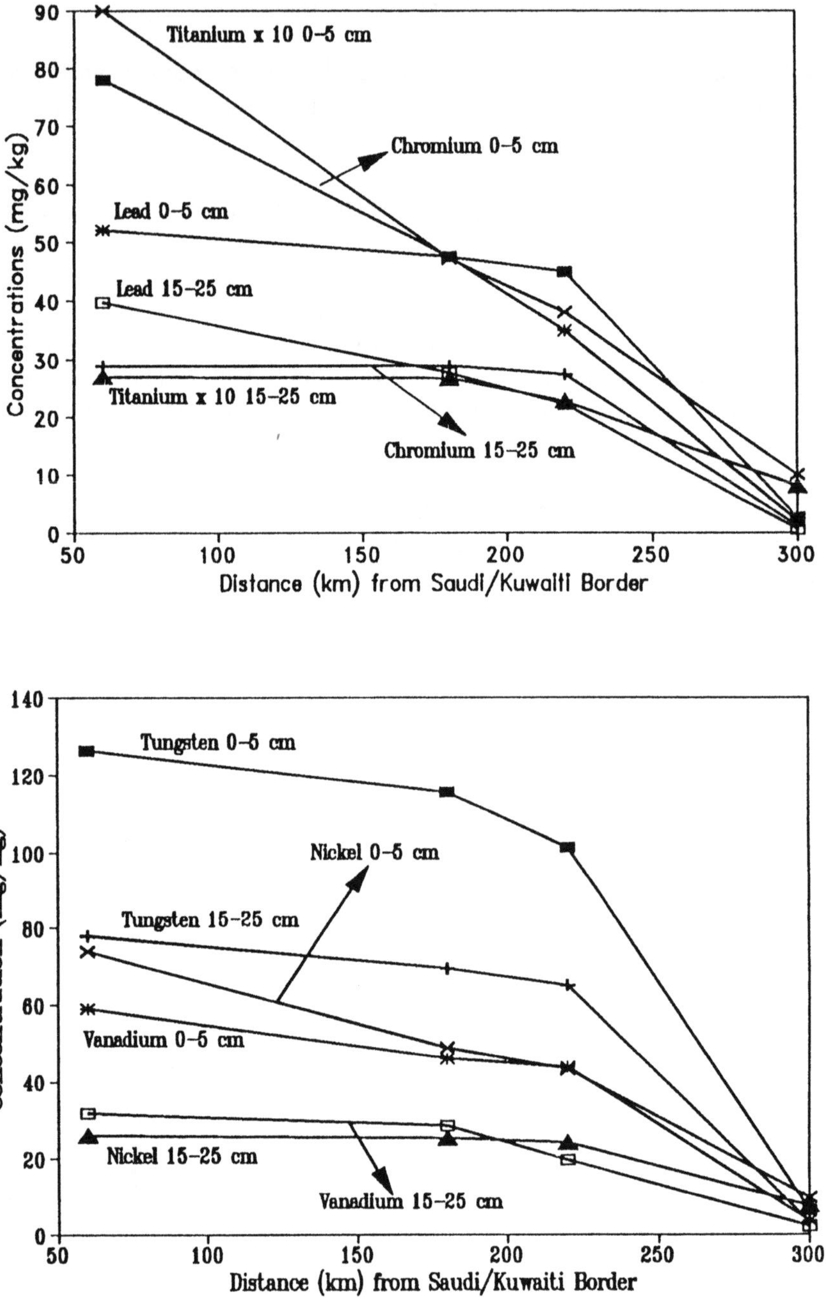

Figure 6.10. A comparison of metal contamination of soil samples collected from different locations in the northeastern Saudi Arabia.

are plotted in Figure 6.9. Soil samples were also collected perpendicular to the Kuwait border. Metal concentrations decreased rapidly with an increase in distance from the Kuwait border. Concentration distributions of copper, chromium, and lead in the upper soil surface samples are shown in Figure 6.10. In general, the data in Figures 6.9 and 6.10 suggest that the Gulf war activities might have been responsible for the metal contamination of soils in the areas bordering Kuwait. The soil samples were collected at the end of March 1991 when these areas were briefly covered with smoke from the Kuwait oil fires. Emissions from the Kuwaiti oil fires would probably have a limited effect on soil contamination by that time.

6.5.2.4. Soil contamination from combustion products of fuel consumed

The impact of combustion products from the fuel consumed during the Gulf war will be similar to the Kuwait oil fires. From the soil contamination point of view, metals and organic chemicals would be the major pollutants in the combustion products. Since soil contamination from the fuel is expected to be similar to the Kuwait oil fires, it will not be discussed further.

6.6. Damage to desert vegetation

The desert ecosystem supports a wide range of plant species. Desert vegetation can be divided into three main groups: ephemeral annuals, succulent perennials, and non-succulent perennials.

Ephemeral annual plants are short-lived, i.e. generally a few weeks under favorable weather conditions. This type of vegetation comprised over 50% of desert vegetation on a numeral basis. Because of their small size, ephemeral annual plants may not, on a biomass basis, contribute so significantly. In general, an ephemeral plant completes its life cycle in about 6 to 8 weeks, and thus growth activity is restricted to rainy periods only. If rainfall is low (generally <20 mm), the ephemerals may not germinate (Epstein, 1978). The majority of the desert ephemerals in the Gulf region are seasonal grasses.

Succulent perennial plants have enlarged parenchyma tissue which enables the plant to store water during the rainy season. The plants are generally leafless or have rudimentary leaves, and largely devoid of stomata. There may be several other morphological adaptations that help the plants survive in the harsh desert climate (Epstein, 1978).

Non-succulent perennials can be divided into two groups, geophytes and xerophytes. Geophytes are perennial plants which complete their growth cycle in a favorable season and remain dormant during harsh desert climate periods. Most of the vegetative activities of such plants are restricted to the winter or rainy season. Xerophytes are plants that adapt to desert conditions by changing their morphological and anatomical characteristics (Epstein, 1978). In the Gulf region, xerophytes are the most abundant plants and they survive well in the harsh climate of the Arabian Gulf.

The activities associated with the Gulf crisis that would damage desert vegetation are the deployment of troops in Kuwait, Saudi Arabia, and Iraq and other Gulf war activities, and the Kuwait oil fires. The effect of these activities on desert vegetation in the Gulf region is discussed here.

The most damaging factor to vegetation was the deployment of close to one million troops in Kuwait and northeastern Saudi Arabia. The extent of troop deployment and associated activities has been discussed earlier and will not be repeated here. The major impact of troop deployment occurred in about 40,000 km^2 of the territory of Kuwait, Saudi Arabia , and Iraq. Some desert ecosystem experts suggest that about 50% of desert biomass would have been destroyed in the above areas.

The Kuwait oil fires, and especially the oil lakes, caused physical damage to vegetation. Vegetation was destroyed in areas where oil lakes were formed. Coating with crude oil of desert vegetation throughout Kuwait had adverse effects. The entire vegetation cover around the burning oil fields vanished either by the oil coating, dried due to high temperatures, or fire extinguishing activities.

In spite of expected extensive damage to desert vegetation, no systematic study was undertaken by any of the affected countries. Unfortunately, part of the damage has been covered with dunes and recovery from the damage might have already been initiated. Strictly speaking, there were no data to really quantify the damage and and to study the recovery of the flora in the area. To help our readers to realize the nature of the damage and the slowness of the recovery process of the desert ecosystem, an example from the Mojave desert in California, USA is cited. This Californian desert has been used for many years as training ground for the armed forces of the United States (Lathrop, 1978). With reference to the Gulf situation, the areas of troop deployment can be considered similar to the Mojave desert situation.

Lathrop (1983) investigated two of the sites used for armored maneuvers in the Mojave desert. Results of on-site ground transect comparison of tests and controls after approximately 36 years of military exercises, are shown in Figure 6.11. The number of perennial plants (vegetation density) and ground covered by these plants were measured and are plotted in this figure. It is evident that, even after 36 years, recovery of the desert vegetation was not complete. Vegetation on the road sites and tank tracks showed a reduction of about 70 and 60% even 36 years after of original disturbances. Lathrop (1983) also reported a trend of decreasing community similarity, as compared with the control, with increased disturbance and intensity of use. He observed that vegetation composition was altered as a result of military maneuvers in the California desert. Roadways displayed the greatest reduction in vegetation density, cover, and alteration in their composition. The recovery period for the desert vegetation to pre-disturbance levels required many decades. From the foregoing, it is concluded that military maneuvers had a significant adverse effect on natural vegetation (Lathrop, 1983). It was reported from Australia

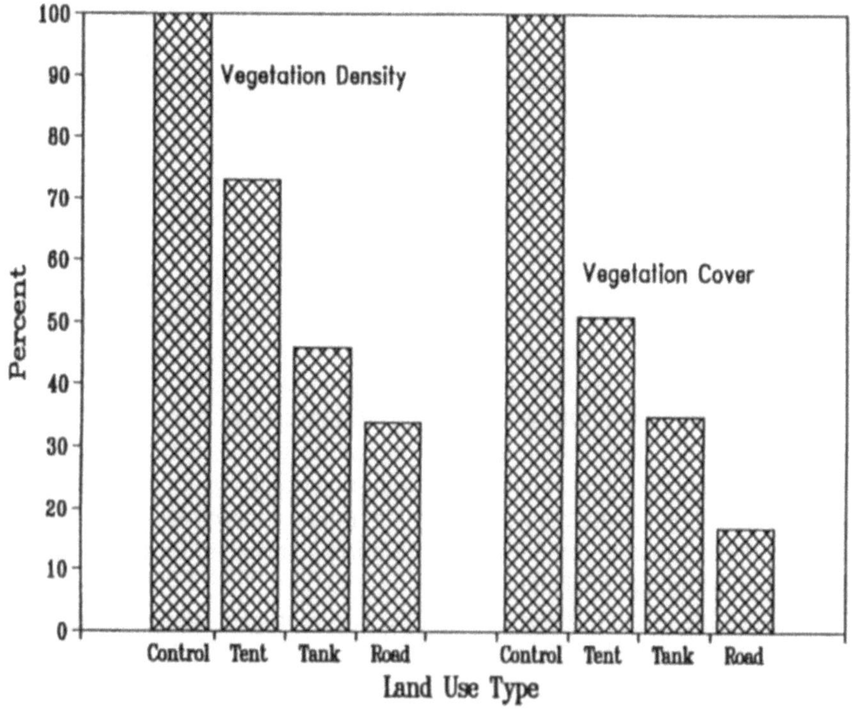

Figure 6.11. Recovery of desert vegetation after military exercises in a Californian desert, USA.

that off-road vehicles significantly damage natural vegetation in arid areas (Gilbertson, 1983). The climatic conditions in the Arabian desert are more severe than those in Mojave desert in California, especially the amount of precipitation. The vegetation in the Gulf region is expected to be under more stress than those in California. It is, therefore, expected that recovery of the damaged vegetation in the Gulf region could be slower than that of California and may require many decades.

6.7. Contamination of desert vegetation

Contamination of vegetation could result from the absorption and accumulation of chemicals in plant parts or deposition of pollutants on the aerial vegetative parts. The Gulf region is arid with very little precipitation. There is not much water to dissolve and transport pollutants from the soil surface to the root zone. It is, therefore, expected that contamination through absorption might require more than one year of exposure. However, deposition of pollution on the aerial vegetative parts is a real possibility.

The Gulf crisis generated a spectrum of pollutants that can cause injury or contaminate desert vegetation. Of importance are gaseous pollutants largely produced as combustion products from the fuel consumed during the Gulf war and Kuwait oil fires. This class of pollutants will be discussed separately. In addition to gases, the Gulf war and Kuwait oil fires also produced large quantities of organic and inorganic pollutants, most of which was absorbed on the soot particles. The contaminated particles were scavenged by dry deposition and fallout might have been deposited on the vegetation. In this section, contamination from metals and organic compounds will be discussed. In addition, oil mist from Kuwait oil fires will also be addressed.

Metal concentrations in the atmosphere were increased due to the Gulf war and the Kuwait oil fires (Sadiq et al., 1992; Sadiq and Mian 1993a, 1993b). Mitchell and Al Sunaid (1992) reported that vanadium and nickel contributions of the Kuwait oil fires to air particulates decreased from north to south. Previously, Sadiq et al. (1992) showed that concentrations of metals in the soil surface layer decreased with an increase in distance from the Saudi/Kuwaiti borders. In conclusion, the Gulf crisis has contributed to metal pollution of ambient air particulates. Thus vegetation on which such particles were deposited, will be contaminated. Data on metal concentrations of desert vegetation were not collected by any organization involved in the monitoring of environmental impact of the Gulf war.

Like metals, concentrations of aliphatic and aromatic hydrocarbons were enhanced because of the Kuwait oil fires (Fayad et al., 1992, 1993). Dust collected from wheat plants in the Eastern Province of Saudi Arabia indicated polycyclic aromatic hydrocarbons. A few polycyclic aromatic compounds were also detected in these samples (personal communication). These reports suggest that desert vegetation could be contaminated by dry deposited dust high in organic chemicals.

Information on the bioaccumulation of organics by plants is limited. Travis and Arms (1988) observed that the bioaccumulation of organic compounds in animals and milk is directly proportional to the octanol-water partition coefficient. The octanol-water partition coefficient of organic chemicals have also been significantly correlated to their bioaccumulation in fish (Chiou et al., 1977; Kenaga, 1980) and beef (Kenaga, 1980, Travis and Arms, 1988). The food chain is the primary source of human exposure to a large class of organic chemicals (Travis and Arms, 1988; 1987). Therefore, assessment of the magnitude of exposure and resulting impact largely depends on our ability to understand bioaccumulation of these chemicals in the aquatic and terrestrial human food chains. No information on the bioaccumulation of organic chemicals by the vegetation in the Gulf region is available.

One of the very common and widely reported problems of the Kuwait oil fires was the rain of oil droplets in Kuwait. Recalling his first impression of the oil fires, Dauod (1991) stated, " A thick spray of sticky, tarry droplets covered our bodies" in the city of Ahmadi in Kuwait. The rain of oil mist was also reported by Al Hassan (1991, 1992), EPC (1991), Hobbs and Radke (1991), and fire fighting teams. Everything outside, including the vegetation, was covered with oil. Al Hassan (1991, 1992) has shown photographs where plants were heavily coated with oil. The problem of oil mist was local, i.e.

confined to Kuwait. Contamination from oil mist was dependent on the wind direction and distance from the fires. The oil mist spray was more severe in the downwind direction and near the fire source. Quantitative evaluation of oil contamination of the desert vegetation was not undertaken by Kuwait scientists or other organizations. At present, it seems too late to collect data on oil contamination of desert vegetation.

From the foregoing, it was concluded that there is indirect evidence suggesting that vegetation in the War impacted area was contaminated by dry deposition, especially with metals and organic chemicals. However, no direct data could be collected to substantiate vegetation contamination.

6.8. Effect of gaseous pollutants on vegetation

Many gaseous pollutants adversely effect plant growth, and important amongst these are sulfur dioxide nitrogen oxides, carbon dioxide and monoxide, and ozone. Exposure assessment of natural terrestrial vegetation to gaseous pollutants is very complex. Terrestrial vegetation is exposed to gaseous air pollutants intermittently or continuously, and generally exposure concentrations are variable. Similarly, the response of vegetation is variable and depends on many morphological and physiological factors. Under some conditions, the effect may be beneficial to the exposed vegetation. The effect of exposure can be acute and/or chronic, reversible and/or irreversible, and direct or indirect. Exposure assessment becomes even more difficult when considered in terms of the Kuwait crisis. The effect of sulfur dioxide, carbon dioxide, nitrogen oxides, acid aerosols, and ozone are discussed here.

6.8.1. Effect of sulfur dioxide on vegetation

Sulfur dioxide gas was produced from the combustion of fuel during the crisis and burning of oil in Kuwait. The total production potential of the Gulf crisis was estimated at about 2.3 million tonnes of sulfur dioxide (Chapter 4). Ground level monitoring observations ranged between below detection to 5214 $\mu g/m^3$. Relatively high concentrations were found near the fires. An appreciable increase in the yearly mean concentrations of sulfur dioxide was noticed in Dhahran, Rahima, and Tanajib of Saudi Arabia during the Kuwait oil fires. Except for few occasions in the vicinity of oil fields in Kuwait, concentrations of sulfur dioxide was found below air quality guidelines. The World Health Organization (WHO) indicate that sulfur dioxide concentrations above 11 ppb (about 29 $\mu g/m^3$) can adversely effect plant growth (Harrison, 1991). In Germany, 50 $\mu g/m^3$ has been proposed as a guideline to safeguard crops and forest plants (Bialobok, 1984). Except for those observations taken in the burning oil fields, the mean concentrations of sulfur dioxide in Kuwait, Saudi Arabia, and other Gulf countries were below the above mentioned plant injury limits (Table 4.15, Chapter 4). It is expected that sulfur dioxide concentrations in ambient air were not high enough to adversely effect vegetation in Kuwait, Saudi Arabia, Iraq, and other countries in the region.

6.8.2. *Effect of carbon dioxide on vegetation*

Very large amounts of carbon dioxide were generated by the Gulf crisis. A conservative estimate suggests a total production of about 200 million tonnes (71 million tonnes from the fuel combustion and 133 million tonnes from the Kuwait oil fires). Carbon dioxide is not toxic. It is in fact rather beneficial to plants, especially to field crops. The photosynthetic activities in desert plants are lower as compared with agricultural plants. It is anticipated that carbon dioxide did not effect desert vegetation.

6.8.3. *Effect of nitrogen oxides on vegetation*

Nitrogen oxides can cause injury to plants (Bialobok, 1984). Nitrogen oxides have shown strong synergism to sulfur dioxide and ozone. The total amount of nitrogen oxide produced during the Gulf crisis was estimated to be 187,000 tonnes (107,000 tonnes from the fuel consumption and 80,000 tonnes from the Kuwait oil fires). Ambient concentrations of nitrogen oxide were monitored in Saudi Arabia and Kuwait and ranged between below detection limit to 69,771 $\mu g/m^3$ with a mean of less than 51 $\mu g/m^3$ (Table 4.7, Chapter 4). The highest concentrations were found inside the burning oil fields in Kuwait. The injury guidelines to plants for this pollutant are not well established. In general, it is believed that nitrogen oxide concentrations of 100 $\mu g/m^3$ as a yearly mean and 244 $\mu g/m^3$ as a 24 hour mean are below the threshold for visible effects on vegetation (Bialobok, 1984). WHO recommended guidelines for nitrogen oxide exposure to plants of 66 $\mu g/m^3$ (33 ppb Harrison, 1991). Considering the data given in Table 4.7 and the plant injury threshold discussed above, it is concluded that the ambient concentration of nitrogen oxide in Kuwait, Saudi Arabia, and other Gulf countries during the Gulf crisis were below the plant injury threshold. Therefore, no adverse effect on desert vegetation is expected from the concentrations of nitrogen oxides monitored.

6.8.4. *Effect of ozone on vegetation*

Concentrations of ozone were not effected by the Gulf crisis. Therefore, it is expected that this pollutant did not enhance injury to desert vegetation.

From the foregoing, it was concluded that gaseous air pollutants were largely below their plant injury thresholds during the Gulf crisis. Therefore, no adverse effect is expected to desert vegetation of Kuwait, Saudi Arabia, and other countries in the Gulf region.

6.9. Impact of the Gulf crisis on agriculture

The Gulf crisis had the potential to disrupt agriculture in the Gulf region and cause a deterioration the quality of crops. Agricultural activities could be influenced by troop deployment in Kuwait and Saudi Arabia, the embargo imposed on Iraq, enhanced pollution due to the Gulf crisis, and effects on the range resources. Before discussing the impact of the above listed factors, it seems imperative to review the agricultural potential of the Gulf region.

6.9.1. Agricultural potential of the Gulf region

The war impacted area includes Kuwait, southern Iraq, Saudi Arabia, and northwestern Iran. Kuwait has about one percent arable land and the remaining 99% is non-arable desert land (Cutter, 1991). Agriculture in Kuwait is more of a hobby than a national resource. Before the occupation, most of the irrigated agriculture was concentrated in the coastal areas south of the Kuwait City. The primary crops were dates, vegetables, and fodder. Most of the irrigation water came from desalination plants and a minor part from groundwater and treated wastewater. Iraq has only 13% arable land and about one third of this is irrigated. The alluvial plains of the Tigris and Euphrates are the major agricultural areas of Iraq. Iraqi farmers grow almost all crops, especially dates, rice, cotton, and wheat. Iraq is the only country in the Arabian Gulf region that utilizes runoff freshwater for irrigation.

The Eastern Province of Saudi Arabia is mostly desert with a few scattered small and large farms. Irrigation is mainly from the groundwater. Wheat, fodder, and vegetables are the main crops. Dates are the major fruit crop. Before the Gulf war, appreciable animal husbandry resources in the Eastern Province existed. The Allied troops were deployed largely in the Eastern Province. Therefore, to give an idea, some data on crops and animal husbandry resources for year 1987 of the Eastern Province of Saudi Arabia are listed in Tables 6.2 and 6.3.

In Qatar, about 2.5% of the land is cultivable. There are three main types of farms in Qatar. Vegetable farming is common in northern Qatar, vegetable-orchard-forage farming is generally practiced in central and northern Qatar, and orchard-forage farming can largely be found in southern and central Qatar (Batanouny, 1981). Information on other countries in the Gulf region was not collected.

6.9.2. Effect of troop deployment on agriculture

Troop deployment in the northeastern Saudi Arabia and Kuwait adversely effected agriculture. Iraq systematically destroyed Kuwait agriculture. Many of the desalination plants that supply water for agriculture were damaged or closed either due to the war or the oil spill. The agriculture sector in Kuwait was largely abandoned, neglected, or vandalized by the Iraqis. Agriculture

Table 6.2. An estimate of agricultural productivity in the Eastern Province of Saudi Arabia for 1987 (source: Statistical Yearbook, 1988).

Items	Area in winter (donum)*	Production in winter (tonnes)	Area in summer (donum)*	Production in summer (tonnes)
Cereal				
Wheat	170,102	75,452		
Vegetable Crops				
Tomato	4383	8580	3306	7306
Potato	3320	4794		
Squash	1967	1509	1306	1055
Eggplant	2090	805	1338	1251
Carrot	945	745		
Onion	1663	1120		
Cucumber	234	1562	418	1729
Other vegetables	25,197		20,139	
Fodder Crops				
Alfalfa			18,207	
Fodder crops	19,203		15,896	
Melons			1137	2009
Watermelon			8641	10,121
Total	267,033		75,173	

* Donum is the Saudi Arabian measure of agricultural land and is equal to 0.1 hectare.

Table 6.3. Fruit production and animal husbandry resources of the Eastern Province of Saudi Arabia for 1987 (source: Statistical Yearbook, 1988).

Item	Area (donum)	Production
Fruits		
Dates (tonnes)	103,060	24,613
Citrus (tonnes)	952	912
Grapes (tonnes)	357	1049
Other fruits (tonnes)	571	
Animal husbandry		
Broilers (head)		12,994,131
Eggs (number)		275,698,948
Goats (head)		20,541
Sheep (head)		67,506
Cows (head)		11,122
Camels (head)		8810
Milk produced (liters)		38,867,966

sustained severe, almost complete, elimination in Kuwait during the Gulf crisis. After the liberation of Kuwait, government priority was aimed towards getting the population settled, improving security, repairing the damage to the oil industry, etc. Agriculture seemed less important at this point, and therefore, it may require several years to come back to the pre-war levels.

In Saudi Arabia, most agricultural work is carried out by expatriate labor. At the time of invasion of Kuwait in August 1990, everyone was searching for ways to get out of Saudi Arabia. This resulted in less attention to the farms. Unofficial estimates are that productivity in the Eastern Province was dropped below 50% due to the Gulf crisis. In addition to the labor problem, many farm storehouses and buildings were rented by the army on such attractive terms that farmers abandoned agriculture. The magnitude of this adverse effect cannot be ascertained.

Information on agriculture in Iraq was not available. It is believed that the impact of troop deployment on Iraqi agriculture was minimal. Few Troops were deployed in other Gulf countries.

6.9.3. Agricultural pollution due to the Gulf crisis and agriculture

Pollution can effect crops either by reducing their productivity or causing deterioration of their quality. The reduction in crop productivity can result from pollutants that cause acute effects to plant growth. These are largely gaseous pollutants. Important among these are sulfur dioxide, ozone, nitrogen oxides, and acidic aerosol. Beneficial effects can occur from enhanced concentrations of carbon dioxide. The impact of the gaseous pollutants on desert vegetation have already been discussed. It is concluded that their ambient concentrations were not sufficiently high to adversely effect vegetation. It is anticipated that agricultural crops were not effected.

It has been shown in Chapter 4 that phytotoxic metals and organic compounds in air particulates in Kuwait and Saudi Arabia were enhanced due to the Kuwait oil fires (Mitchell and Al-Sunaid, 1992; Sadiq and Mian, 1993a). To induce toxicity, metals have to be dissolved in soil solutions. Most agricultural soil in the Gulf region are calcareous and alkaline. Alkaline soil limits metal concentrations to lower levels as compared with acidic soil. It is therefore expected that a reduction in crop productivity due to metal and organic contamination from the Gulf war is a remote possibility. However, metal concentrations may be enhanced in plant tissue and ultimately may contaminate the food chain. Long-term effects of the Gulf war should be investigated.

From the foregoing, it is concluded that contamination of food can be expected due to the Kuwait oil fires and other activities associated with the Gulf crisis. However, data on food contamination were not available. A small study involving less than 100 samples of vegetables, fruits, and other agricultural products was completed in Kuwait. The collected samples were analyzed for petroleum hydrocarbons and toxic metals. However, results of this study were not available.

6.9.4. Effects of the Gulf crisis on range resources

There are many nomad tribes that utilize the range resources of the Gulf region. Many of these tribes raise sheep, goats, and camels to earn their living. In 1987 in the Eastern Province of Saudi Arabia, there were about 20,541 goats, 67,506 sheep, and 8810 camels (Table 6.2). Most of these animals were grazing mainly in the northern area of the Eastern Province. Because of troop deployment, range lands were occupied by the military and grasses and other vegetation was adversely effected. Many herds of sheep, goats, and camels were forced to move out of the deployment area. Displacement of animals from the north put pressure on the range lands in the southern part of Saudi Arabia. Over-grazing probably occurred, adversely affecting the southern area. There were some animal losses and inconvenience that were associated with the moving of the herds.

Some of the herdsmen insisted on staying in the northern area. It was later reported that they complained to the Saudi government of unusual circumstances of the death of their animals with altered morphological characters, such as hair color. These reports suggest that, in general, sheep were more prone to these conditions than goats or camels. These herdsmen blamed these events on the Gulf war and the Kuwait oil fires. The Saudi government investigated these claims but the report is classified.

As stated above, the Kuwait crisis probably contaminated natural vegetation in the war zone. Animals grazing upon this vegetation can ingest these contaminants. An appreciable amount of contaminants can also come from the direct ingestion of soil along with the grasses. Information on the contamination of range animals and their meat and milk in troop deployment areas is not available.

In Kuwait, the situation is even worst. About 60% of the Kuwait area was mined by the Iraqi military and it was very risky to graze animals in such a hazardous area. It was reported that many animals were killed in Kuwait by these mines (Al Hassan, 1992). Al Hassan (1992) reported that many domesticated and experimental animals were also killed by the Iraqis during their occupation. The pollution created by the Kuwait oil fires was reported to have killed several sheep. From these reports it can be concluded that air pollution in Kuwait was sufficiently bad so as to induce toxicity in sheep in Kuwait. Although several Kuwaiti scientists documented Iraqi atrocities, losses to the range resources were not quantified.

From the foregoing, it can be concluded that range resources in Kuwait and in the Eastern Province of Saudi Arabia were adversely affected by the Gulf crisis. There is substantial evidence suggesting that animals and their products were contaminated by toxic chemical produced during the Gulf war and Kuwait oil fires. However, data to substantiate these fears are not available.

6.9.5. Indirect effects of gulf crisis on agriculture

Agricultural productivity is the outcome of many continuous and/or discrete factors. Many of these factors which effect productivity independently and/or simultaneously have already been discussed. In this section, the effects of the United Nations (UN) embargo on Iraqi agriculture, disruption of solar radiation, decrease in ambient temperature, effect on monsoon rains, and quality of natural precipitation in the Gulf region will be discussed.

6.9.5.1. Effects of embargo on Iraqi agriculture

The United Nations (UN) tried hard to convince Iraq to leave Kuwait peacefully. When such pleas were turned down, an international embargo was imposed on Iraq. The Gulf war and the UN embargo had numerous effects on Iraqi agriculture. Iraq has enough agricultural resources to sustain its population. However, for good productivity, crops require water, fertilizers, and pesticides. Because of the embargo, many of these chemicals were barred from entering Iraq. Iraq has limited capacity to produce fertilizers and pesticides, certainly not enough to fulfill the country's requirements. Without fertilizer and pesticide applications crop yield can be reduced drastically. The shortages in fertilizers and pesticides would have adversely effected Iraqi agriculture.

Because of the embargo, the fuel supply was rationed in Iraq. In a state of war, military needs superseded those of the civilians, and agriculture was placed at the bottom of the civilian list of priorities. Without fuel, agricultural machinery cannot be utilized. In addition to fuel, spare parts of agricultural machinery was in short supply and a major portion of this machinery would be out of order, and adversely effected agricultural productivity in Iraq.

The Gulf crisis resulted in labor shortages into the agricultural sector. The majority of the male population between the age of 15 to 45 years were drafted into the army. This left a severe shortage of human resources available for agricultural activities. Lack of human resources deteriorated agricultural productivity in Iraq. It was reported that because of the Iran-Iraq War and now the Gulf war, more than 50% of Iraq's population is below 15 years of age. In addition to the shortages of agricultural work force, Allied bombardment left many electricity plants damaged. This might have effected some of the irrigation pumping stations.

Data on the current status of Iraqi agriculture are not available. However, all the above factors indicate that the agricultural capacity of Iraq was adversely affected by the UN embargo – probably more severely than anticipated. There are no reports of mass starvation in Iraq. However, many of those who visited Iraq, reported that malnutrition reached alarming proportion, especially among children (Chapter 7).

6.9.5.2. Effect of decreased temperature on agriculture

One of the early warnings regarding the catastrophic environmental impact of the Gulf war was based on the possibility of a significant decrease in the atmospheric temperatures and an increase in the stratospheric temperatures due to the large quantity of soot produced if the Kuwait oil fields were to be ignited. Numerical modeling investigations assumed many different scenarios to predict the effect of soot on temperature and other meteorological phenomena on a global and regional basis. Rao (1991) predicted a cooling of 8°C in the Indian Subcontinent if 600 tonnes per day of smoke from the Kuwait fires reached the stratosphere. Manins and Bouma (1991) assumed that if 10,000 tonnes of soot is produced per day and remains suspended for 24 hours, the daytime temperatures underneath the plume will drop about 10°C. Browning et al. (1991) reported a significant decrease in solar radiation underneath the smoke plume and as a result, predicted a decrease of 10°C in daytime temperature within about 200 km of the fire sources. Graf (1991) predicted a significant cooling effect of about 4°C for the summer of 1991 over the Gulf region (about $500 \times 500\ km^2$). Blanchet and Vupputuri (1991) reported the results of the Canadian Climate Center General Circulation Model (CCC/GCM) and noted that the model predicted a cooling of 5 to 10°C in the Kuwait area. They suggested that beyond 100 km from the fire source, natural variability in radiation will dominate.

Several investigators measured actual temperatures in the Gulf region. Alnasr et al. (1992) noted a decrease in ambient temperature in Bahrain. The differences were enhanced when the relative prevalence of smoke increased and peaked in July 1991 at a difference of 4°C. The differences between 1990 and 1991 narrowed with the extinction of fires after August 1991. Shaw (1992) reported that a reduction of ambient temperatures between 0.8 to 2.4°C from March to September. Similar results were reported by MEPA in Dhahran, Saudi Arabia. Because of smoke over Bahrain, several new meteorological records were set. For example, March 1991 was the fourth coolest on record, May–June 1991 was the second coolest, and July–August 1991 the coolest on record at the Bahrain Airport. The data of Shaw (1992) are plotted in Figure 6.12 to show differences in the mean monthly ambient temperature. The decrease in the ambient air temperature in Kuwait was of the order of 3 to 4°C (EPC, 1991) as compared with 0.8 to 2.4°C in Bahrain from March to September 1991. From the foregoing, it was concluded that the decrease in temperature was more prominent in Kuwait and narrowed with distance. Beyond about 1000 km, the decrease in the ambient temperature was insignificant.

Decrease in ambient temperature occurred during February to September 1991. The smoke plume was largely visible in Kuwait, northeastern Saudi Arabia, Bahrain, and Qatar. In February–March, cereal crops such as wheat and barley, flower and grain starts formation in these countries. There are reports of decreased crop productivity in the Gulf region. Agriculture in Kuwait

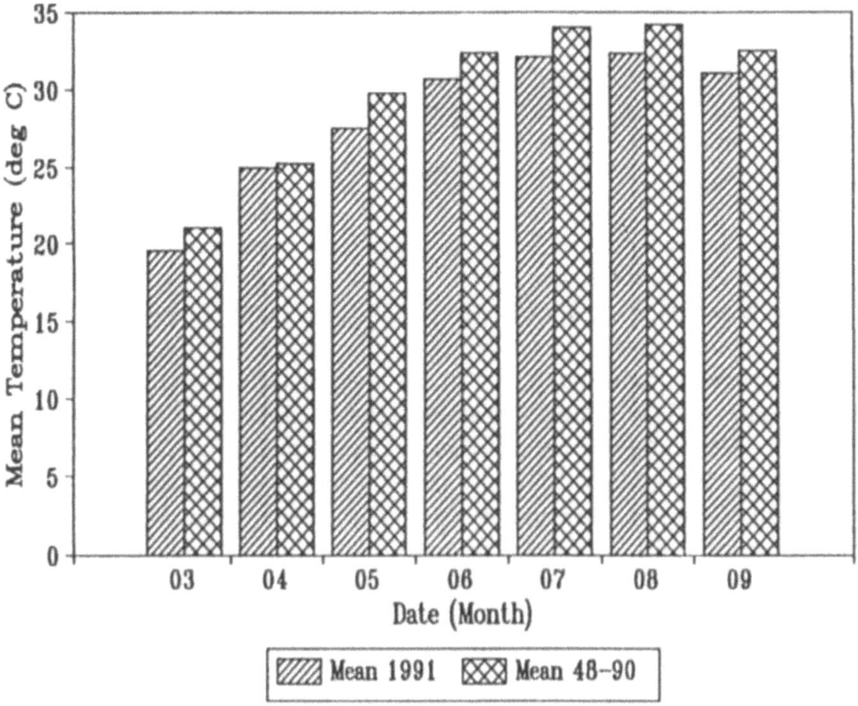

Figure 6.12. A comparison of mean monthly temperatures observed during 1948–90 and 1991 in Bahrain.

completely vanished due to the Iraqi occupation. A large farm in the Eastern Province of Saudi Arabia reported a significant decrease in wheat yield during 1990–91 season as compared with 1989–90 and 1991–92 seasons. However, it is difficult to determine if this decrease was due to temperature variation, or excessive weeds and lack of timely application of fertilizers, pesticides, and irrigation during the 1990–91 season. Agriculture in Bahrain is insignificant anyway. In Qatar and other countries in the region, the decrease in ambient temperature was not significant. Smoke from the Kuwait oil fires was occasionally pushed over Iran and Iraq, especially during March 1991 because of an unusual weather pattern. However, information from these countries was not available to assess the impact of temperature on agriculture. From the foregoing, it was believed that effect of the temperature decrease on crop productivity would be insignificant. Rather, cold temperature in summer might have helped some crops and desert vegetation.

6.9.5.3. Effect of decreased solar radiation on agriculture

In addition to temperature, solar radiation was also expected to decrease due to Kuwait oil fires. During the early days of the Gulf conflict, it was suspected that if 10,000 tonnes of soot per day remained suspended for 24 hours, about 25% of the light will reach to the ground underneath this plume (Manins and Bouma,

1991). Small (1991) argued that smoke, being a good absorber of solar radiation, will cause a reduction in solar radiation at the ground levels, warm the upper atmosphere and cause a cooling of the lower atmosphere. He predicted that smoke from the Kuwait oil fires would be produced in such a large amount that the ecology in the Gulf region would be adversely affected by it.

Many investigators collected data on solar radiation during the Kuwait oil fires. Considering the data collected during airborne investigation of Kuwait oil fire plume, Hobbs and Radke (1992) reported the absorption of over 90% of the solar radiation in the thickest part of the smoke plume and about 75–80% in the super-composite region of the smoke plume. They suspected that reduction of solar radiation to such an extent would certainly effected ambient temperature. In Saudi Arabia, it was noted that solar radiation was reduced as much as 36% during the period from January through June 1991 in Jubail (Riley et al., 1991). Shaw (1992) also reported a decrease in solar radiation on the Island of Bahrain. He found below average sunshine from March through September 1991. The significant reduction in sunshine was due to the smoke cloud cover originating from the Kuwait oil fires. Some of the data were extracted from Shaw and are plotted in Figure 6.13. The decrease in solar radiation from March to May 1991 was mainly due to smoke cover from the

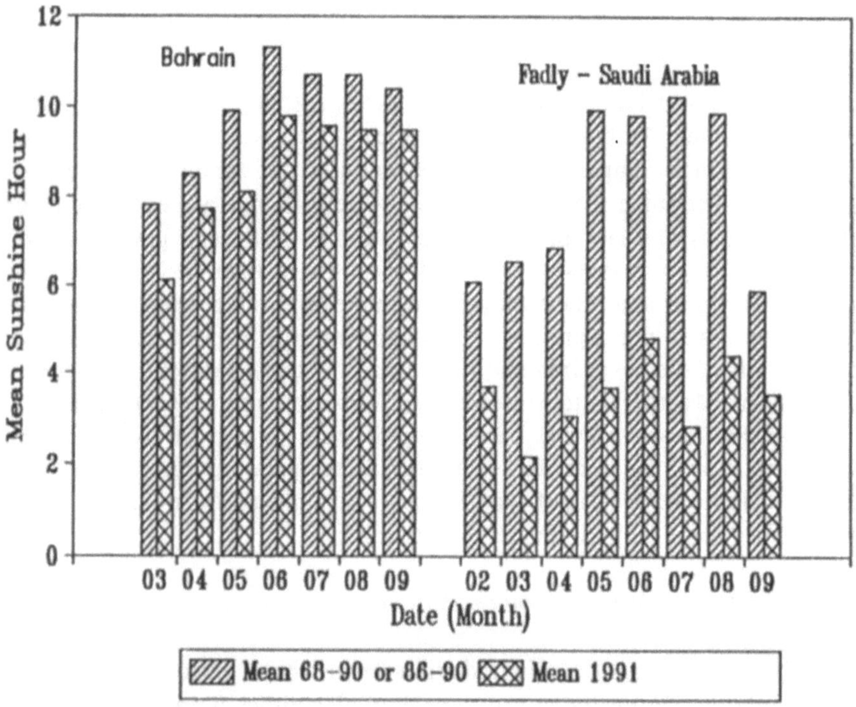

Figure 6.13. A comparison of mean monthly sunshine hours observed during 1968–90 and 1991 in Bahrain and northeastern Saudi Arabia.

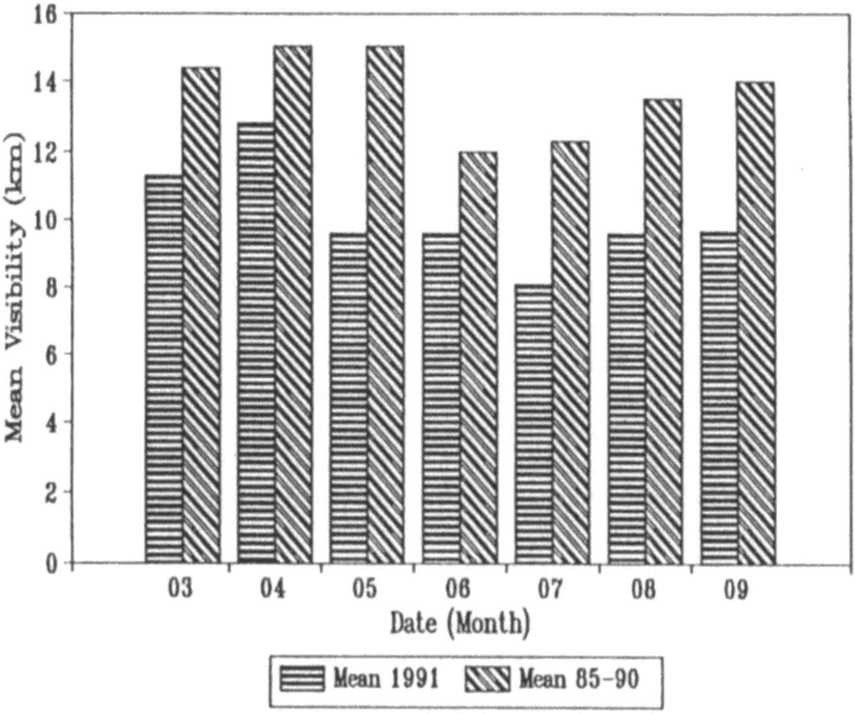

Figure 6.14. Mean monthly visibility during 1985-90 and Kuwait oil fires (1991).

Kuwait fires, while in June to August 1991 it was due to combined effects of dust and smoke, and for September 1991 the difference was mainly due to dust. Smoke also effected visibility – it decreased significantly when smoke was present. The data on the mean monthly visibility from March to September 1991 were adopted from Shaw (1991) and are plotted in Figure 6.14. This figure also shows a comparison between visibility data from 1991 and those collected during the 1985–1990 period.

Solar radiation is critical for photosynthetic activity in plants. A significant decrease in solar radiation for prolonged periods can reduce crop productivity. With reference to the Gulf crisis, solar radiation was reduced over Kuwait, the northeastern part of Saudi Arabia, and Bahrain during March to August 1991. As stated earlier, investigation of agricultural activities was undertaken only in Saudi Arabia during the Gulf Crisis. Decreased wheat productivity at one of commercial farm in the smoke-covered area was reported. However, it is difficult to assess the contribution of solar radiation to the observed decrease.

6.9.5.4. Effects of disruption in rainfall on agriculture

Eearly pessimistic predictions of environmental impact suggested that the smoke and soot produced by Kuwait oil fires would alter the monsoon in the Indian Subcontinent and cause famine in the area. Rao (1991) predicted a cooling of 8°C in the Indian Subcontinent and argued that a decrease of more than 1 to 1.5°C above the normal range of temperature would significantly effect the monsoon in India and Pakistan. However, Jayaraman (1991), another Indian scientist, disagreed with Rao's assessment of the effects of the Kuwait oil fires on the Indian monsoon. He pointed out that the monsoon depends largely what happens over the Tibetan plateau and Bay of Bengal rather than what happens over the Arabian Sea. He cited an Indian study that concluded that the Kuwait oil well fires will have no adverse effect on Indian agriculture or weather including monsoon. The UK Meteorological Office (1991) suggested that if the cooling resulted from the smoke plume extends to the Indian Subcontinent, the behavior of the monsoon will be changed. However, later Manins and Bouma (1991) discounted this possibility. Small (1991) was of the opinion that the monsoon in India could be affected. Browning et al. (1991) reported that their calculations showed an insignificant effect on the Indian monsoon and commented that previously established data indicated higher annual variation than those which would have been seen from the impact of the Kuwait oil fires. Graf (1991) suspected a weakening of the monsoon in the Indian Subcontinent as a result of the Kuwait oil fires (WMO, 1991). It has been proven now that the monsoon in the Indian Subcontinent was not significantly affected by the smoke from the Kuwait oil fires. This was mainly due to the fact that the smoke plume was not elevated to the stratosphere in sufficiently large amounts to effect the monsoon, rather it remained suspended between 2 to 5 km above the ground.

6.9.5.5. Effects of Kuwait oil fires on precipitation

Natural precipitation in the Gulf region is not sufficient to effect agriculture in any appreciable way, especially quality. There was unusual weather during March–April 1991, and Kuwait and Saudi Arabia received unusually high precipitation during this period. Precipitation samples were collected and analyzed. The analytical data are given in Table 6.4 along with data for 1987–88 and 1992.

6.10. Impact of Gulf crisis on wildlife

It was suspected that the deployment of over one million troops in the desert of Kuwait, Saudi Arabia, and Iraq would change the desert ecosystem, including wildlife, in many ways. The adverse impact on wildlife can be due to damage to food resources, trampling of the animals and their burrows, unintentional killing, destructive effects due to enhanced noise, effects of air pollution, the impact of the oil lakes and trenches, and heat generated from the Kuwait oil fires. A discussion of these factors follows here.

Table 6.4. Chemical composition of rain water samples collected in Dhahran, Saudi Arabia. All values measured in μg/l.

Parameter	1991[a]	1992[b]		1987-88[c]	
		Range	Mean	Range	Mean
pH	6.7	6.4–7.9	7.4	5.1–7.2	5.48
Calcium	734.6	14,400.0–91,600.0	32,366.7	76.0–2540.0	925.8
Sodium	424.9	1800–26,000.0	86166.6	28.3–1026	130.0
Arsenic	0.01				
Cadmium	0.56				
Cobalt	0.42				
Copper	0.56				
Iron	8.32				
Molybdenum	2.00				
Nickel	0.80				
Lead	3.20				
Vanadium	2.80				
Zinc	10.20				
Magnesium	570.3	600.0–3800.0	1632.0	18.2–1493.0	179.0
Potassium	432.6	400.0–2500.0	1150.0	0.8–48.7	11.5
Strontium	42.7			0.4–8.6	2.4
Ammonium				27.8–125.0	45.6
Sulfate		7200.0–31,200.0	15,800.0	75.0–1825.0	466.0
Nitrate		0.0–7800.0		83.8–1600.0	287.4
Chloride		1200.0–28,900.0	10,066.7	16.9–1219.7	144.2
Bicarbonate				75.9–392.5	151.0

a = Data extracted from KFUPM/RI (1991b).
b = Data extracted from Al Sunaid (1992).
c = Data extracted from Azhari et al. (1990).

6.10.1. Wildlife of the Gulf region

Deserts are inhabited by many animals. Wildlife in the Arabian desert (that includes desert areas of Iraq, Kuwait, Saudi Arabia, Bahrain, Qatar, the United Arab Emirates, and Oman) are similar to the wildlife in the desert area influenced by the Gulf war, both in number and types as well as in the ways animals have adapted to the desert climate. Like all deserts, the Saudi desert has a large population of rodents. The principal rodent in Saudi and other desert areas of the Gulf region is the jerboa of which there are at least two species present in the area. There are also six species of gerbil or jird, three species of mice, four species of rates and one species of hedgehog – sixteen species in all each filling a different environmental niche in the desert. The most amazing thing about jerboa is that it lives without ever drinking water. It relies totally on dew which collects on his food during the night and the moisture present in its food. Other animals in Saudi Arabia are the gazelle, camel, Arabian hare, wild ass, porcupines, Syrian hyena, Asiatic jackal, Persian and Arabian wolves, and wildcats. There are numerous invertebrates, lizards including the Turkish and fan-footed gecko, the desert monitor, dice, horned, sand and Montpelier snakes, and the Levant viper.

6.10.2. Effects of troop deployment on wildlife

Desert animals are not accustomed to the presence of humans. During deployment, large desert areas were occupied by military activities. Many of the animals (gazelle, camel, the Arabian hare, wild ass, porcupines, the Syrian hyena, Asiatic jackal, Persian and Arabian wolves, wildcats, etc.) that were living in these areas before troop deployment would probably have migrated to other more peaceful areas. Many rodents such as jerboa, mice, hedgehog, snakes, and lizards were not able to leave the area of deployment. Troop deployment was so swift and extensive, especially that of the Iraqi occupation of Kuwait and the Allied ground operation, that many of gazelle, camel, the animals would not have had the chance to leave the War zone. Trigger-happy troopers might have killed many animals. It is clear that wildlife was stressed because of troop deployment, but quantitative data on the wildlife sufferings impacts were not collected. Iraqi troops are known to have caused severe damage to Kuwaits zoological park. They shot and maimed animals.

6.10.3. Destructive effects of noise on wildlife

Both Allied and the Iraqi committed very large number of troops and military hardware to the Gulf war (Chapter 5). As a result, the wildlife in the War zone was exposed to very high levels. Such noise might have caused a severe impact on the wildlife in the area. Some studies conducted in USA suggest aggressive behavior changes in kangaroo rats in California desert due to increased noise levels. Damage to the acoustic sensitivity of other desert rodents was also reported (Bondell, 1976). In these reports, it was concluded that high intensity sounds in formerly quite ecosystems could adversely effect wildlife in the impacted area. Information on sound intensity before and during the Gulf war was not available. However, to gain a general indication of noise problem for the wildlife during the Gulf crisis, data on the sound pressure levels in the California desert were extracted from Brattstrom and Bondello (1983). These data are given in Figure 6.15. Many of the activities undertaken in the war zone would produce noise of more than 100 dB, as shown in Figure 6.15.

The adverse effects of sound from motorcycles and sand dune buggies on the selected desert wildlife were investigated by Brattstrom and Bondello (1983). In an experiment with dune buggies, they found that the hearing capacity of the kangaroo rat was severely impaired by the sound but recovered gradually. Exposure to the sound also reduced the distance from which the rat can

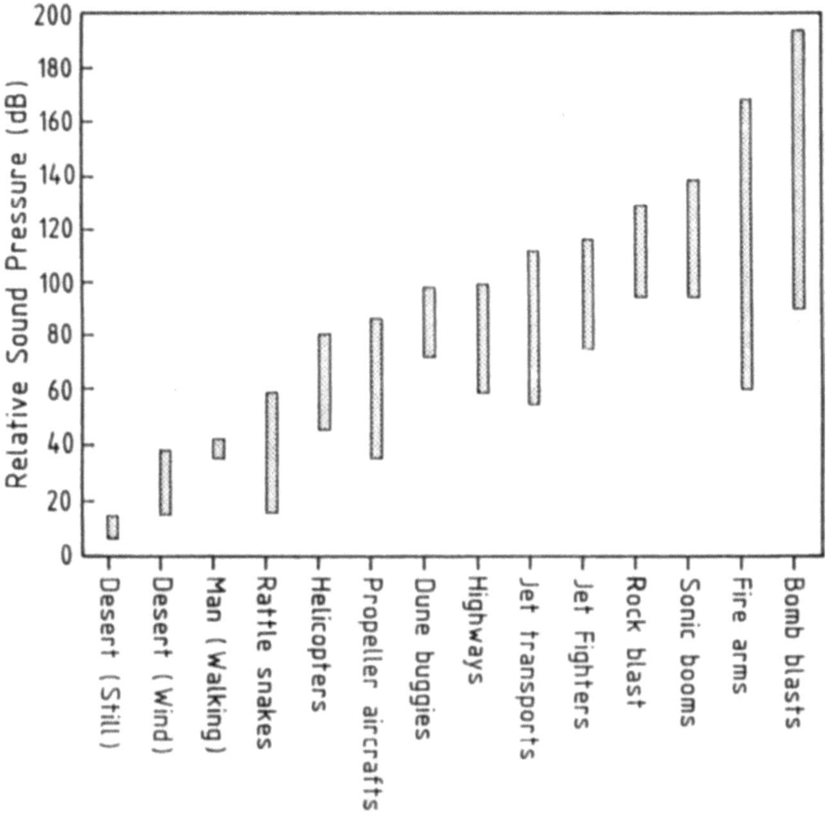

Figure 6.15. General noise levels to which the wildlife in the war zone was exposed.

recognize the sound. Effective warning distance impairment and its gradual recovery are shown in Figure 6.16. Similar behavioral changes were noticed when the rats were exposed to motorcycle sound. They also reviewed the results of other investigations and concluded that increased sound intensity would have disruptive and often destructive effects on native desert wildlife. The noise from the motorcycles and dune buggies definitely caused hearing loss in animals with little or no recovery, interferred with their ability to detect danger, and caused unnatural behavioral changes. Considering these reports, it can be argued that the anthropogenic activities associated with the Gulf war probably produced enough high intensity sounds to disturb native wildlife in the war area. Information on the sound intensity before and during the Gulf war are not available.

The area affected by Gulf war activities is shown in Figure 6.1. In this area, about one million military and civilian personnel were deployed. The military personnel used heavy armor during exercises before and during the Gulf war. There were very intensive air operations starting with the invasion of Kuwait which peaked in January–February 1991. Kuwait and Iraqi territories were

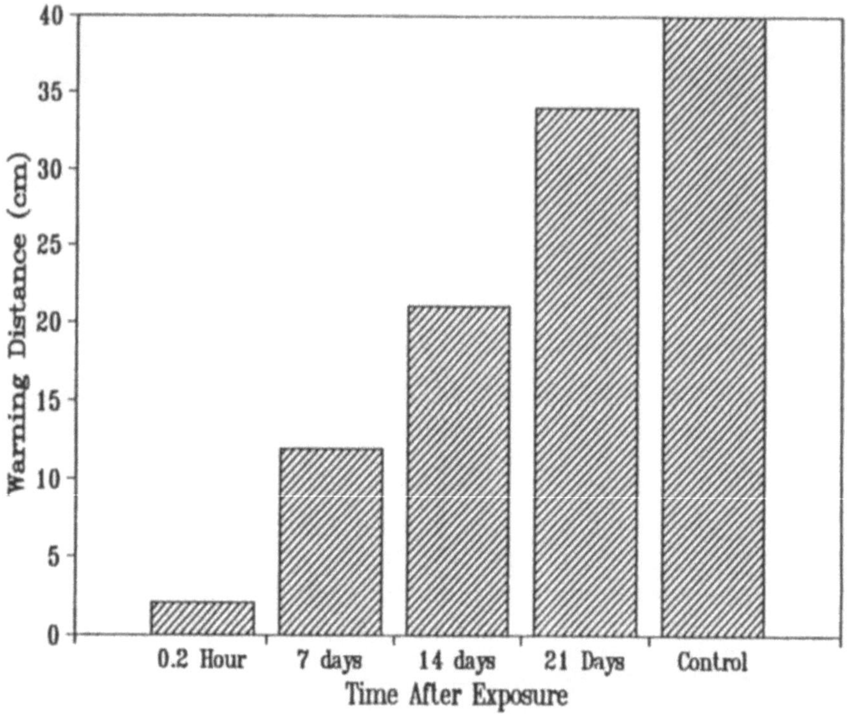

Figure 6.16. Impairment and recovery of hearing capacity in a desert rate species exposed to nois. from dune buggies.

bombed very heavily. Cannons, missile, rockets, and fire arms were discharged before and during the Gulf war. Looking at the data presented in Figure 6.14, one could conclude that the sound intensity in the area of impact was higher during the Gulf Crisis than for the dune buggies mentioned by Brattstrom and Bondello (1983). This level of sound would certainly have effected the hearing ability of many native animals. Unfortunately, no attempt was made to investigate the impact of sound on native wildlife of this area. Our main emphasis was focused on events which directly affected humans – probably a selfish but justifiable attitude. However, we humans were responsible for the impacts to native wildlife during the crisis. The least one should do is to sympathetically recognize the hardships on wildlife in the war zone.

6.10.4. Crushing of animals and their burrows

The principal rodents in Gulf areas are the jerboa, gerbil, mice, rat, and hedgehog. In addition to the rodents, there are numerous invertebrates, lizards, Turkish and fan-footed gecko, the desert monitor, dice, horned, and snakes. All these animals live on or under the ground and make their burrows near the surface. These animals are suseptable to adverse impact due to military vehicles. As has been shown above, noise from dune buggies caused hearing loss in rodents with little or no recovery, interfered with their ability to detect danger, and caused unnatural behavioral changes. These adverse effects would certainly make wildlife more prone to the killing by trampling and running over them with vehicles. Unfortunately, no data could be found to quantitatively show the plight of burrowing animals in the War zone.

6.10.5. *Accidental killing of wildlife*

In a state of war situation, the military has to be vigilant, patrolling its territory day and night and testing the enemy's readiness by spying. Driving in the desert with limited headlights caused many accidental deaths of wildlife. Many rodents and other vertebrates are blinded by a vehicle's headlights and thus increase their chance of being killed. There is no account of such accidental killing. A few camels were killed daily by vehicles on the desert roads (personal communication). No data could be found on the accidental killing of the wildlife in the War region.

6.10.6. *Effects of air pollution on wildlife*

Air pollution effects all types of life, including desert wildlife. The pollution potential of the Gulf war was appreciable. Pollution can induce acute as well as chronic effects on wildlife. Chronic effects will be manifested by increased concentrations of air particulates, sulfur dioxide, nitrogen oxides, acidic aerosol, and ozone. It has been shown in Chapter 4 that concentrations of air particulates were certainly enhanced due to the Gulf crisis, particularly by the Kuwait oil fires. Concentrations of sulfur dioxide were found to be very high in or near the burning oil fields in downwind areas. Wildlife in such highly contaminated niches was adversely affected by the synergistically acting particulate matter and sulfur dioxide. A Kuwaiti scientist reported that many sheep in Kuwait City died due to inhalation of the polluted air (Al Hassan 1991, 1992). However, the extent of damage to wildlife is unknown. It has been shown that other pollutants that can cause acute poisoning of wildlife were below the permissible limits for humans. It is anticipated that such concentrations did not alter wildlife dynamics.

Chronic effects could arise by the ingestion of toxic chemicals which might be deposited on food. Many organic and inorganic chemicals that were produced during the Gulf crisis are known to be biotoxic. Bioaccumulation and depuration mechanisms for these toxic chemicals in wildlife cannot be established so far. Because of the lack of information on many aspects of the chronic poisoning of wildlife by chemicals generated during the Gulf war, this topic cannot be discussed further.

6.10.7. *Effects of oil lakes and trenches on wildlife*

Many oil lakes were formed during the burning oil fields in Kuwait. All life underneath these lakes vanished forever. Wild animals in Kuwait must have inhaled oil mist during the Kuwait oil fires. Reports from Kuwait have shown that many rodents, lizards, snakes, and birds fell into the oil lakes and trenches and never came out alive. The number of animals killed by falling into the lakes may have exceeded several thousand. These are preliminary estimates of the number of animals killed and recovered from the oil lakes and trenches; however, the real number of animals killed by falling into the oil lakes and trenches will never be known.

6.10.8. Effects of heat from the Kuwait oil fires on wildlife

There were more than 600 oil fires in the Kuwait oil fields with a temperature around 1000°C at the well head. The fire fighting crews reported temperatures of about 130° to 150°C in the vicinity of the fires. Heat could be felt several kilometers away from the fires. Wildlife could have been trapped and killed when the wells were suddenly sabotaged by Iraq. The fire flames may have burned any nearby animal. More animals would have been killed because of the intense heat if they come out of their burrows or would have starved to death underground. It is expected that the wildlife in the burning oil fields was devastated. Recovery of the wildlife will probably require several decades or it may never be restored.

6.11. Contamination of water due to the Gulf crisis

Most groundwater in the Gulf region is saline and is primarily used in agriculture and as a feedstock for some inland desalination plants. The groundwater which is utilized for irrigation and is normally drawn from the deep aquifers of several hundred meters depth. The catchment areas of these aquifers are located several hundreds kilometers away, generally in the northern part of the Kingdom. As stated earlier, natural precipitation in the Gulf region is very low and sparsely distributed. Whatever rain falls on the ground, is evaporated back to the atmosphere before it has a chance to seep to the groundwater aquifer. This results in limited local recharge of the deeper aquifers. At present, primarily deeper aquifers are exploited. In addition to the deeper aquifers, there are secondary shallow aquifers. The shallow aquifers occur a few meters below the surface. Groundwater in the shallow aquifers is generally very salty. These aquifers are not currently exploited and thus are of no economic/social significance. These are the aquifers which might be threatened by the Gulf crisis.

Groundwater contamination could come from ground disposal of military waste and seepage of oil from the lakes and trenches. As stated above, sanitary and other solid wastes were collected and burned and the residue after burning, was buried. The groundwater table may raise sufficiently or water may move upwards via capillaries so as high to soak the residual wastes. Organic and inorganic contaminants in the waste residue can dissolve in the groundwater and be transported to other shallow aquifers due to diffusion processes. It has been shown that the Gulf crisis resulted in the contamination of surface soil with both organic and inorganic chemicals. With precipitation, these chemicals can be transported to the deeper soil layers and ultimately to the shallow groundwater aquifers. Once in the groundwater, these chemicals can be absorbed and accumulated in vegetation and may be transferred to consumers.

There were more than 100 small and large oil lakes in the Kuwait oil fields (Al-Besharah, 1991). These lakes were formed from oil spillage from the damaged oil wells. Kramer (1991) reported that about 35 million cubic meters of soil have been contaminated by the oil lakes. In addition, the Iraqi troops dug several hundreds of kilometers of trenches and filled them with oil as a part of their defensive strategy (Al Hassan, 1991). Seepage from the lakes and trenches can contaminate shallow groundwater aquifers with petroleum hydrocarbons. Later investigations reported that oil soaking was found to be within 0.5 meter of the soil surface (Al Besharah, 1991; Al Besharah and Saad, 1992). These observations suggest no serious contamination of groundwater with petroleum hydrocarbons; however, the biodegradation product of petroleum hydrocarbons and inorganic chemicals may move further down and contaminate groundwater in the future.

From the foregoing, it is concluded that contamination of deep groundwater aquifers due to the Gulf crisis is a possibility in the distant future. However, the probability of contamination of shallow groundwater, especially in the northeastern Saudi Arabia and Kuwait, remains high.

Freshwater resources of the Gulf region, except for Iraq, are negligible. There is no natural river in Saudi Arabia, Qatar, Bahrain, and the United Arab Emirates. During the air campaign, the Allied Forces bombed many electrical and civilian facilities that stopped water treatment and pumping stations. The people of Baghdad were forced to drink from the river which was used for bathing, washing clothes, and other domestic needs. Sewage was also flowing to the river. It is suspected that unhealthy drinking water may cause medical problems such as enteritis, etc. Other information on surface water contamination in Iraq could not be found.

6.12. Future response and research needs

Many investigators speculated about widespread damage to the land resources due to the Gulf crisis (Abdal, et al., 1992; Barnaby, 1991; Dauod, 1991; El-Baz, 1992; EPC, 1991; Khuraibet, 1992), while others presented pictorial overviews of the adverse effects of the Gulf war (Al Hassan, 1992; Sepa Press, 1992). By December 1992, there were more than 30 books, mostly narrative accounts of different aspects of the Gulf crisis. So far, no serious attempt has been made to study the damage sustained by the land resources of the Gulf region during the Gulf crisis. Therefore, many factors effecting natural resources remain to be investigated.

The desert soil crust is very crucial in maintaining an equilibrium within a desert ecosystem. Substantial evidence and indirect data indicate that the desert crust was seriously disturbed from the activities related to the Gulf war. In a preliminary investigation, it was shown that soil erosion was enhanced during 1991. Because of the adverse social, economic, and health consequences of enhanced erosion, a long-term program should be initiated to stabilize the disturbed desert crust. Off-road driving should be regulated, especially in sensitive areas. Recovery rates of the disturbed desert crust should also be investigated.

Among the land resources, soil was the most frequently disturbed component during the Gulf crisis. In addition to disturbances to the surface, preliminary reports indicated soil contamination from organic and inorganic chemicals produced during the Gulf crisis. Soil contamination is the vehicle for transporting these toxic substances to the human food chain, particularly via vegetation and groundwater contamination. Therefore, an international soil research program to assess contamination and monitor recovery should be started as soon as possible. This program should include soil from Kuwait, northeastern Saudi Arabia, and southern Iraq.

After burning sanitary wastes, the residue was buried. It is suspected that this waste management practice might have resulted in chemical and biological contamination of desert soil and groundwater in the areas of troop concentration in Saudi Arabia. These burial sites should be identified, documented, and soil and groundwater should be monitored on a long-term basis.

Vegetation was badly affected by the Gulf crisis. Recovery of natural vegetation in severely impacted areas should be monitored. Moreover, chemical contamination of the desert vegetation should also be investigated.

Like soil and vegetation, wildlife in the war zone was adversely affected. So far, no data have been collected on the damage sustained by the desert animals. Population dynamics of representative animal species should be studied on a long-term basis.

In the best interest of the regional land resources, information collected during any of the above research and monitoring programs, should be shared with the scientific community. Therefore, there is a genuine need to develop a Gulf war database. The international community should benefit from the outcomes of above research programs, e.g., combating desertification in Africa.

References

Abdal, M., Al Houty, W., and Zaman, S. 1992. Gulf war impact on desert ecosystem of Kuwait. Paper presented at the Third World Academy of Sciences, Fourth General Conference. 23–26 November, Kuwait City, Kuwait.

Al Basharah, J. and Al Saad, A. 1992. Oil lake formation and means of remediation. Paper presented at the Third World Academy of Sciences, Fourth General Conference. 23–26 November, Kuwait City, Kuwait.

Al Hassan, J.M. 1991. The Iraqi Invasion – Environmental Catastrophe. In Environmental and

Health Impact of the Kuwaiti Oil Fires. Proc. of an International Symp. held at the University of Birmingham, 17th October, pp. 6–11.

Al Hassan, J.M. 1992. The Iraqi Invasion of Kuwait, An Environmental Catastrophe. Fahad Al Marzouk Printing Press, Kuwait.

Al Sunaid, A.A. 1992a. Saudi Aramco AMMNET historical air quality and Kuwait fire SO_2 removal mechanism. Paper presented at the Second WMO Meeting of Experts to Assess the Response to and Atmospheric Effects of the Kuwait Oil field Fires, Geneva, 25–29 May.

Al Sunaid, A.A. 1992a. Saudi Aramco AMMNET historical air quality and Kuwait fire SO_2 removal mechanism. Paper presented at the Second WMO Meeting of Experts to Assess the Response to and Atmospheric Effects of the Kuwait Oilfield Fires, Geneva, 25–29 May.

Aldhous, P. 1991. Big test for bioremediation. Nature 349:447.

Alnaser, W.E., Awadalla, N.S., and Almortaza, L.A. 1992. Use of solar radiation measurement to detect the pollution of Bahrain's sky caused by the Gulf war. Proc.

Arab News, 1992a. Kuwait continues battle to clear Gulf war mines. p. 3, April 30, 1992, Arab News, Saudi Arabia.

Arab News, 1992b. Kuwait recover 6.5 million oil. p. 6, December 26, 1992, Arab News, Saudi Arabia.

Azhari, F.M., Singh, R.P., and ElMubarak, A.H. 1990. Chemistry of atmospheric precipitation at the western Arabian Gulf coast. Atmos. Environ. 24A:2927–2934.

Bakan, S., Chlond, A., Cubasch, U., Feichter, J., Graf, H., Grassel, H., Hasselman, K., Kirchner, I., Latif, M. Roeckner, E., Sausen, R., Schlese, U. Schriever, D. Schult, I., Schumann, U., Sielmann, F., and Welke, W. 1991. Climate response to smoke from the burning of oil wells in Kuwait. Nature 351:367–371.

Barnaby, F. 1991. The environmental impact of the Gulf war. Ecologist 21:166–172.

Batanouny, K.H. 1981. Ecology and Flora of Qatar. Alden Press Limited, Oxford on behalf of the Center for Scientific and Applied Research, University of Qatar.

Bialobok, S. 1984. Controlling atmospheric pollution. pp. 451–478. In Treshow, M. (ed.), Air Pollution and Plant Life. John Wiley and Sons, New York.

Bondello, M.C. 1976. The effects of high-intensity motorcycle sounds on the acoustical sensitivity of the desert iguana, Dipsosaurus dorsalis. M.S. thesis, California State University, Fullerton, USA.

Brattstorm, B.H. and Bondello, M.C. 1983. Effects of off-road vehicles on desert vertebrates. pp. 167–206. In Webb, R.H. and Wilshire, H.G. (eds.), Environmental Impacts of Off-road Vehicles. Impact and Management in Air Regions. Springer-Verlag, New York.

Browning, K.A., Allam, R.J., Ballard, S.P., Barnes, R.T.H., Bennetts, D.A., Maryon, R.H., Mason, P.J., McKenna, D., Mitchell, J.F.B., Senior, C.A., Slingo, A., and Smith, F.B. 1991. Environmental effects from the burning oil wells in Kuwait. Nature, 351:363–367.

Chiou, C.T., Freed, V.H., Schmedding, D.W., and Kohnert, R.L. 1977. Environ. Sci. Technol. 11:476.

Church, G.J. 1991. The battleground. The 100 hours. Time, March 11, pp. 26–34.

Cooke, R.U. and Warren, A. 1973. Geomorphology in Deserts. University of California Press, Berkeley, USA, pp. 374.

Cutter, L.S. 1991. Ecocide in Babylonian. Focus, Summer Issue 41:26–31.

Daoud, M.W. 1991. Ahmadi City under the smoke. In Environmental and Health Impact of the Kuwaiti Oil Fires. Proc. of an International Symp. held at the University of Birmingham, 17th October, pp. 27–31.

Daum, P.H., Al-Sunaid, A., Busness, K.M., Hales, J.M., and Mazurek, M. 1992. Studies of the Kuwait oil fires during mid-summer 1991. Brookhaven National Laboratory - USA, Battelle Pacific Northwest Laboratory - USA, and Saudi Arabian Oil Company - Saudi Arabia, Final Report.

Dregne, H.E. 1966. Surface materials of deserts environments. pp. 290–377, In McGinnes, W.G., Goldman, B.J., and Payrole, P. (ed.), Desert of the World. The University of Arizona Press, Tucson, USA.

El Desouky, M. and Abdulraheem, M.Y. 1991. The impact of oil well fires on the air quality in Kuwait. In Environmental and Health Impact of the Kuwaiti Oil Fires. Proc. of an International Symp. held at the University of Birmingham, 17th October, pp. 16–26.

Emerson, S. 1991. When earth takes the hit. International Wildlife, July–August, pp. 38–41.

Epstein, J. 1978. Techniques of desert reclamation. pp. 40–47. In Ali, C.T. and Brown, J. (ed.) Landscape Design for the Middle East. RIBA Publications Limited, London.

Fayad, N.B., El Mubarek, A.H., Edora, R.L., Al Gamdi, A., and Polancos, A.B. 1993. Polycyclic aromatic hydrocarbons in ambient air following the burning of Kuwait oil wells. Arabian J. Sci. Eng. (accepted).

Garwin, R.L. and Kendall, H.W. 1991. Quenching the wild wells of Kuwait. Nature 354:11–14.

Gilbertson, D. 1983. The impact of off-road vehicles in the Coorong dune and lake complex of south Australia. pp. 356–373. In Webb, R.H. and Wilshire, H.G. (eds.), Environmental Impacts of Off-road Vehicles. Impact and Management in Air Regions. Springer-Verlag, New York.

Hahn, J. 1991. Environmental effects of the Kuwaiti oil field fires. Environ. Sci. Technol. 25:1530–1532.

Harrison, R.M. 1991. Environmental effects of the Kuwaiti oil fires. In Environmental and Health Impact of the Kuwaiti Oil Fires. Proc. of an International Symp. held at the University of Birmingham, 17th October, pp. 66–68.

Hobbs, P.V. and Radke, L.F. 1992. Airborne studies of the smoke from the Kuwait Oil Fires. Sci. 256:987–991.

Holden, C. 1991. Kuwait's unjust deserts: damage to the desert. Sci. March 19, pp. 1176.

Horgan, J. 1991. Up in flames. Kuwait burning oil wells are a sad test of theories. Scientific Am. May, pp. 7–9.

HPI Consultants Inc. 1987. Crude oil assay handbook, First Edition, Austin, Texas, USA.

Hunt, C.B. 1966. Hydrologic Basin, Death Valley, California. Geochemistry of the saltpan. U.S. Geol. Sur. Professional Pap. 494-B:B40–B103.

Jayaraman, K.S. 1991. Impact of Gulf war. Nature 349:447.

Jayaraman, K.S. 1991. Impact of Gulf war. Nature 349:447.

Jenkins, G.J., Johnson, D.W., McKenna, D.S., and Saunders, R.W. 1992. Aircraft measurements of the Gulf smoke plume. Weather 47(6):212–220.

Johnson, D.W., Kilsby, C.G., McKenna, D.S., Saunders, R.W., Jenkins, G.J., Smith, F.B., and Foot, J.S. 1991. Airborne observations of the physical and chemical characteristics of the Kuwait oil smoke plume. Nature 353:617–621.

Kelly, K. and Schnadelbach, R.T. 1976. Landscaping the Saudi Arabian Desert. pp. 182. The Delancey Press, Philadelphia, USA.

Kenaga, E.E. 1980. Environ. Sci. Technol. 14:553.

KFUPM/RI, 1991a. Physical properties of oil slick and metal concentrations in oil slick and water samples. Ninth Progress Report, Gulf Oil Spill 1991. The Research Institute, King Fahd University of Petroleum and Minerals, Dhahran.

KFUPM/RI, 1991b. Metal Concentrations in Rain Water and total suspended air particulates in the Eastern Province. Twenty third Progress Report, Gulf Oil Spill 1991. The Research Institute, King Fahd University of Petroleum and Minerals, Dhahran.

Khuraibet, A.M. 1992. The Iraqi left over ordinance in Kuwait and their health effects. Paper presented at the Third World Academy of Sciences, Fourth General Conference. 23–26 November, Kuwait City, Kuwait.

Kovda, A.A., Samoilova, E.M., Charley, J.L., and Skujins, J.J. 1979. Soil processes in arid lands. pp. 439–470. In Goodall, D.W., Perry, R.A., and Howes, K.M.W. (ed.), Arid-Land Ecosystems: Structure, Functioning, and Management. Cambridge University Press, Cambridge.

Kramer, M. 1991. Kuwait: back to the past. Time, August 5, 12–18.

Lagrone, S.F. 1991. Potential community exposure to toxic chemicals. Environ. Sci. Technol. 25:366–368.

Lameloise, P., Thibaut, G., Masse, R., Lafuma, J., Person, A., and Pasquereau, M. 1991. Measurement Campaign of the Regional Mobile Laboratory for Measurements of Air Quality in Kuwait. Final Report, March 27 to April 4, Paris, France.

Lane, C. 1991. What just lies ahead. pp. 28–29. March 4, Newsweek.

Lathrop, E.W. 1983. Recovery of perennial vegetation in military maneuver areas. pp. 265–277. In Webb, R.H. and Wilshire, H.G. (eds.), Environmental Impacts of Off-road Vehicles. Impact and Management in Air Regions. Springer-Verlag, New York.

Linden, E. 1991. Getting blacker everyday. Times, May 27.

MacLeod, S. 1991. Iraq. With his county in ruins, how long can Saddam hang on. Time, March 11, pp. 37–38.

Manins, P. and Bouma, W. 1991. Burning oil fields in the Middle East. Clean Air 25:47–48.

McCauley, J.F., Breed, C.S., Grolier, M., and Mackinnon, D. 1980. The US dust storm of February 1977. In Pewe, T.L. (ed.), Desert Dust: Origin, Characteristics, and Effects on Man. Geol. Soc. Am. Special Pap. 186:123–147.

Mitchell, D.P. and Al-Sunaid, A.A. 1992. AMMNET air particulate chemical characterization and metal Analysis. Proc. the International Conference on Chemistry in Industry. November 14–16, Bahrain.

National Review, 1991. Black mischief. National Review, February 26.

Okada, K., Ikegami, M. Uchino, O., Nikaidou, Y., Zaizen, Y., Tsutsumi, Y., and Makino, Y. 1992. Kuwait soot over Japan. Nature 355:120.

Okita, T. 1991. Review of the survey by Japanese team. Paper presented at the Boston, USA.

Parmelee, J. 1991. Kuwaiti Emir snuffs out last Iraqi-lit oil fire. Washington Post, pp. A1, A47, 7 November.

Raloff, J. and Monastersky, R. 1991. Gulf oil threatens ecology, maybe climate. Sci. News, 139:71–73.

Rao, U.R. 1991. Gulf Oil spill and the monsoon. Nature 351:704.

Riley, J.J., Alick, N.G., and Thompson, T.L. 1991. Effects of Kuwait oil fields fires on human comfort and environment in Jubail. Cited by Alnaser et al. (1992).

Sadiq, M., AlThagafi, K.M., and Mian, A.A. 1992. A Preliminary Evaluation of Metal Contamination of Soils From the Gulf war Activities. Bull. Environ. Contam. Toxicol. 49:663–639.

Sandia, 1991. Operation quick response predicting environmental effects of oil sabotage. Energy & Environment, July 00, Sandia National Laboratories.

Schroeder, W.H. and Lane, D.A. 1988. The fate of toxic airborne pollutants. Environ. Sci. Technol. 22:240–246.

Shaw, W.S. 1992. Smoke at Bahrain during the Kuwait oil field fires. Weather 47(6):220–226.

Sipa Press, 1992. Kuwait: War in the Giulf. Sipa Press, The Kuwait Bookshops Co. Ltd.

Small, D. 1991. Environmental impact of fires in Kuwait. Nature 350:11–12.

Statistical Year Book, 1988. Kingdom of Saudi Arabia, Ministry of Finance and National Economy, Central Department of Statistics. Statistical Yearbook 24th Issue, pp. 587–622.

Suthers, J.R. 1991. Air and water pollution in Jubail. In Environmental and Health Impact of the Kuwaiti Oil Fires. Proc. of an International Symp. held at the University of Birmingham, 17th October, pp. 32–39.

Tandy, C.R.V. 1978. Background. pp. 6–12. In Ali, C.T. and Brown, J. (ed.) Landscape Design for the Middle East. RIBA Publications Limited, London.

Tawfiq, N.I. 1991. Response by Saudi Arabia to the environmental crisis caused by the Gulf war. In Environmental and Health Impact of the Kuwaiti Oil Fires. Proc. of an International Symp. held at the University of Birmingham, 17th October, pp. 40–43.

The Kuwaiti oil fires and oil lakes – facts and numbers. In Environmental and Health Impact of the Kuwaiti Oil Fires. Proc. of an International Symp. held at the University of Birmingham, 17th October, pp. 12–16.

The Middle East, 1992. The Gulf war. Blind man's buff. The Middle East January, p. 23.

Travis, C.C. and Arms, A.D. 1987. Toxic Chemicals, Health and the Environment. The Johns

Hopkins University Press, Baltimore, USA.
Travis, C.C. and Arms, A.D. 1988. Bioconcentrations of organics in beef, milk, and vegetation. Environ. Sci. Technol. 22:271–274.
UK Meteorological Office, 1991. Possible environmental impact of burning oil wells in Kuwait, January 17.
USIAAT, 1991. Results of air sampling and monitoring by U.S. Interagency Air Assessment Team between March 13–20.
Webb, R.H. and Wilshire, H.G. (eds.), 1983. Environmental Impacts of Off-road Vehicles. Impact and Management in Air Regions. Springer-Verlag, New York.
Williams, R.S., Heckman, J., and Schneeberger, J. 1991. Environmental Consequences of the Persian Gulf war 1990–91. National Geographic Res. & Exploration 7 (Special Issue):1–48.
Wilshire, H.G., Nakata, J.K., and Hallet, B. 1980. Field observations of the December 1977 windstorm, San Joaquin Valley, California. In Pewe, T.L. (ed.), Desert Dust: Origin, Characteristics, and Effects on Man. Geol. Soc. Am. Special Pap. 186:233–251.
WMO, 1991. Report on the WMO meeting of experts on the atmospheric part of the Joint UN. Response to the Kuwait Oilfield Fires. Geneva, April. 27–30.
World Oil, May 1991. Kuwait: the mother of all blowouts. World Oil, May, pp. 38–48.

CHAPTER 7

Human health and the Gulf crisis

7.1. Summary

- The assessment of health implications of the Gulf crisis encountered many problems, especially inadequacies in health statistics and technical limitations of the available data. Any human health impact assessment would not be realistic, but speculative at best.
- The literature suggests that smoke from the Kuwait oil fires and burning of diesel fuel was not toxic but those individuals who were exposed to the repeated smoky episodes for prolonged periods could be adversely affected. The smoke situation in Kuwait, Saudi Arabia, and other countries in the region has improved dramatically since November 1991 when the last oil fire was extinguished.
- Ambient concentrations of sulfur dioxide (SO_2) were found to be below air quality guidelines throughout the Gulf region. It is speculated that such low SO_2 levels would not effect human health adversely, even in Kuwait. Concentrations of SO_2 might have been elevated to above air quality guidelines but only in the burning oil fields. Therefore, firefighters could have been exposed to the elevated SO_2 levels but these are technical people and presumably took all precautions to minimize exposure against any such mishap. The Kuwait oil fires could potentially produce very large quantities of SO_2 but luckily most of the SO_2 was oxidized immediately to sulfate and then to calcium sulfate. In our opinion, this phenomenon probably saved many lives in Kuwait, Saudi Arabia, Iraq, Iran, and many other neighboring countries of Kuwait.
- The health effects of the ammunition and explosives used in the Gulf war can never be assessed quantitatively. Substantial evidence, however, suggests that the local population and combat personnel in the war zone of Kuwait and Iraq were exposed to an unknown quantity of organic and inorganic chemicals, many of which could induce adverse health effects.
- Various evidence suggests that military-related activities would have enhanced ambient concentrations of inhalable particulates (generally referred as PM10). It was found that concentrations of particulates were

higher near the Kuwait oil fires. The data collected in Kuwait and Saudi Arabia indicate that PM10 concentrations violated air quality guidelines several times over while the oil fires were burning. The available data on the health implication of the PM10 data suggest that the enhanced concentrations of PM10s in Kuwait, Saudi Arabia, and other countries in the region would have adverse health effects. However, no data are available (probably not existing) to suggest that any excess mortality is related to the PM10 load in the air during the Gulf crisis.

- Concentrations of nitrogen oxides (NO_x) in the smoke plume and on the ground were mostly below the air quality guidelines and certainly below the toxicity limit necessary to induce adverse health effects. It is, therefore, concluded that NO_x produced by the Kuwait oil fires and combustion of diesel fuel was too low to pose any human health hazard in Kuwait or anywhere in the Gulf region.
- Carbon monoxide (CO) concentrations at ground level posed no health hazard in Kuwait and Saudi Arabia during the Gulf war. All the measurements were found to be below air quality guidelines and certainly below the toxicity limit for humans. It may be concluded that the Gulf crisis did not elevated CO concentrations to hazardous levels in any of the countries in the region, including Iraq.
- Concentrations of Ozone (O_3) in Kuwait, Saudi Arabia, and probably in other countries in the region were within air quality limits, and therefore, no adverse health effects are expected.
- Ambient concentrations of volatile, alipahtic, and polycyclic hydrocarbons were below the permissible limits. Blood concentrations of volatile organic chemicals in humans who stayed in Kuwait during the Gulf war were lower than that found in the many urban centers of the world. It is believed that acute health effects associated with organic chemicals will be minimal. The chronic effects of these chemicals is difficult to estimate.
- Concentrations of lead, cadmium, and cobalt in inhalable particles were found to be higher than their safe limits as suggested in the literature, whereas nickel, vanadium, and copper were within safe limits. A comparison of metal concentrations worldwide and those found in the Gulf region indicated higher concentrations of nickel, lead, vanadium, and chromium. Inhalation of metal-contaminated dust particles may cause the accumulation of metals in specific human organs and induce adverse chronic health effects in supersensitive subjects. Therefore, continuous monitoring of both air and humans for metal concentrations is recommended.
- Many teams of scientists investigated the acute (short-term) health effects of the Kuwait oil fires. These studies may be questioned on their scientific and statistical merits, but all indicated that respiratory and cardiovascular diseases were enhanced and these responses were dependent on distance from the burning oil fields. These cases were only seen in Kuwait.
- Limited available data suggests that there may be some chronic (long-term) health effects, especially from metals and organic chemicals, that could be

associated with the Gulf war. These are speculations at best and largely based on extrapolation. These chronic effects are expected to be due to exposure to contaminated inhalable particulates, known and unknown organic and inorganic chemicals, and contamination of the food chain.

- Iraq did not use chemical weapons during the Gulf war. However, the Allied air force claims that it destroyed several facilities allegedly containing toxic chemicals. No information could be collected regarding the extent of damage to Iraq's chemical weapon plants and stockpiles, whether toxic chemicals were released as a result of bombing, and if the local population was exposed, etc. Considering the extent of bombing, it is suspected that damage to chemical weapon facilities and stockpiles might be extensive.
- Allied casualties in the Gulf war are well known (149 killed and 513 wounded – Church, 1991). However, information on Iraqi casualties is at best speculative. The majority of the estimates reported more than 100,000 Iraqi military personnel dead during the 43 days of the War. It is suspected that about 100,000 persons were wounded, the majority of whom may have died due to lack of medical facilities and medicine.
- As a result of the war and the UN embargo, Iraqi medical services and stocks of medicine are said to be in a desperate short supply. No one has estimated civilian causalities. However, considering the pitiful plight of many Iraqis, it is believed that as many as half a million people may have died (including military personnel) due to the causes associated with the Gulf crisis. Kuwaiti causalities during the Iraqi occupation has been estimated at about 2,000 dead.
- It has been reported that Kuwait was left littered with live and spent ammunition and anti-personnel mines. By April 1992, about 4,000 civilian and 85 explosive experts had been killed or injured by the mines and live munition. Not all mines have yet been cleared.
- The distress of refugees from Iraq and Kuwait was enormous. Whole populations in Kuwait, Iraq, Bahrain, and northern Saudi Arabia were tormented. In brief, the social cost of the Gulf crisis was colossal.
- Many health related questions are yet to be resolved. Concentrations of inhalable air particulates, many organic compounds, and metals in them, were found to be higher than proposed health limits. The chronic effects of these exceedences should be investigated on a long-term basis.

7.2. Introduction

The environment is always a victim of war and no single component of the environment suffers more than humans themselve. Environmental victimization may either be associated with the land, vegetation, air, or water, but ultimately the negative effects lead to mankind. The Gulf war was no exception to this rule. When the Iraqi leader declared his intentions to use oil as a weapon, many scientists predicted dramatic consequences of an unprecedented magnitude. The world was shocked by the news of the vast oil spill and deliberate ignition of Kuwait's oil wells. The international community responded quickly to reduce adverse effects of these actions. Many international scientific teams visited Kuwait, Saudi Arabia, Bahrain, and other countries in the region. To assess the environmental damage from the Kuwait oil fires, the teams collected scientific data on air quality in the region. Concerted efforts were made to contain the oil spill. One of the often asked, but always remaining unanswered questions was what impact would these events have on human health. The demand of the general public for accurate answers has not been satisfied by the scientific community which could only state that they did not know for sure and that adverse effects may show up later.

Health risk assessment is very a complex task and requires a large body of accurate data. Characterization of the source, exposure mode and duration, pollutant concentration, interaction of pollutants with other components of the environment, background information on the environment, social and physiological conditions of the population involved, and many other aspects need to be considered in a health risk assessment. Information on many of these aspects was not available. Before discussing the health consequences of the Gulf war, some problems faced in this regard will be discussed.

7.3. Problems associated with health risk assessment

It has been difficult to definitively assess the health consequences of the Gulf war. Some of the social and technical problems that compounded the assessment of risks to human health are briefly discussed here.

1. Iraq invaded Kuwait on August 2, 1990. There was a mass exodus of both Kuwaitis and expatriates in the following months. Many of the foreigners were not allowed back into Kuwait and many opted voluntarily to stay out. The situation in Saudi Arabia and Iraq was not much better than Kuwait. In Saudi Arabia, Bahrain, and other Gulf countries in the region, the human health issue is considered sensitive. Background information on the pre-war human health status, even if available, is not easily accessible.
2. In 1990, more than 61% of the total Kuwait population were expatriates (Al Shatti, 1991). The majority of expatriates moved out of the country during the Iraqi occupation or immediately after the liberation. How long they were exposed and to what conditions are very hard to document. A large proportion of Kuwaitis either escaped or were out of the country at the time of occupation. These individuals returned to Kuwait after the liberation and thus were partially exposed to pollutants. Iraq imposed travel restrictions on expatriates but ultimately many of these were allowed to leave. Those who left Iraq did not go back. Before the start of the air compaign, a major portion of the population, including Saudis and expatriates, from the Eastern Province and Riyadh region were moved to relatively safe places. A large majority of this population came back in March–April 1991. Many

people also came to the Gulf region for the first time. It is difficult to assess the health implications of the Gulf war in a population that was so dynamic and differentially exposed to the polluted environment.

3. The Kuwaitis and expatriates that remained in Kuwait were severely stressed and depressed. These individuals may respond differently to environmental conditions as compared with a normal individual. In addition, environmental conditions in Kuwait leading up to the end of the ground and air operation are unknown. Because of the very complex socio-environmental conditions to which these individuals were subjected, it makes the evaluation of adverse health effects that could be associated with the Gulf war very difficult. A similar case but of lesser magnitude could be argued for those who stayed in Iraq, Saudi Arabia, and Bahrain.
4. Health care systems in Kuwait and other Gulf countries were well established before the Gulf crisis. Records on patients were kept in all health clinics and hospitals. However, statistics on different diseases were not maintained properly. Therefore, there is no readily available historical baseline data that can be compared and form a basis for the evaluation of health risks in the Gulf region.
5. As a result of Iraqi occupation, government and private infrastructures in Kuwait were completely disrupted. It took considerable time to restore government services after the liberation of Kuwait. In the early days of the Kuwait oil fires, difficulties were encountered by pioneer international scientists collecting health-related scientific data in Kuwait. Iraq was almost an out-of-bound territory for the world and it still remains that way.
6. The majority of the medical staff working in Kuwaiti hospitals were expatriates and left Kuwait during the crisis. For example, at Al Amiri Hospital, there were 140 physicians on August 2, and that number dropped to about 24 by the liberation of Kuwait. As a result, there were acute shortages of technical personnel, at least in the beginning of the liberation (McGinn, 1991). Also, as Kuwait claims, much of their equipment was either vandalized or taken to Iraq. Lack of equipment and trained staff severely curtailed the capabilities of the Kuwait national institutes to gather necessary data on the nation's health and to extend technical support to visiting scientists.
7. Many of foreign scientific teams were motivated by their invested interests. The scientific teams were generally given instructions to concentrate on areas where their nationals were stationed. For example, the United States Environmental Protection Agency was rightfully more concerned about US soldiers. Therefore, the team concentrated more on the locations where there was heavy US troop concentrations. French, British, and the Norwegian measurements were also biased towards areas containing their own nationalities. As a result of this, data on air pollutant levels to which the local population might have been exposed are fragmented both in time and geography.
8. The embargo on The Iraq, travel restrictions imposed by the Iraqi govern-

ment, and other political realities made the collection of any data in Iraq nearly impossible. The plight of the Iraqi people is a complete mystery to the world. Even if information becomes available, it will be impossible to differentiate between health deterioration associated with the embargo or environmental pollution. It is suspected that no data on the environmental conditions in Iraq were collected because the country was in a state of emergency.

9. In certain cases, foreign scientists were overwhelmed by political realities. The people and environment in Iran and Iraq were completely ignored by the international scientific community. Information from areas other than Kuwait and Saudi Arabia is scarce.
10. Major efforts of the scientific community concentrated on the oil spill and the Kuwait oil fires. The environmental and human health impact of the Allied bombardment and the outfall of explosives were completely ignored. It is our understanding that United States medical personnel collected a wealth of health-related data but these are not generally accessible to the scientific community.
11. The United Nations Environmental Program (UNEP) failed to coordinate scientific activities worldwide on the health implications of the Gulf war. Many teams came to Kuwait independently and did what the team felt was right. Most of the United Nations' efforts were focused on the meteorological and modeling aspects of the Kuwait oil fires (WMO, 1991, 1992). Actual monitoring was completely ignored. This approach on the part of the United Nations is regrettable.
12. Human health is a very sensitive issue worldwide and the Arabian Gulf region is no exception. People do not like to be guinea pigs for scientists. Collecting the desired samples and data is extremely difficult. Therefore, most of our knowledge of environmental implications of the Gulf crisis on human health can be derived from analogous situations and statistically valid associations.

Because of the above problems, the real impact of the Gulf crisis on human health may never be known. The available information on the health and environment is synthesized in this chapter.

7.4. Air quality guidelines of most probable pollutants

Information on air quality guidelines (may be referred to as health standards) of pollutants suspected to be emitted in large amounts as a result of the Gulf crisis were collected and summarized in Table 7.1. These guidlines will form the basis for the evaluation of health risks of various pollutants.

Table 7.1. Air quality (health) guidelines proposed in the literature.

Parameter	Exposure Time	Maximum Concentration	Reference
Sulfur dioxide (SO_2)	1 hour	730 $\mu g/m^3$	MEPA(1402H)
	24 hours	365 $\mu g/m^3$	MEPA (1402H)/USEPA (1978)
	1 year	80 $\mu g/m^3$	MEPA (1402H)/USEPA (1978)
Inhalable particles	24 hours	150 $\mu g/m^3$	MEPA (1402H)
	1 year	50 $\mu g/m^3$	USEPA (1978)
	24 hours	340 $\mu g/m^3$	MEPA (1402H)
	1 year	80 $\mu g/m^3$	MEPA (1402H)
Ozone (O_3)	1 hour	235 $\mu g/m^3$	USEPA (1978)
	1 hour	295 $\mu g/m^3$	MEPA (1402H)
Nitrogen oxides (as NO_2)	1 hour	660 $\mu g/m^3$	MEPA (1402H)
	1 year	100 $\mu g/m^3$	MEPA (1402H)/USEPA (1978)
Carbon monoxide	1 hour	40 mg/m^3	MEPA (1402H)/USEPA (1978)
(CO)	8 hours	10 mg/m^3	MEPA (1402H)/USEPA (1978)
Hydrogen sulfide (H_2S)	1 hour	195 $\mu g/m^3$	MEPA (1402H)
	24 hours	40 $\mu g/m^3$	MEPA (1402H)
Oil mist	-	5 mg/m^3	CONCAWE (1986)
Lead (Pb)	Quarterly	1.5 $\mu g/m^3$	USEPA (1978)
Chromium (Cr)	-	2.5 ng/m^3 as Cr(VI)	Sloof et al. (1990)
Nickel (Ni)	-	50 $\mu g/m^3$	Fairhurst and Illing (1987)
Vanadium (V)	-	1 $\mu g/m^3$	Lameloise et al. (1991)
TNT	-	0.5 mg/m^3	Federal Register (1976)
RDX	-	1.5 $mg/^m$	Federal Register (1976)
PETN	-	15 $mg/^m$	Federal Register (1976)
Tetryl	-	1.5 mg/m^3	Federal Register (1976)

7.5. Health impact assessment strategy

Proper projections of health risks resulting from the exposure to pollutants emitted during the Gulf crisis require information on both the levels and toxicity thresholds of airborne contaminants. The available information on the above aspects has already been discussed and, in this chapter, this data is summarized wherever appropriate.

The assessment of human health impact of a given pollution episode is intriguingly complex. In addition to data on pollutant levels and their toxicities, interactions between several factors should be considered simultaneously in evaluating the effects of a pollution episode or pollutants. The most important of these are ambient concentrations and human exposure time (dose and duration) of the concerned pollutant. Prevailing weather conditions, especially temperature and humidity, can modify the impact of a pollutant. The presence of co-pollutants may provide an additive effect. For example, low concentrations of ozone can cause greater bronco-constrictions if associated with low concentrations of SO_2 or other allergens (Koening, et al. 1990; Molfino, et al. 1991). In addition, other factors may aggrevate adverse effects such as breathing cooler air, exercise, dust storms, etc. (Ayres, 1991). Finally, one has to consider the age, bronchial hyper-reponsiveness, cardiovascular, and other health conditions of the individuals exposed to the pollutants.

Many war-related events and activities had profound effects on humans. Some of these are the invasion of Kuwait, the total embargo imposed on Iraq, the air compaign and ground operations of the Allied Forces, the oil spill in the Arabian Gulf, and the Kuwait oil fires. Considering the Gulf crisis, it is expected that the human health risks would mainly be associated with exposure to petroleum mist, smoke, inhalable particulates, sulfur dioxides, acidic aerosol, nitrogen oxides, carbon monoxide, ozone, hydrocarbons (both polycyclic aromatic and volatile hydrocarbons), toxic metals, radiation, and chemicals related to warfare. As stated in Section 7.3, there are many problems that limit realistic assessment of the impact of the Gulf crisis on humans. Discussion of these problems will not be repeated here. The available data on the source strength of pollutants, ambient concentrations of pollutants, toxicity of contaminants, area of impact, etc., have been collected to help predict the health effects. Available health statistics in Kuwait will also be reviewed.

7.6. Health impact of air pollutants

Any air pollution episode can cause either acute (short-term or immediate) or chronic (long-term) health effects, the most feared and conspicous being acute health effects. With reference to the Gulf, the international community was more concerned with the acute health consequences. The most obvious cause of acute health effects was the Kuwait oil fires. The general pollutants that could induce such effects are the priority air pollutants (gaseous pollutants, oil mist, smoke/soot, and air particulates). The toxic effects and toxicity threshold limits of these and other pollutants are discussed below.

7.6.1. Effects of crude oil mist on human health

It has been reported that crude oil mist fell on the ground in areas surrounding the oil fires, including Kuwait City (Al Besharah, 1991; Dauod, 1991; El Desouky and Adbulraheem, 1991; Al Hassan, 1991, 1992; EPC, 1991). These reports suggest that almost all of Kuwait was drenched with an oil mist from the fires. The magnitude of human exposure to the petroleum mist was dependent on the wind direction (population on the downwind side was more exposed), distance from the fire source (the nearer to the souce, the greater the mist), and type of activity undertaken during the episode (person involved in outdoor activities would have been exposed to more petroleum mist than those who stayed inside their homes).

All literature reports (Al Besharah, 1991; Dauod, 1991; El Desouky and

Adbulraheem, 1991; Al Hassan, 1991, 1992; EPC, 1991) considered oil mist in the most qualitative terms. No attempt was made to quantify oil mist. Human health effects should, therefore, be considered in the form of generalized petroleum toxicity. The toxicity of oil mist has been reviewed by the CONCAWE Task Force (1986) which examined a large number of acute, short-term and long-term studies involving both animals and humans. It was concluded in the short-term studies that concentrations of oil mist less than 50 mg/m^3 did not produce toxic effects. Long-term exposure studies indicated that repeated exposures to high levels of oil mist (>100 mg/m^3) may result in the accumulation of oil in the lungs. It was found that concentrations lower than 100 mg/m^3 of oil mist, equivalent to those found in work places, generally produced no adverse health effects. The CONCAWE Task Force (1986) suggested that oil mist of less than 5 mg/m^3 would have no adverse effect on human health. (CONCAWE, 1986).

Ambient concentrations of oil mist were not measured in Kuwait or anywhere in the Gulf region. Because of the lack of data on ambient concentrations of oil mist during the Kuwait oil fires, the health implications of oil mist can not be evaluated realistically. However, considering the high levels of oil mist required to produce acute and chronic health effects (CONCAWE, 1986), it is anticipated that the oil mist would have had limited effects, if any, on populations in the vicinity of burning oil fires. Fire fighting personnel were probably exposed to more oil mist than any other group of the Kuwait population. However, firefighters were supposed to take all necessary precautions to avoid the direct inhalation of oil. In an international conference on the Kuwait oil fires, a Kuwaiti scientist reported that liver tissue from a few animals that were exposed to the Kuwait oil fires were recently dissected by Kuwait scientists. They reported that these livers were blackened, probably due to the inhalation of crude oil mist and soot particles. Because of the lack of research facilities, these scientists could not undertake a systematic and comprehensive study. These observations suggest that the health of super-sensitive individuals (asthmatics) in Kuwait, especially during February–April 1991 may have been adversely affected because of high concentrations of oil mist in ambient air. In general, however, it is hoped that the general population in Kuwait was not exposed to excessive oil mist from the Kuwait oil fires.

7.6.2. Effects of smoke/soot on human health

With reference to the Gulf war, there were two major sources of smoke/soot particles, smoke emissions from military-related activities and the Kuwait oil fires. It is estimated that approximately 6 billion gallons of diesel fuel were consumed during the Gulf crisis, mainly in the month of February 1991, that consequently generated about 7,000 tonnes of soot/smoke particles. The Kuwait oil fires were the major source of smoke/soot. The oil fires peaked in the last week of February 1991, emitting approximately 6 million barrels of crude per day. Soot production from these fires was estimated at about 10 million tonnes. The smoke/soot produced by diesel burning was about 0.05% of that produced by the Kuwait oil fires. Therefore, for all practical purposes, smoke produced due to diesel combustion can be neglected.

Assessment of the toxicological effects of smoke requires the precise identification of its constituents. Physical and chemical characteristics of the smoke from the Kuwait oil fires are discussed in Chapters 3 and 4. Here we briefly summarize those observations. All airborne studies confirmed that smoke particles were spherical and combined together to form a chain aggregate commonly referred to as soot. The morphological characteristics of smoke particles seemed to be altered by the type of smoke, age, and source of the plume (Jenkins et al., 1991; Johnson et al., 1991; Hobbs and Radke, 1992). Smoke density decreased with distance and, in some instances, smoke was identified approximately 2,000 km from the source. Information on the physical and chemical properties of smoke from the military- related activities is scarce. The major portion of solid emissions, in the case of military operations, is believed to be elemental carbon.

Composition of smoke/soot is further complicated by the fact that soot is a good absorbent for organic combustion products. The toxicity of soot particles, therefore, depends on the chemical composition of the emission, especially on concentrations of polycyclic aromatic hydrocarbons, tetrachlorobezo-p-dioxin, tetrachlorodibezofuran, phenol, and nitrated phenol. Information on concentrations of organic chemicals in soot particles in the smoke plume is limited. Hobbs and Radke (1992) reported that the smoke particles collected from the super-composite plume (smoke plume where fires from the individual wells combine) consisted of approximately 30% salt, 15–20% soot, 8% sulfate crystals, and approximately 30% organic chemicals. Jenkins et al. (1991) measured concentrations of non-methane hydrocarbons (NMHC) in three smoke samples taken from the thickest part of the plume on March 30, 1991. They observed a decrease in concentrations of all eight hydrocarbons (ethane, ethene, propane, propene, benzene, toluene, n-hexane, and NMHC). Ethane and propane were found at the highest concentrations. Hobbs and Radke (1992) reported that the concentrations of non-methane gaseous hydrocarbons (NMHC) in the plume ranged from 55 to 827 ppb of carbon. The NMHC were dominated by alkanes (65 to 82% NMHC on a per carbon basis); alkenes or olefins made up 7 to 23% of the NMHC, aromatic hydrocarbons 5 to 19%, and alkynes 1 to 7 percent.

The chemical composition of the smoke was also variable at the ground. Spengler (1991) reported that carbon in the particulates varied between 2 to 60% elemental carbon, and organic compounds ranged between 30–70 percent. Okita (1991) found that, in the particulate samples collected at different locations in Kuwait, carbon comprised approximately 50% of the particles and elemental carbon approximately 45 percent. Lameloise et al. (1991) observed that high molecular weight organic compounds such paraffins, were relatively abundant, particularly in areas close to the burning oil fields. Concentrations of soot particles or their chemical composition were not determined during the

Gulf war. From the literature, it was found that appreciable concentrations of toluene, anthracene, fluorathene, and pyrene could be found in the emissions of heavy-duty diesel engines. From these observations, it is concluded that the chemical composition of smoke was highly variable.

From the foregoing, it is obvious that, with such high variability in smoke characteristics, its chemical composition cannot be used to assess the health hazards of the Kuwait oil fires. Therefore, an alternative approach was adopted, i.e. extrapolation of the results and conclusions of analogous toxicological studies and apply them to the Kuwait oil fires. Unfortunately, toxicological data on smoke inhalation are scarce as oil fires have been uncommon. Because of the lack of information, assessment of hazards of the smoke to which the populations in the Gulf region were exposed, cannot be based on toxicological studies (Sharratt and Butler, 1991). The argument is also true for the soot/smoke produced due to the consumption of diesel fuel during the Gulf crisis.

Toxicological information on some individual compounds are reviewed for the benefit of the reader. Reviews on the individual hydrocarbons present in the light fraction of the crude, e.g. toluene (Bell et al., 1989), xylene (Bell et al., 1991), and n-hexane (Battershill et al., 1987) conclude that the levels which show no effect in animals and man are in the range of 100 ppm. Another component of the light fraction of crude is benzene which has been reported to be carcinogenic above 100 ppm level (Sharratt and Butler, 1991).

Bitumen is another component of crude oil. The principal concern regarding the inhalation of bitumen fumes is its carcinogenicity (IARC, 1985). However, one study carried out on workers exposed to bitumen provided no evidence of health hazards (Baylor and Weaver, 1968). The evidence that exposure to bitumen fumes leads to cancer was reviewed by IARC (1985) which suggested that in the studies that show bitumen carcinogenicity, the individuals might have been exposed to other cancer causing substances along with bitumen. The carcinogenicity of bitumen at present remains controversial.

Elemental carbon was the major constituent of smoke particulates generated by both combustion of diesel fuel and emissions from the Kuwait oil fires. A number of studies has been carried out on workers exposed to carbon black. In the rubber and printing industries exposure occurs to carbon black in conjunction with mineral oils containing PAHs. Epidemiological studies suggest an increase in stomach cancer when workers are exposed to various dusts containing carbon black (Blum et al., 1979; Parks et al., 1982). Similar but less consistent results were obtained in studies of Pressmen (IARC, 1984b). The toxic effects of exposure to carbon were also reviewed by IARC (1984a). Epidemiological evidence reviewed by IARC suggests diminished respiratory function, particularly in smokers. Long-term exposure to about 7 mg/m^3 carbon black reduced the forced vital capacity and forced expiratory rate of the lungs. Pneumoconiosis has also been reported in a group of workers exposed to carbon black.

The particulates present in the exhaust emissions of diesel engines may have some similarity to emissions from the Kuwait oil fires. Grasso et al. (1990) reviewed the health hazards of diesel particulates. They considered carcinogenicity based on evidence from animal studies and epidemiological data. From different studies, they drew the general conclusion that diesel emissions could present a carcinogenic hazard. Sharratt and Butler (1991) cautioned that this generalized conclusion may not be valid. Long-term inhalation studies on diesel particulates at atmospheric levels (4–6 mg/m^3) produced a low incidence (10 percent) of squamous cell carcinoma in rat's lungs. In relation to the epidemiological investigations, Grasso et al. (1990) commented that several studies have indicated that lung cancer incidence may be slightly higher in certain occupations in which exposure to diesel exhaust is likely to be greater than the general population, e.g. truck drivers.

The above evidence suggests that smoke from the oil fires and burning of diesel fuel was not very toxic. Populations, such as Kuwaitis and Saudis living in the Eastern Province of Saudi Arabia, who were exposed to smoky episodes more often and for prolonged periods could, however, be adversely affected. The situation in Kuwait and in other countries in the region has improved significantly since November 1991 when the last oil fire was extinguished.

7.6.3. Health effects of inhalable air particulates

Suspended air particulates represent a broad class of chemically and physically diverse substances that exist as discrete particles in the atmosphere. The majority of these particles range between 0.005 to 100 μm in size. Originally, the United States Environmental Protection Agency (USEPA) measured particles smaller than 25 μm and termed these as "Total Suspended Particles" commonly referred to as TSP. In 1987, the USEPA replaced TSP with an indicator that includes only those particles that are smaller than 10 μm and referred to these as PM10 or the inhalable particle fraction. PM10 particles can reach the tracheobronchial and deeper regions of the lung and are thus thought to be a more suitable as a representative indicator for most of the adverse health effects that are related to air particulates. PM10 has been adopted universally as an indicator of the inhalable fraction of air particulates and the monitoring of PM10 levels is increasing worldwide. The primary quantitative basis for establishing the PM10 standard has been community epidemiological research, with mechanistic support from toxicological controlled human exposure investigations. For short-term exposures, epidemiological studies showed markedly increased daily mortality within 24 hours with concentrations above 1000 $\mu g/m^3$, with some small increased risk of mortality at concentrations above 600 $\mu g/m^3$. Short-term epidemiological studies also indicate aggravation of bronchitis with PM10 levels in the range of 250–350 $\mu g/m^3$ while small but statistically significant decreases in lung functions have been documented in children exposed to 24-hour peak PM10 levels of 140–250 $\mu g/m^3$. The decrement persisted for 2–3 weeks following exposures. Long-term epidemiological studies provide evidence of likely increased respiratory symptoms and illnesses with PM10 levels above 80–90 $\mu g/m^3$. Recent studies indicate the possibility of increased respiratory symptoms and illnesses in children experiencing multi-year exposure levels of PM10 across a range of 40–58 $\mu g/m^3$.

Data on air particulates were not collected in Kuwait, Saudi Arabia, and Iraq during the war period (i.e. before February 1991). The infrastructure in Kuwait and to some extent in Saudi Arabia was so disrupted that measurements could not be started immediately. Systematic monitoring of PM10 at five locations [Salmiya, Fahahel, Ahmadi, Farwania, and Andalus given in Figure 7.1] in Kuwait was started in May 1991 by the Kuwait Environmental Protection Council. A monthly mean concentration of PM10 at these locations during May–September 1991 were reported as 508, 330, 322, 269, and 238 $\mu g/m^3$ respectively (EPC, 1991). These values are much higher than the proposed air quality limit in Table 7.1. Considering the above toxicological data, one could conclude that inhalable air particulates in Kuwait were alarmingly high. If the foregoing toxicity thresholds are applied to Kuwait, one should expect additional mortality due to high PM10 concentrations. Unfortunately this type of data was not available in any of the countries in the region. Therefore, the question whether or not high concentrations of PM10 resulted in extra mortality in Kuwait remains unanswered.

The local environment effects air particulate concentrations substantially. It

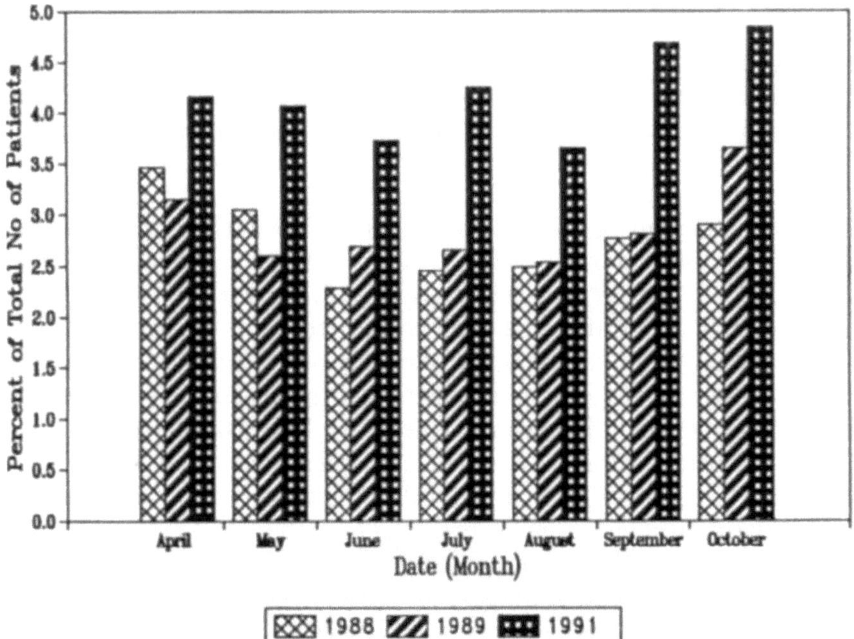

Figure 7.1. Effect of the Kuwait oil fires on respiratory illness in Al Ahamadi City, Kuwait.

was noticed that the location of measurement and wind direction influenced PM10 concentrations. The EPC (1991) found higher mean concentrations of particulates in Al Ahmadi (338 $\mu g/m^3$) and Fahaheel (310 $\mu g/m^3$) as compared with 250 $\mu g/m^3$ recorded at Salmiya-Kuwait City. The Al Ahmadi and Fahahel locations were nearer to the source, i.e., the oil fires (Figure 7.1). Locations in a downwind direction from the fires had a higher concentration of particulates as compared with those located in the upwind direction. In general, relatively lower levels of particulates were measured with the westerly, north-westerly, and easterly winds at Mansoria. North-westerly winds increased both PM10 and TSP in Kuwait as well in eastern Saudi Arabia, Bahrain, and Qatar. North-easterly winds pushed the smoke plume over Iran and Iraq (Al Besharah) and, as a result, concentrations of particulates would have increased. The NILU was the only organization that measured particulates inside Iraq (Sivertsen and Berg, 1991). Concentrations of air particulates were relatively lower at this location than that reported from Kuwait.

From the foregoing, it is concluded that, on many occasions, concentrations of inhalable and total suspended particulates were higher in Kuwait, Bahrain, and Saudi Arabia than the air quality guideline given in Table 7.1 (150 $\mu g/m^3$ for a mean of 24 hours, MEPA, 1402H). Comparing the particulate measurements with the toxicological threshold levels of air particulates, one may expect extra mortality in Kuwait, Saudi Arabia, Bahrain, and Qatar. Health statistics could not be collected in the impacted areas for the period of the Gulf war or immediately after the cessation of hostalities. During March–April 1991 the population of both Kuwait and northeastern Saudi Arabia was very dynamic (people who stayed during the war moved out of the countries and many of those who were out of the countries came back after the war). This means that the population of these areas was not equally exposed to high concentrations of inhalable air particulates. In such a dynamic situation, it would be difficult to make any sense out of health statictics even if these were available. Ozkaynak (1992) addressed the potential for excess mortality impacts associated with short-term exposures to particles emitted during the Kuwait oil well fires. This assessment was performed using relationships between variations in the daily pollution levels and daily mortality rates derived in previous studies, mainly in the United States. Although there were differences in available pollutant measurements, aerosol compositions, concentration ranges, demographic and health profiles of the exposure subjects, and population size between the Kuwait situation and the studies considered by Ozkaynak (1992), the author argued that particles from the Kuwait oil fires would be as toxic as that of urban aerosol. He concluded, therefore, that projection of risks using historic mortality risk coefficients can be considered in establishing a reasonable lower bound on the toxic effects of exposures to particles emitted by the burning oil wells in Kuwait. He selected a typical concentration of PM10 particulates, and not minimum or maximum levels of pollutant concentrations in his calculations. From his calculations and reasoning, Ozkaynak (1992) estimated a 10% increase in mortality in a one month period if PM10 levels increased from 50 to 250 μg/m³. The data reported above and in Chapter 4 certainly show that PM10 concentrations in Kuwait and Saudi Arabia were higher than 250 μg/m³. Data on whether there was excess mortality because of high levels of PM10 in the Gulf region could not be obtained, and probably do not exist.

The Kuwait Environmental Protection Department collected dust deposited on air conditioner filters in the Al Ahmadi region of Kuwait and conducted experiments in cooperation with the Dubai Institute of Environmental Research (Beg, 1992). It was found that the dust caused hemolysis of human erythrocytes in the experimental test systems. A higher lipid peroxidation reaction was observed in erythocytes cells treated with this dust. In a follow-up study, the Environmental Protection Department and Dubai Institute of Environmental Research collected samples of old filters (filters that were not cleaned for two months) and fresh filters (filters that were cleaned within one month of the date of sampling). It was found that old deposited dust was more effective in inducing hemolysis than freshly deposited dust from the same location. By comparing these results, it was shown that the toxicity of freshly deposited dust was about one-sixth that of the old deposited dust. A repeat of these experiments produced similar results (Beg, 1992). Comparing the geographic origin of the dust, the investigator argued that the dust from Kuwait was more toxic than those collected from Dubai during the same period. He attributed the enhanced dust toxicity to the emissions from the Kuwait oil fires.

From the foregoing, it is concluded that the Kuwait oil fires and the military-related activities would have enhanced ambient concentrations of inhalable air particulates (PM10). Concentrations of particulates were higher near the fire source and decreased with distance. PM10 particulates violated air quality guidelines by several orders of magnitude during the Gulf war and the Kuwait oil fires. Health statictics from the impacted areas were not available to directly correlate them with PM10 air loads. However, the literature indicated that enhanced concentrations of PM10 in Kuwait, Saudi Arabia and other countries in the region could cause adverse health effects. From the analogous studies in New York and London, Ozkaynak (1992) suggested a 10% increase in mortality within a one month period if PM10 levels increased from 50 to 250 μg/m³. However, no data were collected to show that excess mortality occurred due to the PM10 load caused by the Gulf crisis.

7.6.4. Effects of sulfur dioxide on human health

Sulfur dioxide (SO_2), being a very soluble gas, is absorbed by the nose and upper respiratory pathways while breathing. This, in turn, produces increased airway resistance in the lung. High doses of SO_2 (i.e., 2–5 ppm or 5000–13,000 μg/m³) may cause severe impairment of lung functions. In some cases, a healthy person without any previous history of asthma may develop a moderately to severe attack of the disease. Among the healthy people studied, about one in ten is susceptible to such an exaggerated reaction. The effect at low concentrations (i.e., 0.5 ppm or 1,300 μg/m³) includes distasteful odor and irritation leading to difficulty in breathing. A chronic cough and mucus secretion may result from the repeated exposures. Sulfur dioxide is one of the few air pollutants whose adverse human health effects have been documented in the general population. An example of SO_2 toxic effects are the epidemiological evidence related to the London fog incidence in 1952. Exposure to combined SO_2 and fog/smoke caused 4000 excess deaths in one week (Logan, 1952). Increased asthma attacks have been found to correlate with increasing SO_2 levels in Yokkaichi, Japan (Yoshida et al., 1966).

The air quality guidelines for ambient SO_2 concentrations are given in Table 7.1. The basis for the 24 hour air quality standard stems from epidemiological studies. The principal effects associated with 24-hour levels of SO_2 at or above 0.19 ppm include increased mortality rates and an increase in bronchitis. High levels of particulates are complimentary in these studies. A number of long-term epidemiological studies of chronic effects has found that populations living in areas with high particulate matter and SO_2 tend to have a higher prevalence of respiratory illnesses and retardation in lower lung function as compared with other groups living in areas with lower pollution levels.

Kuwaiti crude contain sulfur between 2.52% to 4.55% with an average of 3.61% (HPI, 1987). This sulfur would be converted to SO_2 if Kuwaiti crude is burned. It was calculated that total SO_2 produced by the Kuwait oil fires was 2.3 million metric tonnes. Contributions from the burning of diesel fuel during the Gulf war were found to be negligible when compared with the Kuwait oil fires. Therefore, only SO_2 produced by the Kuwait oil fires will be considered.

Many national and international organizations monitored concentrations of SO_2 in Kuwait, Saudi Arabia, Bahrain, and other countries in the Gulf region. Concentrations of SO_2 in the smoke plume approximately 2 to 5 km above ground level were generally below the air quality limits proposed by many countries (Daum et al 1992; Hobbs and Rydake, 1992; Gerges, 1992; Jenkins et al., 1992; Johnson et al., 1992). Ground level measurements by many scientists (Table 4.5, Chapter 4) reported SO_2 ambient concentrations that range between 1.3 and 5214 $\mu g/m^3$. From the available data, the EPC (1991) concluded that, in the residential areas of Kuwait, SO_2 levels were below air quality guidelines, especially when the wind direction was away from the oil fires. Relatively high concentrations of SO_2 were reported in and around the burning oil fields. For example, the USEPA (1991) detected the highest level of SO_2 (0.68 ppm) in the Burgan Oil Field and it was above the maximum 24 hour limit. Similarly, a Japanese team (Okita, 1991) found very high levels of SO_2 in one of the oil fields during an episode of smoke touch down between April 30 and May 8, 1991. Similar results were reported by a French team (Lameloise et al., 1991). These observations suggest that, although the majority of measurements were found below air quality limits, there may have been incidents when SO_2 concentrations were above air quality guidelines, especially in and around the fire sources. Considering the previously mentioned toxicity threshold, irritation leading to airway resistance, chronic coughs (if exposed to repeated high concentrations), and severe impairment of lung functions during the smoke touchdown episodes can be expected in populations living around the burning oil fields. All collected data in Saudi Arabia reported that SO_2 concentrations were below the air quality limit. Considering the Saudi monitoring results, it was concluded that levels of SO_2 in Iraq, Iran, and other Gulf countries were also below air quality guidelines. The low levels of SO_2 in most parts of Kuwait, Saudi Arabia, Bahrain, and other countries in the Gulf region should not pose any health complications that can be associated with this pollutant.

It was concluded in Chapter 3 that SO_2 generated in the oil fires was rapidly oxidized to sulfate and scavenged by soot particulates. The oxidation was so rapid that about 50% conversion took place within one hour of its generation (Hobbs and Radke, 1992; Johnson et al., 1992). The oxidation of SO_2 to sulfate shielded the Kuwait population from its toxic effects but generated particles which were enriched with calcium sulfate (Chapters 3 and 4). Exposure studies of healthy people and people with asthma, however, show no detectable effects at maximum concentrations of many sulfates such as ammonium sulfate and ferric sulfate. It is anticipated that sulfate produced due to oxidation will not be a health hazard to populations within 100 km of fire sources.

During oxidation, sulfuric acid aerosol may have been produced. A recent study has shown that SO_2 as sulfate does not necessarily equate with strong acid aerosols (Speizer, 1989). It was found that symptoms such as cough and sputum correlate better with acid aerosol than with SO_2. The USEPA is considering an acid aerosol as a criterion for this air pollutant (Ayers, 1991). The USEPA (1991) team determined ambient concentrations of sulfuric acid at several locations in Kuwait and two in Saudi Arabia. In Kuwait, the team found sulfuric acid concentrations ranged from below the detection limit to 31 ppb. Sulfuric acid concentrations in Dhahran and Riyadh, Saudi Arabia, were appreciablly lower than those found in Kuwait. The concentrations of acid found in this monitoring campaign were typical of urban concentrations in industrial countries. These observations suggest that concentrations of sulfuric acid aerosol were not high enough to be of a health hazard in Kuwait, Saudi Arabia, and probably in any other country in the region.

7.6.5. Effect of nitrogen oxides on human health

Of the several oxides of nitrogen present in ambient air, only two are known to effect human health. These are nitric oxide (NO) and nitrogen dioxide (NO_2). These oxides, if transferred across the lung-blood barrier, can produce inactive forms of hemoglobin. Eye and nasal irritation will develop after exposure to about 15 ppm nitrogen dioxide and pulmonary discomfort will commence after a brief exposure to 25 ppm. It is likely that pathological changes can be detected on the basis of exposure to 25–50 ppm for a short period. Exposure to 150–200 ppm may lead to the gradual development of fatal pulmonary fibrosis. The evidence for NO_x having a direct effect in inducing the narrowing of airways is conflicting, although the balance of reports suggest that NO_2 will cause broncho-constriction in exercising asthmatics (Ayres, 1991). In normal subjects, levels in excess of 2 ppm are needed to produce an effect (Koren, 1989). There is no evidence on the effect of long-term exposure to high ambient concentrations, although hospital admissions for asthma have been shown to relate to peak NO_x levels (Takafugi et al., 1987). Short-term peak exposure at levels above 1 ppm of NO_2 have been observed to produce effects of concern in healthy subjects. In some controlled human studies, significant pulmonary function changes were noticed in subjects exposed to 1–2 ppm for a short duration of 1 to 3 hours. Data suggest that asthmatics may be hypersensitive to exposure. However, lack of a concentration-response curve and the difficulty in replicating studies of acute pulmonary effects precludes definitive conclusions at this time. Chronic effects of exposures to NO_x are unknown and yet controversial. In brief, a decrease in the lung function at levels above 1 ppm in normal subjects, and increased bronchial reactivity in the range of 0.2 to 0.5 PPM in asthmatics are expected. Age has shown minimal effects on exposure responsiveness to NO_x.

Nitrogen, ranging from below the detection to 0.24 percent, is present in Kuwaiti crude oils (HPI, 1987). The oil fires were, therefore, expected to produce large amounts of NO_x (this refers to all types of nitrogen oxides including nitric acid). Assuming an emission rate of 0.05% for NO_x, a total emission of 80,000 metric tonnes of NO_x was calculated for the burning oil wells in Kuwait (Chapter 3). The other major source of NO_x was emissions from the heavy-duty diesel engines (bombers, fighters, tanks, and other armored vehicles). From the oil consumed during the Gulf war, the total emission of NO_x was 107,086 metric tonnes (Chapter 5). Nitrogen oxides may have been produced by explosives and munitions (Chapter 5). Most of the chemicals in high explosives contain nitrogen (Fordham, 1980). The total amount of munitions and explosives used in the Gulf war was about 120,000 metric tonnes (excluding those Iraqi stockpiles that were destroyed by Allied bombardment). No available information was found that could help assess the amount of NO_x generated from high explosives. It is suspected that this quantity might be substantial. Moreover, in the vicinity of the explosion, temperature and pressure are elevated significantly. We could not find any information that suggests that atmospheric nitrogen can be converted to NO_x during an explosion. In brief, it is concluded that during the war period, concentrations of NO_x were elevated. Understandably, no attempt was made to monitor NO_x concentrations in Kuwait territory during the Gulf war.

Concentrations of NO_x were monitored after the Gulf war both in the smoke plume (Daum et al. 1992; Hobbs and Rydake, 1992; Gerges, 1992; Jenkins et al., 1992; Johnson et al., 1992) and at ground level (Table 4.7, Chapter 4). Concentrations of a NO_x measured about 80 km from the fires were approximately 9 ppb on May 30, 1991 (Hobbs and Radke, 1992). This level was below the air quality guideline (Table 7.1). Hobbs and Radke (1992) estimated a NO_x emission rate of 0.05% of the fuel burned. Their observations also suggested a rapid depletion of NO_x in the plume at a rate of 60% per hour.

This means that elevated levels of NO_X could be expected in a small area around the fires. Ground level data on NO_X concentrations in Kuwait and Saudi Arabia ranged between 1 and 6977 $\mu g/m^3$, with an average of 17 to 51 $\mu g/m^3$ (Table 4.7, Chapter, 4). Except for a few observations in the burning oil fields (Lameloise et al., 1991; USEPA, 1991), all values of NO_X were within the air quality guidelines given in Table 7.1. Previously established data collected at Dhahran and Abqaiq by Saudi Aramco suggest that mean concentrations of NO_2 remained unaffected by the Kuwait oil fires (Al Sunaid (1992a, 1992b). Similarly, the EPC (1991) found that the yearly mean concentrations of NO_X at Mansoria in Kuwait was higher in 1990 as compared with the 1991 levels. Considering the observations in Chapter 4, it can be concluded that the Kuwait oil fires did not elevate the concentrations of NO_X in other countries including Iraq.

The above concentrations of NO_X were, in general, below air quality guidelines, except for a few observations in and around the burning oil fileds in Kuwait. Even the elevated concentrations reported in Chapter 4 were below the toxicity threshold discussed above. From the foregoing, it was concluded that concentrations of NO_X in all the Gulf region were low enough to pose no health hazard.

As stated above, a depletion of NO_X was noted in the immediate vicinity of the flames (Hobbs and Rydake, 1992; Jenkins et al., 1992; Johnson et al., 1992). Nitrogen oxides can form nitric acid when oxidized in air in the presence of water droplets. Therefore, concentrations of nitric acid in the ambient air were determined by a USEPA (1991) team during their visit to the area. Measurements were taken in Kuwait and Saudi Arabia. Nitric acid concentrations in the Burgan Oil Field ranged from below detection limit to 32 ppb. Ambient levels of nitric acid found by the USEPA (1991) were typical of any urban center. It is concluded that acidic aerosol produced from the Kuwait oil fires did not adversely effect human health in Kuwait or Saudi Arabia.

7.6.6. Effects of carbon monoxide on human health

Carbon monoxide is a non-irritating gas that is readily absorbed into the bloodstream from the lungs. Upon entering the bloodstream from the respiratory system, CO combines with blood hemoglobin to form carboxyhemoglobin (COHb). This reduces the capability of hemoglobin to carry oxygen to body tissues. It has been noted that a CO level of 50 ppm leads to an equilibrium value of 7% COHb and of 100 ppm to about 14% COHb. After exposure, CO is slowly released from the blood with a clearance half-life of 3 to 4 hours. At COHb levels of 2 to 5%, the effects are noticed in the central nervous system. At levels greater than 5 percent, there are changes in cardiac and pulmonary functions. The air quality guidelines (Table 7.1) are intended to keep this compound below 2.1% in the bloodstream.

Carbon monoxide is generated during the incomplete combustion of fossil fuels. Combustion of diesel fuel and the Kuwait oil fires were the major sources of CO production associated with the Gulf crisis. It was calculated that about 42,160 metric tonnes of CO would have been produced from the combustion of diesel fuel (Chapter 5). Carbon monoxide was found to be a minor component of total carbon emissions from the Kuwait oil fires. From the emission rate data, it was estimated that about 0.8 million metric tonnes of CO was generated by the Kuwait oil fires. Concentrations of CO measured at ground level in Kuwait and Saudi Arabia ranged from below the detection limit to 32,370 $\mu g/m^3$ and the mean concentrations of different sets of observations were reported between 21 to 1058 $\mu g/m^3$ (Table 4.4 Chapter 4). All the data collected on CO concentrations in Kuwait and Saudi Arabia were below the permissible limit set in the air quality guidelines given in Table 7.1. Local conditions, such as traffic load, seems to have more effect on CO levels than the Kuwait oil fires (El Desouky and Adbulraheem, 1991; EPC, 1991). Data of CO concentrations collected in other countries in the region are not available. However, it can be argued that if CO concentrations were found to be below air quality limits in Kuwait, certainly its concentration would have been lower elsewhere.

From the foregoing, it is concluded that concentrations of CO at ground level posed no health hazard in Kuwait and Saudi Arabia during the Gulf crisis. All measurements were found to be below the air quality guidelines and certainly below the toxicity limit discussed above. Considering the data collected on CO concentrations in Kuwait and Saudi Arabia, it can be argued that the Gulf crisis did not elevate CO concentrations to hazardous levels in any of the countries in the Gulf region including Iraq.

7.6.7. Health effects of ozone exposure

Ozone (O_3) is a respiratory tract irritant that reacts rapidly with the tissue and fluid lining in the airways the lungs. Human responsiveness to O_3 concentrations vary widely. Controlled human studies indicated that acute, reversible decrements in lung function such as coughing and shortness of breath occur in some individuals exposed to 0.12 to 0.16 ppm of O_3 for 1 to 3 hours while exercising. With prolonged exposure, similar effects can be seen at lower concentrations. O_3 is implicated in mucosal damage at levels as low as 80 ppb (Bates and Sizto, 1987). It can increase bronchial reactivity in normal subjects (Thurston et al., 1989). O_3 has been shown to be associated with increased hospital admission of asthmatics in Ontario but not in Vancouver (Speizer, 1989).

O_3 is a secondary photochemical pollutant and a powerful oxidant. The basic requirements for O_3 production are sunlight acting on NO_x in the presence of hydrocarbons. The ingredients for the production of O_3 existed during the Kuwait oil fires. Concentrations of O_3 were measured in the smoke plume and on the ground both in Kuwait and Saudi Arabia. All the data collected were below the air quality limit for this pollutant. No effect of the Kuwait oil fires could be attributed to O_3 concentrations in Kuwait. From the foregoing it is concluded that chronic health effects from O_3 exposure in humans is difficult to establish and is controversial. Concentrations of O_3 in Kuwait, Saudi Arabia, and probably in other countries in the region, were within the air quality limits suggested in Table 7.1. During the Gulf crisis, O_3 concentrations were not enhanced to levels that can be considered hazardous to health.

7.7. Effects of emissions from munition use on human health

It is estimated that approximately 120,000 tonnes of various types of munition were used by the Allied Forces in the Gulf war. These munitions varied widely in their chemical composition. Data on the type and chemical composition of various ammunition used in Kuwait and Iraq are a closely guarded military secret. Because of this limitation, the actual contamination and subsequent effects of the use of ammunition and explosives may never be known. Under the circumstances, only a qualitative assessment of the impact of emissions from ammunition use on human health should be expected in this section.

The most commonly used inorganic and organic chemicals in military ammunition are lead azide, lead styphnate, lead mononitroresorcinate, potassium dinitrobenzofurozan, brass, steel, barium styphnate, zirconium potassium perchlorate, mercury azide, potassium chlorate, mercury fulminate, nitroglycerin, nitrocellulose, dinitrotoluene, trimethylol ethane trinitrate, and triethylene glycol dinitrate. Some of the common high explosive chemicals are PETN (penthrite, pentaerythrite tetranitrate), TNT (2,4,6-trinitrotoluene), RDX (cyclotrinethylene trinitramine), hexanitrodiphenyl oxide, tetranitromethane, Tetryl (trinitrophenylmethylnitriamine), hexanitro diphenylaminoethylnitrate, and many others. In general, explosive formulations contain more than one of these chemicals in combination with other chemicals. Considering the multinational nature of the coalition that ousted Iraq from Kuwait, it is anticipated that many types of high explosives would have been used during the Gulf war. In addition to the above chemicals, there are many other organic and inorganic compounds that are used in different parts of the explosive device. A detailed discussion of all of these chemicals is beyond the scope of this book.

Health effects of any given pollutant is a direct function of its concentration and duration of time to which a subject is exposed. With reference to military operations during the Gulf war, data on the concentrations of pollutants and duration of exposure are not available. Therefore, the effects of pollutants on humans can only be discussed in general terms. Metals are used extensively in many different military devices. A portion of the metals in ammunition and explosives may become airborne on impact. Once airborne, metals can be inhaled and transported to the blood stream. Data on the concentrations of metals in air particulates during the War were not collected. However, in a preliminary study, Sadiq et al. (1992a) reported that concentrations of many metals decreased with an increase in distance from the Kuwait border. This investigation suggested indirectly that the Gulf war probably produced large amounts of toxic metals. No direct data on metal concentrations in air particulates were collected during the Gulf war period. Therefore, health effects of metal exposure, in reality, cannot be assessed. Information on the toxicological effects of some of these metals is listed in another section.

Organic chemicals are another class of pollutant emitted due to the use of explosives and ammunition. High explosives for military applications are generally unstable organics. There are many hundreds of such chemicals designed for a specific application or need. As already mentioned, data on the type and amount of explosives used in the Gulf war are classified. Therefore, assessment of their effects on human health is very speculative. This situation is further complicated by the fact that organic chemicals used in explosives and ammunition are unstable. As the ammunition explodes, high temperatures and pressures are generated. Since the chemicals in explosives are unstable, these undergo chemical changes during the explosion and new chemicals are produced. It is beyond the scope of this book to discuss the chemistry of these explosives. However, byproducts of high explosive chemicals may be more dangerous to humans than the original chemicals.

From the foregoing, it is anticipated that a multitude of organic chemicals would have been produced during the military operation in Kuwait and Iraq. The exact nature and quantities of organics emitted, and their exposure effects on human health will probably never be known. One can simply conclude that a large but unknown amount of both unstable and stable chemicals were generated because of the use of explosives and ammunition. These chemicals would have affected air quality adversely but how much and for how long is unknown. In fact, the effect of ammunition and explosives used in the Gulf war may never be assessed quantitatively. All evidence, suggests that the local population and combat personnel in the war zone of Kuwait and Iraq were exposed to a number of organic and inorganic chemicals, and many of these could induce adverse health effects.

It is not possible to speculate on the type and possible degree of exposure to organic chemicals generated by the explosives. However, toxicity information on some of the high explosives is summarized below. TNT is the most commonly used explosive in military and other applications. TNT, if inhaled, may irritate mucous membranes and cause sneezing, coughing, and a sore throat. TNT may produce toxic hepatitis and aplastic anemia. This compound yellows the skin, hair, and nails on contact. Dermatitis, erythema, papules, and itchy eczema can be severe. Ingestion of 1–2 grams of TNT is considered to be an acute fatal dose of TNT (Buck and Wilson, 1975). The suggested maximum airborne dust concentration is 0.5 mg/m^3(Federal Register, 1976). Another high explosive with military application is RDX. Workers who inhaled RDX for several months have become unconscious and suffered loss of reflexes. The suggested maximum permissible airborne concentration of RDX is 1.5 mg/m^3 (Federal Register, 1976). Cyclotol is generally used as an explosive filling in military applications. Its toxicity is similar to that of TNT and RDX. Toxicity data on DATB (a high explosive used in military applications) are not available. Nitroguanidine (NQ) is commonly used in gun powder but data on its toxicity are not available. PETN is another high explosive of interest and was found to be slightly toxic. The recommended maximum atmospheric concentration for an 8 hour period is 15 mg/m^3(Federal Register, 1976). Tetryl yellows human skin and sometimes causes dermatitis. Repeated exposures may induce irritation of eyes and nose which can lead to excessive sneezing and nose bleeding. The suggestive maximum concentration of Tetryl dust is 1.5 mg/m^3.

In summary, the effect of emissions from the use of ammunition and explosives in the Gulf war can probably never be assessed quantitatively. Broadly, it is concluded that the local population and combat personnel in the war zone of Kuwait and Iraq would probably have been exposed to a number of organic and inorganic chemicals and, all evidence suggests that many of them could induce adverse health effects. The United States military may have collected related information but such data would be classified.

7.8. Airborne organics and human health

Organic pollutants such as benzo(a)pyrene and benzene are carcinogenic (Santodonato et al., 1979; Albert, 1978), and hexachlorocyclopentadiene and carbontetrachloride can produce liver damage, narcotic effects, headache, nausea, dizziness, stupor, and loss of consciousness (Sax and Lewis, 1973). With reference to the Gulf crisis, there are several sources of organic pollutants. The most important of which were emissions from the Kuwait oil fires, evaporation from oil lakes, trenches filled with oil, the oil spill, combustion products from military hardware, and fallout from explosives. Broadly speaking, the organics in the oil fire emissions were considered to be polycyclic aromatic hydrocarbons (PAH), aliphatic hydrocarbons, and volatile organics.

Many volatile organic chemicals, if present at high concentrations in the ambient atmosphere and inhaled for longer periods, can adversely effect human health. The USEPA (1991) determined volatile organics at many locations in Kuwait and Saudi Arabia from March 12 to April 4, 1991. A French air monitoring team (Lameloise et al. 1991) also determined volatile organics in Kuwait. The mean concentrations of volatile hydrocarbons were typical of an industrial/urban situation. However, the maximum concentrations were generally found to be higher in and around the burning Oil Fields (Burgan and Ahmadi Oil Fields). According to the French team, long-term adverse effects from the exposure to the volatile organic chemicals in the Gulf region should be very minimal, if any. Aliphatic hydrocarbons are largely non-toxic. Fayad et al. (1993) reported that PM10 samples collected at Dhahran were contaminated with aliphatic hydrocarbons.

Several PAH compounds have both mutagenic and carcinogenic properties. The proposed allowable limit for benzo(a)pyrene in air was set by OSHA at 0.2

$\mu g/m^3$. Recently the proposed total PAHs criterium for human health protection was set at a value of 0.0109 $\mu g/m^3$ in air. However, this criterium is still controversial (Eisler, 1987). Concentrations of PAH in ambient air particulates were determined in PM10 samples collected from Kuwait, Bahrain, and Saudi Arabia by several investigators and these are summarized in Table 4.14 (Chapter 4). Concentrations of total PAHs in PM10 air particulates, as reported in Table 4.14, ranged from 7 to 387 ng/m^3 and that of benzo(a)pyrene in air varied between 0.1 to 31 ng/m^3. Concentrations of benzo(a)pyrene and other PAH compounds were lower than the proposed criteria for PAHs set by international organizations (Eisler, 1978). Moreover, PAH concentrations found in the Gulf region were comparable to those found in urban areas in several parts of the world. It was anticipated that acute health hazards from organic chemicals in air would be minimal during the Gulf crisis. However, chronic health effects could not be determined with the limited, highly variable, and questionable data on the concentrations of organic pollutants in ambient air.

7.9. Airborne metals and human health

Crude oils contain different metals, particularly nickel and vanadium. From the available information, total metal emissions in tonnes from the Kuwait oil fires were: vanadium 7,224, nickel 2,084, aluminum 917, iron 747, zinc 432, molybdenum 125, lead 114, and copper 60. In addition to the Kuwait fires, metals, especially lead, would have been generated by burning diesel fuel during the Gulf crisis. The emissions from the ammunition and explosives use during the war are expected to enhance ambient metal concentrations, although their quantities are unknown (Sadiq, et al. 1992). It can be seen from the above discussion that the Gulf crisis has the potential to greatly enhance metal concentrations in the Gulf region.

Metals can enter the human body via different routes, of which air, food, and drinking water are the most important. Drinking water requirements in Kuwait, Saudi Arabia, and other GCC countries are largely met with desalinated water. The Gulf war probably had no effect on this source. However, river water may constitute a major source of drinking water in Iraq and thus could be contaminated from dust fallout. Because of the censorship and other hardships, no data related to Iraq could be obtained. Desalination plants in Kuwait were vandalized during the war but have been repaired. Metal introduction to humans via a food route requires more time. Metals may pass the lung/bloodstream barrier more rapidly if inhaled. Most of the metals that were generated during the Gulf crisis would have been airborne. Therefore, air particulates, especially PM10, if contaminated could elevate metal concentrations in the human body. Because of the probability of contamination, our attention will largely be focused on metals in air particulates.

As already stated, no data could be directly related to metal contamination potentials of air and ground operations of the Gulf war. However, Sadiq et al. (1992a) have published some work that indirectly indicates that large amounts of metals were released during the war. Many investigators determined metal concentrations in air particulates collected during the Kuwait oil fires. Data reported by Sadiq and Mian (1993a, 1993b, 1993c, 1993d) suggest that the Kuwait oil fires enhanced metal concentrations in air particulates. Similar observations were noted by Mitchell and Sunaid (1992). These data were discussed in Chapter 4. Metal concentration ranges extracted from these results are given in Table 7.2.

Vanadium, if occurring at high concentrations in air particulates (as vanadium pentaoxide) could produce acute irritation of the eyes and respiratory tract. Exposure to vanadium particles in excess of 0.5 mg/m^3 of a respirable size has caused conjunctivitis, nasopharyngitis, and signs of pulmonary irritation and bronchospasm, lasting in some cases for several days. There is some evidence to suggest that asthma may be an effect of chronic vanadium exposure (Sharratt and Butler, 1991; Read, 1992). Longterm exposure to vanadium concentrations higher than 1 $\mu g/m^3$ air may induce adverse health effects (WHO Air Quality Guidelines as cited by Lameloise et al., 1991). Concentrations of vanadium were generally found below this threshold (Table 7.2). It is concluded that vanadium in air particulates was not a health concern.

The health effects of airborne copper in humans results primarily from industrial exposure and are manifested predominantly by dermatological and respiratory symptoms (USEPA, 1987; Williams, 1982). Copper intoxication has been reported in persons exposed to 0.1 mg $copper/m^3$ (USEPA, 1987). In general, about 1% of the total daily intake of copper may come from direct inhalation. The health effects of airborne copper in humans result primarily

Table 7.2. Metal concentrations of inhalable air particulates (PM10).

Metals	Range (ng/m^3air)*
Arsenic	0.1–15.4
Barium	8.2–598
Cadmium	1.0–167
Cobalt	1.7–30.1
Chromium	3.0–1155
Copper	2.7–594
Manganese	7.6–630
Molybdenum	2.1–32.4
Nickel	1.5–1300
Lead1	3.9–1222
Titanium	7.0–1159
Vanadium	1.8–1165
Zinc	3.0–2020

* Data extracted from Table 4.12 Chapter 4.

from industrial exposure and are manifested predominantly by dermatological and respiratory symptoms (USEPA, 1987; Williams, 1982). Copper concentrations in air particulates in the war area were found to be below the limit for copper intoxication (Table 7.2). It is concluded that copper concentrations in air particulates were not a health hazard to the population in Kuwait, Saudi Arabia, Iraq, and other countries in the region.

The WHO air quality guidelines (as cited by Lameloise et al., 1991) suggest that exposure to cadmium in air should be kept below 5 ng/m^3 in rural areas and 20 ng/m^3 in urban and industrial areas. Adverse health effects may be induced in populations exposed to higher cadmium levels for over a year. As shown above, concentrations of cadmium in the inhalable air particulates were enhanced by the Gulf crisis. The data in Table 7.2 suggest that populations in the Gulf region may have been exposed to relatively high cadmium concentrations, at least for the better part of a year. Contemporaneous data on metal concentrations in the human body were not collected, and therefore, the exact extent of human exposure to cadmium in this area cannot be assessed. If high concentrations in air particulates persisted for years, cadmium may pose a health problem. It is anticipated that with the end of the Gulf war and dousing the oil fires in Kuwait, concentrations of cadmium will decrease.

Chromium toxicity is dependent on its chemical form, e.g., chromium(VI) is considered to be a genotoxic carcinogen and chromium(III) is not (Slooff et al. 1990). Chromium(III) at a concentration of 5 ng/m^3 constitutes only a negligible portion of total daily intake and should not pose any health effects. Using the procedure proposed by WHO (1987) for estimating carcinogenic potency, Slooff et al. (1990) estimated a risk factor of 10 – 6 (one extra case of lung cancer per million population) for life-long exposure by inhalation of 0.025 ng chromium(VI)/m^3 air. The permissible limit for chromium in the Netherlands is 2.5 ng chromium(VI)/m^3 air. The available data on ambient concentrations of chromium in the Gulf region is total chromium. Information on chromium speciation in air is not available. Slooff et al. (1990) estimated the fraction of chromium(VI) in air ranging between 0.01 to 30 percent. It is assumed that about 30% of the total chromium measured in the Gulf region was chromium(VI) produced by oxidation during the burning process. An annual mean of total chromium of 35 $\pm$ 27 ng/m^3 was found in the inhalable particulates collected in 1991 in Dhahran, Saudi Arabia (Sadiq and Mian, 1993a). From the above information, a yearly mean concentration of chromium(VI) as 10 $\pm$ 8 ng/m^3 Dhahran was calculated. This concentration is about four times higher than the limit proposed in the Netherlands. High concentrations of chromium in PM10s could adversely effect human health in the Gulf region. However, the exact impact of human exposure to high concentrations of ambient chromium is difficult to assess as the daily total intake and contribution of ambient chromium to this total are unknown.

The toxicity of nickel has been reviewed by Fairhurst and Illing (1987). Repeated exposure to nickel oxide above 4 g/m^3 produces pulmonary inflammation and fibrosis. In general, the permissible level of exposure to insoluble nickel compounds is set at 50 $\mu g/m^3$. Nickel concentrations in the Gulf region ranged from 1.5 to 1300 ng/m^3 (Table 7.2), i.e. several times lower than the toxic limit. It is concluded that nickel concentrations in the air particulates in the Gulf region do not pose a health hazard.

Lead is a toxic metal to which a population can be exposed through air. Analysis by USEPA of the relationship of blood lead levels and the amount of lead inhaled shows that the slope of the blood lead and air lead relationship is approximately linear in the range of normal ambient exposures (0.1 to 2 μg lead/m^3, quarterly average) among individuals whose blood lead levels do not exceed 30 μg lead/dl (10 ml or cc). The median slope of the blood lead and air lead relationship in three key studies on children was 1.92 $\mu g/dl$ per $\mu g/m^3$ (i.e., for each increment of 1 $\mu g/m^3$ in air lead, there will be an increase of 1.92 $\mu g/dl$ in blood lead).

Readily apparent disturbances of important physiological functions and organ systems occur only at very high lead exposure and blood lead levels. The principal symptoms to lead exposure: impaired mental and physical development and cognitive function of infants and children, impairment in heme synthesis and activity of blood enzymes in all age groups and races, and increase in both diastolic and systolic blood pressure (especially in middle aged subjects). Chronic renal nephropathy (i.e., a form of kidney disease) especially in adults with high occupational lead exposures, has also been reported. Reproductive effects manifesting in premature delivery and possibly other complications in pregnant women and altered spermatogenesis in men were also noted. Impairment in the liver, gastrointestinal tract, and endocrine systems were noted at high lead exposures. Fetus, infants, and children appear to be more vulnerable and sensitive than adults to lead effects. The risk of human exposure, especially infants and children, to high ambient lead concentrations is well documented (Brunekreef, 1985, Scroeder et al., 1987, 1988). Adverse health effects may be apparent in populations exposed to >500 ng/m^3 in air for a year or more (WHO as cited by Lameloise et al., 1991). The safe level of blood lead recommended by NAS (1980) is below 150 ng/m^3 in air. A yearly mean of 282 $\pm$ 144 ng of lead/m^3 air was calculated from the data for inhalable particulates collected in Dhahran, Saudi Arabia (Sadiq and Mian, 1993b). The mean value is almost double than the limit recommended by NAS (1980) for the United States. However, the observed lead concentrations in the Gulf region were below the maximum limit of quarterly lead concentrations of 1500 ng/m^3 as proposed by the USEPA (National Ambient Air Quality Standards). Lead poisoning is a chronic problem and automobile emissions may constitute a siginicant source of lead in air particulates in the Gulf region. Further monitoring of children, especially in the cities, for lead contamination is recommended.

Air quality guidelines for other metals were not available, and therefore, the health implications of the metal concentrations in air particulates cannot be discussed further. Data collected in the Gulf region were compared with metal concentrations found at other locations as discussed in Chapter 4. Relatively higher concentrations of nickel, vanadium, and chromium were found in the inhalable particulates from the Gulf region. Considering the level of industrialization in this region, concentrations of metals in air particulates seems high. Data on metal concentrations in humans exposed to the Gulf crisis were not available on a large scale. In a preliminary study, Sadiq et al. (1992) analyzed over 100 random hair samples collected from five different Saudi cities involved in some way in the Gulf crisis. Higher mean metal concentrations were found hair collected in cities that were closer to the war zone as compared with a city (Jeddah) that was farthest from the Gulf crisis.

From the foregoing, it is concluded that concentrations of lead, cadmium, and chromium were higher in inhalable particulates than the limits suggested in the literature. Concentrations of nickel, vanadium, and copper were found to be within safe limits. Comparing metal concentrations with worldwide data, it was found that nickel, vanadium, and chromium were higher in the air particulates of the Gulf region. It is suspected that some metals may accumulate in humans and induce adverse chronic health effects in certain supersensitive subjects. Therefore, continuous monitoring of both air and humans for metal concentrations is recommended.

7.10. Acute health effects of the Kuwait oil fires

Health effects due to the exposure to a chemical or a mixture of chemicals may be immediate or manifest itself after some period. The initial reaction is referred to as an acute effect whereas those requiring a longer period before reaction is called a chronic effect. Acute effects, being immediate and direct, are easy to recognize. Chronic effects may show after months or years. Due to delayed response and the probability of additional exposure or interaction with other chemicals, chronic effects are very difficult to recognize. Chronic effects represent a response to a collective impact of more than one factor. Because of the very nature of chronic effects, these are generally expressed as speculative. In comparison, acute effects are discrete and discussed in quantitative terms. The impact of gaseous pollutants would be acute and those of chemicals in the inhalable particulates as chronic.

Many people are concerned about the health implications of the Kuwait fires. As discussed above, all gaseous pollutants were found to be within the air quality guidelines, except in areas in or around the burning oil fields. Concentrations of inhalable particulates were found to be very high. It is suspected that these particles might contain substantial amounts of toxic chemicals. In response to the above concerns, many organizations collected health statistics which are discussed below.

Ahmadi was probably the first town to be engulfed by the smoke from the burning oil wells. The population, about 10,000 individuals at that time, was shocked by the closeness and intensity of the fires. The oil mist from the fires blackened the thoubs (garments) of those outside in only a few minutes.

Medical authorities advised the town people, especially children and elderly people, to stay inside and to reduce outdoor activities if possible. The medical authorities planned to monitor the environment and collect health-related statistics. However, this plan was severely hampered by the lack of staff, equipment, and a power supply (Daoud, 1991). Peak flow measurements are the simplest procedure to test for lung function. A preliminary evaluation of 120 individuals in Ahmadi during the initial stages of the oil fires did not show any effect on the peak flow measurements (Dauod, 1991). Published data on respiratory cases, as a percentage of total cases in Al Ahmadi City, were provided by Dauod (1991) and are plotted in Figure 7.2. It is evident that smoke and other pollutants from the Kuwait oil fires have increased the incidence of respiratory illness. Statistical analyses of these data suggest an insignificant effect on patients suffering from respiratory illness (Dauod, 1991). One of the authors (of this book) has some personal experience in using a peak flow meter. It requires training to master the technique and generate reproducible measurements. To our understanding, the reported data by Dauod (1991) were collected in an emergency and had several technical limitations. Therefore, no conclusion should be based on these data.

The EPC (1991) discusses several small studies in Kuwait. In a study, recent health statistics (March 1 to September 25, 1991) in Adnan Hospital were compared with those collected in 1986. Although the comparison is questionable, the EPC (1991) reported that asthma and chest complaints increased in 1991 as compared to 1986. Similarly, complaints related to the gastro-intestinal tract increased and probably reflected a deterioration in drinking water quality since the desalination plants in Kuwait were vandalized. In another investigation, the data collected in two clinics, one in the vicinity of the Burgan Oil Field and the other far away from the oil fires, were compared. The EPC (1991) found higher chest complaints in the areas closer to the fires. At the clinic removed from the fire, the study noted higher chest complaints only on days when the smoke was very intense and pollution levels were higher. This trend was not noted in areas closer to the Burgan Oil Field. In another survey, a total of a 195 asthmatic subjects were investigated. They were asked to fill out a questionnaire about their exposure and their lung ability was tested by peak flow measurements. The data collected from 115 patients were correlated with air quality parameters. A significant positive correlation between aggravation of asthmatic symptoms and ambient concentrations of NOx was found. The PM10 concentrations were positively correlated to asthmatic symptoms but the correlation was not significant.

There are also a few unpublished medical reports from Saudi Arabia that suggest no significant acute effects due to the Kuwait oil fires emissions. The Ministry of Health of Saudi Arabia prepared a report on the health statistics in Al Khafji (the northern most town of Saudi Arabia), covering from April to September 1991 (KFUPM/RI, 1992). The data reported suggest the highest rate of infection of the upper respiratory tract in April 1991 that decreased gradually to September 1991. However, in September 1991 the rate of the respiratory

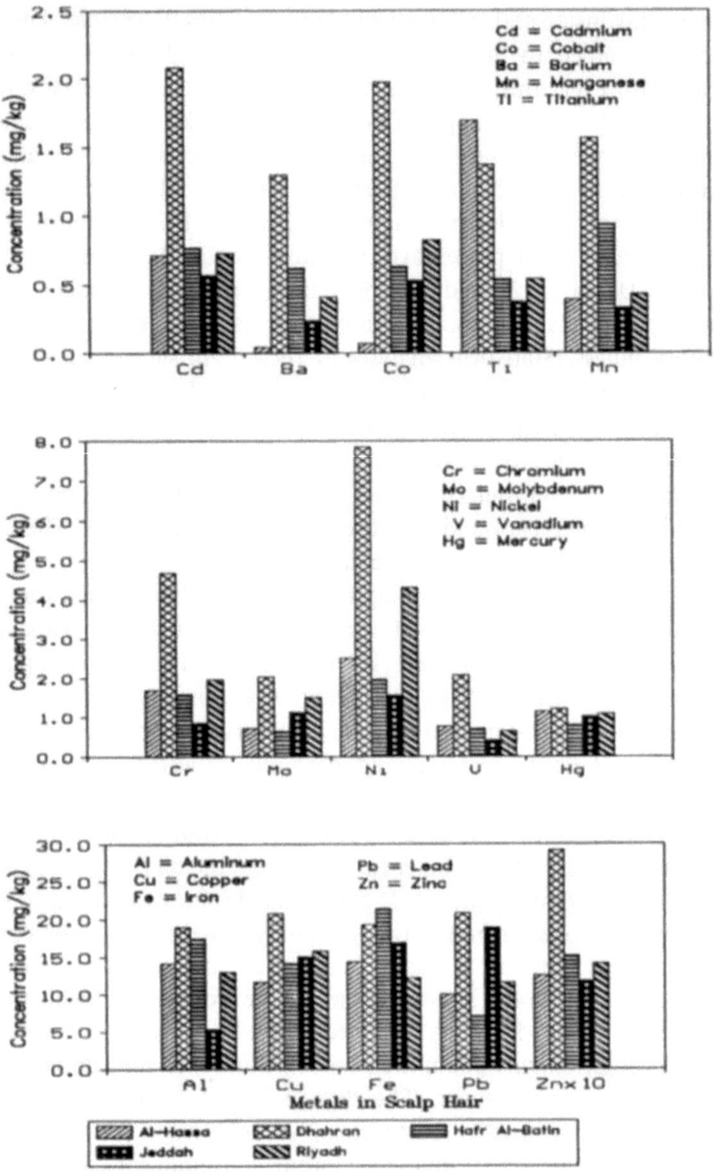

Figure 7.2. Metal concentrations in scalp hair collected during the Gulf crisis from different Saudi cities.

infection was higher than calculated from historical data. In general, the decrease in complaints in Al Khafji coincided with success in extinguishing the Kuwait oil fires. Unlike respiratory complaints, the eye infection cases in Al Khafji were found to increase from April to September 1991. The increase in eye infections may be related to the dust storms. Information from other countries in the region was not available. In summary, the investigations may be questionable as to their scientific and statistical validity, however, they indicate that respiratory and cardiovascular diseases were enhanced and that response was dependent on distance from the burning oil fields.

7.11. Chronic health effects of the Kuwait oil fires

As mentioned above chronic effects are delayed manifestations of an exposure that may manifest itself several months or years later. Under the given circumstances, it is very difficult to predict the chronic effects of the Gulf crisis. However, preliminary results of a few investigations are discussed here.

The Kuwait oil fires generated a number of organic chemicals. These chemicals may cross the lung/bloodstream barrier if inhaled. Blood samples from 14 volunteers, mostly American troops serving in Kuwait, were analyzed for concentrations of a number of volatile organic compounds. The median level of these compounds were found to be in the lower range of concentrations found in the general American population. This suggests that serving in Kuwait has not enhanced the concentrations of volatile compounds in the test subjects (Etzel, 1991). For example, the study found a median value of 0.035 ppb of benzene in blood samples from Kuwait as compared with a median value of 0.064 ppb in the United States. In a similar investigation, blood or urine samples were taken from 32 persons working for Siemens in Kuwait. Concentrations of volatile organic compounds were determined in these samples. The concentrations of the tested organic compounds were found to be comparable with those found in urban centers in industrialized countries (Stockmann, 1991). These investigations suggest that the tested organics were not accumulated in subjects that were residing in Kuwait during the War. These results are not surprising. The USEPA (1991) and Lameloise et al. (1991) determined volatile concentrations of organic compounds in the air in Kuwait and Saudi Arabia. They found low concentrations of these chemicals. If ambient concentrations are low, concentrations of these chemicals would correspondingly be lower in blood samples. Concentrations of other high molecular weight organic chemicals, such as PAHs, should be monitored in Kuwait and Saudi subjects who lived in the area throughout the crisis. Increases in the concentrations of PAHs and aliphatic hydrocarbons in Kuwait and Saudi Arabia have been reported (Fayad et al., 1993; EPC, 1991).

Concentrations of inhalable particulates were found to be above air quality guidelines. High concentration episodes of inhalable particulates in urban areas have been associated with excess mortality (mortality that would not occur if inhalable particulates were not elevated). Ozkaynak (1992) addressed the potential for excess mortality associated with short-term exposures to particles emitted during the oil well fires. From his calculations and reasoning, Ozkaynak (1992) estimated a 10% increase in mortality in a one month period if PM10 levels increased from 50 to 250 $\mu g/m^3$. The air particulates data

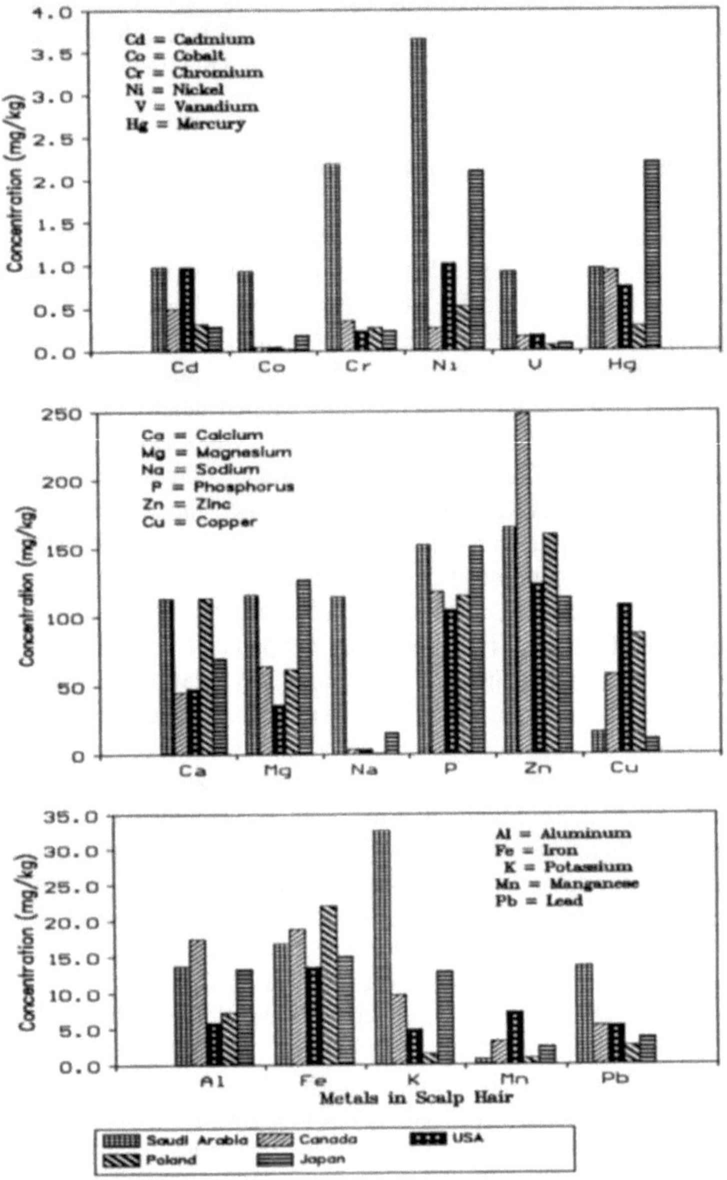

Figure 7.3. A worldwide comparison of metal concentrations in hair samples from Saudi Arabia collected during the Gulf crisis.

collected in Kuwait and Saudi Arabia were higher than 250 $\mu g/m^3$ (Table 7.1). It is anticipated that the adverse effects of enhanced concentrations of PM10 would not be as bad as projected by Ozkaynak (1992). It was found that the dust collected in the city of Al Ahmadi during April, 1991 did cause hemolysis of human erythrocytes in experimental test systems. Later, a comparison showed

Figure 7.4. Map showing locations of health monitoring stations in Kuwait

that the toxicity of freshly deposited dust (dust deposited in September–October 1991) was approximately one-sixth that of the old deposited dust (dust deposited in April 1991 – Beg, 1992).

Metal concentrations in the air particulates were found to be enhanced due to the Kuwait oil fires. Once airborne, the metal can be absorbed into the blood stream via the respiratory tract. Concentrations of a few toxic metals in air particulates were found to be the highest in the world. Realizing the metal contamination potential, Sadiq et al. (1992) collected hair samples from subjects in five different cities variously exposed to the Gulf war and the Kuwait oil fires. It was found that the mean concentration of aluminum, barium, cadmium, cobalt, chromium, copper, nickel, lead, vanadium, titanium, copper, and molybdenum were higher in the hair samples from Dhahran (center of the Gulf crisis in Saudi Arabia) than any other city investigated in this study. Hair samples from Jeddah (not exposed to the Gulf crisis) were found to contain the lowest mean concentration of the above metals (Figure 7.3). The results of this study were compared with those from other countries and are shown in Figure 7.4. Concentrations of lead, cobalt, chromium, nickel, vanadium, and potassium are higher in Saudi hair as compared with other countries. It has been suggested that elemental data from scalp hair can be a first-step approximation of metal contamination in humans (IAEA, 1978). The data presented in Figures 7.3 and 7.4 and by Sadiq et al. (1992) indicated that metal concentrations in the Gulf area may be elevated due to the Gulf war. Long-term monitoring of metal concentrations in humans should be considered in Kuwait, Saudi Arabia, and Iraq.

The above studies suggest that there might be some chronic effects of the Gulf war due to relatively higher ambient concentrations of organic chemicals, toxic metals, and inhalable particulates. The most probable indications may be higher blood pressure, increased respiratory tract complaints, enhanced incidence of cancer, and cardiovascular disorders.

7.12. Health impact of Iraqi chemical weapon

Health implications of chemical weapon arsenals are no longer a secret. Chemical weapons are primarily designed to inflict acute and chronic suffering to humans. Luckily no chemical weapons were used by Iraq in the Gulf war. However, the Allied air force reported destroying several facilities allegedly containing toxic chemicals. No information could be collected regarding the extent of damage to the chemical weapon plants and stockpiles from the Allied bombardment. It is also not known if toxic chemicals were released as a result of the bombardment. If released, it is not known whether the local population was contaminated and if so to what extent. Considering the extent of Allied bombing, it is suspected that damage to chemical weapon facilities and stockpiles may have been extensive. Early reports, referring to United Nations' experts, mentioned that many components of the Iraqi chemical arsenal were leaking or very poorly maintained. It is not clear that this obviously dangerous condition was due to damage sustained by the Allied bombardment or lack of maintenance and concern by Iraqi authorities.

It was reported on September 21, 1992 that the United Nations' experts had destroyed about 800 tonnes of chemical munitions and blown up leaking rockets containing 2.5 tonnes of nerve agent (Read, 1992). United Nations experts involved in blowing up Iraq's stockpile of hazardous nerve gas rockets at the main chemical weapon plant of Muthana reported in September, 1992 that the area was contaminated (Saudi Gazette, 1992). The extent of contamination was not discussed. Although the exact extent of Iraq's chemical arsenal is unknown, it has been estimated that Iraq possessed approximately 45,000 pieces of chemical ordinance (Saudi Gazette, 1992).

7.13. Health implications of UN embargo

Almost two years after the Gulf war, assessing human and environmental costs of the UN embargo on Iraq with any precision is exceedingly difficult if not impossible. We have inadequate data, or rather no data, on the impact of the embargo on the Iraqi population, especially on the state of their health. The UN embargo was initially imposed to persuade Iraqi authorities to end the occupation of Kuwait peacefully. Later it was tightened to a total embargo with almost no imports and exports with the outside world. Iraq has been isolated now for over two years.

Iraq is an oil producing country with substantial agriculture (the impact of embargo on agriculture has been disscussed above). However, its industrial base was not well developed. Having been war with Iran for almost a decade, whatever industrial development was achieved during that decade, was largely dedicated to the military. The civilian sector was severely neglected. During the initial stages of the embargo, items related to the military services were stockpiled. Naturally, medicine was included in the stockpile. As a result, civilian medical services were severely curtailed by the UN embargo. Innocent infants and children were frequent victims of the embargo. After the Gulf war, medical teams from the World Health Organization, the United Nation Arab American Medical Association Children's Fund, Harvard University, International Physicians for the Prevention of Nuclear War, Medicine for Peace, and many others visited Iraq and came to the unanimous conclusion that malnutrition and chronic diarrhea are epidemic among children, and that infectious diseases such as typhoid, cholera, hepatitis, tetanus and meningitis outbreaks are worsening. Medicine, medical equipment, and infant formula are in short supply (Medical News, 1991). The situation in Iraqi hospitals were described as desperate. Many teams strongly pleaded that, if the situation worsened further or remained unchanged, many people, especially infants and children, would soon die. Nothing substantial has been done since then to improve the plight of the Iraqi people. Considering the experts' reports cited above, one can imagine that by now many thousands of childern and infants have died that should have been healthy and alive otherwise.

When supply runs short, the privileged class really are previliged. In Iraq, armed forces are the privileged. Priority would therefore have been given to the treatment of Iraqi military personnel during and after the Gulf war. The civilian population, especially the destitute, were probably left unattended to suffer the dire consequences of the embargo. No one knows how many thousands of people may have died because of the lack of medical supplies. No one can imagine how many infants would have been living and playing today if there was no occupation of Kuwait which ultimately lead to the embargo. There were reports suggesting that infant mortality has increased by over 20% because of the UN embargo (Medical News, 1991).

Sanitary services in Iraq were badly damaged by the Allied bombardment. The UN embargo made the repair and replacement of the damaged marchinary even harder, if not impossible. The population of Baghdad had to drink unsafe water. Outbreaks of contagious diseases would most likely have occurred. Reports from many international medical organizations suggest widespread complaints of chronic diarrhea among children, typhoid, cholera, hepatitis, tetanus and meningitis (Medical News, 1991). It is ironic that these excess mortalities would be associated with the UN embargo on Iraq.

As stated in Chapter 6, agricultural productivity would have been adversely effected by damage to irrigation systems and the scarcity of pesticides, fertilizers, and spare parts. The UN embargo has partially aggrevated this situation. Although there is no report of famine in Iraq, malnutrition of large masses, especially children, have been reported by many international medical teams (Medical News, 1991).

In summary, the UN embargo adversely effected Iraq. It was of no surprise to learn that the number of human causaulities as a result of the UN embargo totalled more than the number killed during the 43 days of war. The people most affected by the UN embargo are infants, children, and the destitute. It is deplorable that instead of punishing those who were responsible for the occupation of Kuwait, the world has unintentionally chosen to victimize innocent Iraqi infants and children. Unfortunately, the innocent infants and children of Iraq are still being sacrificed on the altar of world peace.

7.14. Human cost of the Gulf crisis

Wars always extract a high toll on human health and lives. The Gulf crisis started with the invasion of Kuwait by Iraq on August 2, 1991, and ended with the extinguishing of the last Kuwait oil fire on November 6, 1992, by the Emir of Kuwait (however, for the Iraqi people the crisis continues with no end in sight). The punitive measures against Iraq, such as the UN embargo, are still continuing. In comparison with the past conflicts, the Gulf war was the most intensive and devastating. The bombardment was unprecedented in the history of mankind. In just 43 days, over 84,000 tonnes of bombs were dropped over an area of about 4,000 square miles. It is estimated that about 120,000 tonnes of ammunition were used in the War by the Allied Forces alone. The air and ground offensives were so large that they consumed approximately eight billion gallons of diesel fuel. The Gulf war is also distinguished for unprecedented intentional destruction of the environment. As part of the scorched earth policy, Iraq damaged over 800 oil wells and over 600 of these wells caught fire. The Kuwait oil fires emitted about 1.12 billion barrells of crude oil. In addition to the fires, emission fallouts from ammunition and burning of diesel fuel contaminated the atmosphere. There are indications that the Gulf war, like its ferocity, would surpass its predecessors in environmental destruction and human cost.

Allied Forces were expecting close to 40,000 casualties in the Gulf war. The United states alone was expecting about 30,000 to 40,000 casualties including approximately 15,000 killed in action. To provide necessary emergency medical services to the wounded persons, the United States deployed 23,482 medical personnel to the operation (Medical News, 1991). Fortunately, the above estimates were wrong and Allied Forces suffered casualties of only 149 dead and another 513 wounded (Church, 1991; Khuraibet, 1991). Approximately 20 people died, all Americans, and about 100 wounded in Dhahran, Saudi Arabia as a result of the Iraqi scud attacks.

There is still silence regarding the number of Iraqi deaths. There may never be accurate figures of Iraqi casualties but only very crude estimates. Circumstantial evidence suggests that the number of deaths was large - in the hundreds of thousands. The United States Defense Intelligence Agency has estimated that as many as 100,000 Iraqi military personnel died during the Gulf war (The Middle East, 1992). This estimate specifically pertains to the military casualties. A similar or even higher number, of civilians may have been killed. Other estimates suggest that 100,000 to 140,000 Iraqi army personnel were killed in the 43 days of the Gulf war, of which 50% died in the last 100 hours of the ground operation. The ferocity of bombardment is apparent as examplified by the death of more than 25,000 Iraqi soldiers on "death road" (the road leading from Kuwait City to the Iraqi border) during a mass retreat (Khuraibet, 1991). Arkin et al. (1991) estimated that the Gulf war killed between 100,000 to 120,000 Iraqi soldiers, half of whom died during the four days of the ground war. In comparison, the number of Iraqi soldiers killed during the 43 days of the Gulf war were more than the number killed during the eight-year Iran-Iraq War. In the opinion of some US medical experts, the Iraqi military may have suffered 30,000 to 100,000 killed and 100,000 to 300,000 wounded (Medical News, 1991). Church (1991) also put Iraqi casualties at over 100,000 killed. Most of the above estimates suggest that over 100,000 Iraqi military personnel were killed in action during the 43 days of the Gulf war. The human cost of the Gulf war is compared with other historical conflicts in Table 7.3. The Iraqi casualties was very high as compared with those of the Allies. This difference widens further when the total human cost of Gulf crisis is considered.

Table 7.3. Human fatalities of some conflicts.

Conflict	Year	Nationality	Fatalities
Vietnam[a]	1964–73	USA	47,358
Lebanon[a]	1982–84	USA	264
Grenada[a]	1983	USA	18
Panama[a]	1989–90	USA	23
Gulf war[a]	1991	Allied	149
Gulf war[b]	1991	Iraq	100,000–140,000
Gulf war (this book estimates)[b]	1991	Iraq	100,000–120,000

[a] Data taken from Watson et al. (1991).

[b] The Iraqi estimates include only deaths which occured in the battle field.

It is generally accepted that an accurate account of Iraqi military casualties will never be known. Assessment of the civilian casualties is an even bigger challenge. Information on civilian casualties is scarce. Arkin et al. (1991) suggested that between 7,000 to 12,000 civilians deaths were directly related to the Allied bombardment. Considering the ferocity of the Allied bombardment during which over 84,000 tonnes of bombs were dropped, this estimate seems very conservative. About 90% of the bombs dropped in Iraq were conventional (not smart) bombs with questionable accuracy. It was repeatedly mentioned during the War that many Iraqi targets were in populated areas. Allied Forces claimed to target ordnance factories, military facilities, installations, and communication centers. Many of these facilities were run, partially or wholly, by civilians. Therefore, during the bombardment civilian casualties were unavoidable. It would not be unreasonable to suggest that civilian deaths totaled more than 30 to 50% of that of the military. This would therefore mean that civilian deaths were in the range of 30,000 to 50,000.

The aftermath of the Gulf war presents Iraq with serious adverse problems. The United Nations embargo has severely depleted the stocks of expirable life-saving medicine. Additionally, the Allied bombardment knocked out many of electric, communication, and sanitation facilities, and destroyed Iraq's infrastructure. This damage was bound to adversely effect medical facilities. It has been estimated that 100,000 to 300,000 military personnel were wounded during the 43 days of the Gulf war (Medical News, 1991). The hardships mentioned above would have killed many of the wounded who could have been saved had there been medical supplies and peace. Assuming conservatively that approximately 50% of those wounded died or were permanently handicapped suggests additional Iraqi casualties figures of between 50,000 to 150,0000 with a mean value of 100,000 casualties.

Civil medical services were severely hit by the UN embargo. Innocent infants and children were frequent victims. There were several media reports after the War that suggest that health conditions in Iraq were deplorable. After the Gulf war, medical teams from the World Health Organization, the United Nation Arab American Medical Association Children's Fund, Harvard University, International Physicians for the Prevention of Nuclear War, Medicine for Peace, and many other visited Iraq. As stated above, these medical teams concluded that malnutrition, and chronic diarrhea were epidemic among children, infectious diseases such as typhoid, cholera, hepatitis, tetanus and meningitis outbreaks were worsening and that medicine and infant formula were in short supply (Medical News, 1991). The situation in Iraqi hospitals was said to be desperate. This gloomy picture of Iraqi medical services helps one to realize that many hundreds of thousands of people, including innocent infants and children, may have died due to lack of medicine, hospital supplies, and medical services. In addition to deaths, the suffering of all patients in Iraq has been prolonged because of reasons that can be associated with the Gulf crisis. It is difficult to quantify additional mortality associated with the above factors. It is suspected that more than 200,000 deaths may be associated with the above causes.

On January 23, 1991, it was announced that Allied bombers had destroyed two Iraqi reactors at the country's nuclear research establishment at Al Tuwaitha, some 20 km north of Baghdad. One had a thermal power output of 800 kilowatts and the other 5000 kilowatts. These Iraqi reactors were fueled by highly-enriched uranium. As the fuel in the reactor core was consumed, radioactive fission products were produced. The amount of radioactivity that was present at the time of bombing depends on how long the fuel had been used. As a general analogy, if the fuel had been in the reactor for a year, the amount of radioactivity would have been approximately 0.0033% of that of the Chernobyl accident. In addition to the fission products, radioactive isotopes would have been produced by neutron bombardment of materials in the reactor cores. The amount produced depends upon the age of the reactor. One reactor was in service for over 20 years and is suspected to contain a considerable amount of radioactivity (Barnaby, 1991). The two active radio isotopes in the reactor core of greatest danger to humans are cesium-137 and iodine-131. Once in human organs, these elements give off radiation and exposure to this radiation may induce cancer. The thyroid glands of children, due to their very small size, are particularly at risk to damage from iodine-131 radiation. Like other aspects of the War, information on the possible release of radiation during or after the bombing of these reactors is not available. If some radioactive material or radiation was released, people living near Al Taiwatha may be exposed to radiation deposited on the ground. They may have also inhaled radiation through the atmosphere. Exposure to such radiation can increase the incidence of cancer and the risk of genetic mutation in offspring.

The Allied air force attempted to destroy several facilities allegedly containing toxic chemicals. No information could be found to suggest the extent of damage to chemical plants and stockpiles due to Allied bombardment. It is unknown if toxic chemicals were released due to bombardment, or if the local population was contaminated and to what extent. Considering the extent of Allied bombing, it is suspected that damage to chemical weapon facilities and stockpiles might have been extensive. Earlier reports referring to United Nations' experts, mentioned that many components of the Iraqi chemical arsenal were leaking or very poorly maintained.

During the occupation, Iraqi soldiers persecuted many Kuwaitis. The exact extent of Kuwaiti civilian deaths by Iraqi is unknown. It is suspected that about 2,000 Kuwaiti citizens were killed during the occupation (Arkin et al., 1991).

Several million mines were deployed by Iraq as a defensive measure to slow the advancement of Allied Forces. By April 1992 more than 85 ordnance experts were believed to have been killed by mines, unexploded shells, and bombs. In addition to the experts, about 4,000 civilians have been killed or injured by mines and live ammunition (Arab News, April 30, 1992). Not all mines have been cleared. How many more people will be killed by these mines is unknown.

As stated above, the Gulf war and the Kuwait oil fires caused severe deterioration of air quality in the Gulf region. Populations in Kuwait, Saudi Arabia, Iraq, Iran, and many other countries were exposed to chemicals that were not present previously in their environment. Measurements in Kuwait and Saudi Arabia suggest that all gaseous pollutants were below air quality guidelines. However, concentrations of toxic metals and organics such as PAHs were increased appreciably. Inhalable particulates that adsorb these chemicals increased due to the Gulf war. Ozkaynak (1992) estimated a 10% increase in mortality in a one month period if PM10 levels increased from 50 to 250 $\mu g/m^3$. The data show that PM10 concentrations in Kuwait and Saudi Arabia were higher than 250 $\mu g/m^3$. Because of the desert conditions, inhalable particulates are generally higher in the Gulf region. It is suspected therefore that the adverse effect of enhanced concentrations of PM10 may not be as bad as projected by Ozkaynak (1992). Luckily, acute health effects of the Gulf war proved to be negligible; however, we may not be as lucky with reference to chronic health effects.

From the foregoing, it is concluded that Iraqi casualties may never be known to the world. It will not be surprising if the final toll exceeds half a million deaths as a result of the Gulf war and the punitive measures taken by the international community. In summary, the toll the Iraqi people paid for invading Kuwait is a sickening testamony to the insanity and ruthlessness of the Iraqi leader Saddam Hussain.

7.15. Social stress due to the Gulf war

The psychological cost of the Gulf war is going to be immense especially for Iraq and Kuwait. Both the Iraq and Kuwait social systems have been shaken very badly. Occupation of Kuwait by Iraq was very disastrous for expatriates, especially those from the Third World Countries. Many families lost their life-long earnings. During and after the invasion, many Kuwaiti citizens escaped Iraqi occupation and crossed over to Saudi Arabia. Social suffering of all those living in Kuwait is incalculable.

The Saudi population also paid a high price for the Gulf war. During the war many Saudi families from the Eastern Province and Riyadh region were moved to safe places. Many of the expatriates could not move out of the War zone because of the lack of transportation. Scud-related depression and fears of people in Riyadh, Dhahran, and Tel Aviv cannot be described in words. The psychological stress that followed the Iraqi occupation was not only felt by those who were directly involved but also by those who were several thousand kilometers away, living in their own homes.

The psychological cost of the Gulf war is immense, especially for Iraqis and Kuwaitis. Women in Kuwait were raped. Of the 230,000 Palestinians who fled Kuwait following Iraqi invasion, only a handful of those were allowed to return. Most of the remaining 170,000 Palestinians in Kuwait have been fired from their jobs (Kramer, 1991). The situation in Iraq is even worst. Many children in Iraq are orphans. Most families lost a member or close relative in the war. The suffering is continuing. Everyone in Iraq was severely stressed during the Gulf war, especially because of the relentless Allied bombardment of Iraq.

A wave of unrest struck Iraq immediately after cessation of hostilities between Iraqi and Allied forces. Widespread fighting triggered what had become by early April a vast human calamity. In a matter of days, an estimated 1.5 million Iraqi people fled to Turkey and Iran (UN Chronicle, September, 1991). Barnaby (1991) estimated that during the civil war that followed the Gulf war, 2.5 million Iraqis, about 15% of the total population, fled to Iran and Turkey. The oppression of the Iraqi people is continuing – partly by the international community (because of the UN embargo) and partly by their government. The psychological cost of the Gulf war is immense for all those involved and beyond imagination for the Iraqis.

7.16. Future response and research needs

The following is suggested:

1. One of the critical deficiencies in the evaluation of acute effects of the Gulf crisis on human health is the lack of appropriate historical health statistics. The regional governments should select hospitals to collect relevant health statistics in a format compatible in the region, or at least to a country. Data on the primary and secondary air quality parameters should be collected at or near the selected hospitals. All the collected data (on health and air quality) should be stored in a database and analyzed to investigate inter-associations between health statistics and air quality in a region.
2. On many occasions, concentrations of inhalable air particulates exceeded air quality limits proposed by the Gulf countries. Literature suggests that an increase in the inhalable air particulates, especially in association with high concentrations of atmospheric sulfur dioxide, adversely effects human health. A long-term research program should be initiated in the Gulf region to investigate the effects of the elevated concentrations of inhalable air particulates, especially during the shamal season, on different aspects of human health, especially allergy, cardiovascular complications, and cancer incidents.
3. Concentrations of many organic compounds and metals in inhalable air particulates were found to be higher in the Gulf region. These chemicals can enter blood stream via the lungs. Many of these chemicals are biotoxic and adversely effect human health if accumulate to high concentrations in sensitive organs. Air particulates should, therefore, be monitored on a regular basis for these chemicals. The pollutants found to be in high concentrations in air particulates should also be monitored in human blood.
4. Recent reports suggest a close linkage between concentrations of oil and particulates in the atmosphere and asthma. Allergy is increasing in the Gulf region. Serious efforts should be made to investigate the effects of environmental contaminants on asthmatics.
5. Many problems were associated with the assessment of the impacts of the

Gulf crisis on human health in the Gulf region. Several of these problems have been resolved, but a few of them still remain. For example, because of a large number of expatriates living in the Gulf region (approximately 50% of the total population), population mobility is very high. Determination of the duration and magnitude of exposure to an environmental episode becomes very difficult in such a situation. The regional governments should be collecting statistics on population mobility.

6. No air quality warning system has been officially adopted by any of the countries in the Gulf region. In cooperation with the United States Environmental Protection Agency, Saudi Arabia and Kuwait have proposed a preliminary air quality warning system, but none of these countries has officially adopted it. Efforts should be made to develop and implement such a system. In addition to the warning system, the local population should also be educated regarding the warning system.

References

Al Besharah, J. 1991. The Kuwaiti oil fires and oil lakes – facts and numbers. In Environmental and Health Impact of the Kuwaiti Oil Fires. Proc. of an International Symp. held at the University of Birmingham, 17th October, pp. 12–15.

Al Hassan, J.M. 1991. The Iraqi Invasion – Environmental Catastrophe. In Environmental and Health Impact of the Kuwaiti Oil Fires. Proc. of an International Symp. held at the University of Birmingham, 17th October, pp. 6–11.

Al Hassan, J.M. 1992. The Iraqi Invasion of Kuwait, An Environmental Catastrophe. Fahad Al Marzouk Printing Press, Kuwait.

Al Shatti, A.K.S. 1991. Introduction to Kuwait. In Environmental and Health Impact of the Kuwaiti Oil Fires. Proc. of an International Symp. held at the University of Birmingham, 17th October, pp. 1–5.

Al Sunaid, A.A. 1992a. Saudi Aramco AMMNET historical air quality and Kuwait fire SO_2 removal mechanism. Paper presented at the Second WMO Meeting of Experts to Assess the Response to and Atmospheric Effects of the Kuwait Oil field Fires, Geneva, 25–29 May.

Al Sunaid, A.A. 1992b. Saudi Aramco air quality monitoring program (AMMNET). Proc. Ist Bahrain International Conference on Environment, 24–26 February, pp. 82–93.

Albert, R. E. (1978). Carcinogen Assessment Group's Final Report on the Population Risk Due to Ambient Benzene Exposures. U.S. EPA, Washington, D. C.

Arab News, 1992. Kuwait continues battle to clear Gulf war mines. p. 3, April 30, 1992, Arab News, Saudi Arabia.

Arab News, 1992b. UN. inspector heading for Iraq. September 21, p. 3.

Arkin, W.M., Durrant, D., and Cherni, M. 1991. On Impact: Modern Warfare and the Environment, Greenpeace, Washington, D.C.

Ayres, J.G. 1991. Potential respiratory effects of the Kuwaiti oil fires on the population of Kuwait. In Environmental and Health Impact of the Kuwaiti Oil Fires. Proc. of an International Symp. held at the University of Birmingham, 17th October, pp. 57–58.

Barnaby, F. 1991. The environmental impact of the Gulf war. Ecologist 21:166–172

Bates, D.V. and Sizto, R. 1987. Air pollution and asthma admission in southern Ontario: the acid summer haze effect. Environ. Res. 43:317–331.

Battershill, J.M. Illing, H.,P.A., Shillarker, R.O., and Smith, A.M. 1987. N-hexane. Toxicity Rev. 18, Health and Safety Executive, HMSO, London, U.K.

Baylor, C.H. and Weaver, N.K. 1968. Health survey of petroleum asphalt workers. Archive of Environ. Health, 17:210–214.

Beg, M.U. 1992. Dissipation of toxic factor from the airborne dust in Kuwait. The Proceeding of the 1992 Middle East Environmental Conference, April 7–9, Dubai, UAE.

Bell, G.M., Battershill, J.M., and Shillarker, R.O. 1989. Toluene. Toxicity Rev. 20, Health and Safety Executive, HMSO, London, U.K.

Bell, G.M., Shillarker, R.O., Padgham, M.D.J., and Standring, P. 1991. Xylene. Toxicity Rev. 22, Health and Safety Executive, HMSO, London, U.K.

Blum, S., Arp, E.W., Smith, A.H., and Tyroler, , H.A. 1979. Stomach cancer among rubber workers: an epidemiological investigation. In Dement, J.M. and Lemen, R. (ed.). Dusts and Diseases. Pathotox Publishers, Park Forest South, USA.

Brunekreef, B. 1985. The relationship between the environmental lead and blood lead in children. A study in Environmental Epidemiology. Report 1985–211. Department of Environmental and Tropical Health. Agriculture University of Wagengen, The Netherlands, pp. 1–185.

Buck, C.R. and Wilson, S.E. 1975. US Army Report USEHA-32-049.

Church, G.J. 1991. The battle ground. The 100 hours. pp. 26–34. Time, March 11.

CONCAWE, (1986). Health aspects of worker exposure to oil mists. Report No. 86/69, CONCAWE, Den Haag.

Daum, P.H., Al Sunaid, A., Busness, K.M., Hales, J.M., and Mazurek, M. 1992. Studies of the Kuwait oil-fire plume during mid-summer 1991. Final Report, U.S. Department of Energy, Brookhaven National Laboratory, Battelle Pacific Northwest Laboratory, and Saudi Aramco Oil Company, Saudi Arabia.

Dauod, M.W. 1991. Ahmadi city under the smoke. In Environmental and Health Impact of the Kuwaiti Oil Fires. Proc. of an International Symp. held at the University of Birmingham, 17th October, pp. 27–31.

Eisler, R. (1987). Polycyclic Aromatic Hydrocarbons, Hazards to Fish, Wildlife and Invertebrates. United States Fish and Wildlife Service, PB 87-189825.

El Desouky, M and Adbulraheem, M.Y. 1991. The impact of oil well fires on the air quality in Kuwait. In Environmental and Health Impact of the Kuwaiti Oil Fires. Proc. of an International Symp. held at the University of Birmingham, 17th October, pp. 16–26

EPC, 1991. State of the Environment Report. A case study of crimes against the environment. Environment Protection Council, Kuwait. November, 191. pp. 49.

Etzel, R. 1991. personnel communication as reported by EPC (1991).

Fairhurst, S. and Illing, H.P.A. 1987. The toxicity of nickel and its inorganic compounds. Toxicity Rev. 19, Health and Safety Executive, HMSO, London, U.K.

Fayad, N.B., El Mubarek, A.H., Edora, R.L., Al Gamdi, A., and Polancos, A.B. 1993. Polycyclic aromatic hydrocarbons in ambient air following the burning of Kuwait oil wells. Arabian J. Sci. Eng. (accepted).

Federal Register, 1976. Code of Federal Regulations, 49, Transportation Parts 100–199, Office of the Federal Register, General Services Administration, Washington D.C.

Fordham, S. 1980. High Explosives and Propellants. 2nd Edition, Pergmon Press, Oxford.

Gerges, M.A. 1992. Preliminary results of observations on atmospheric pollution resulting from Kuwait oil well fires, collected during the UN Inter-Agency Plan of Action (April–August 1991), United Nations Environment Program, May 15, pp. 1–12.

Grasso, P., Mann, A., and Irvine, D. 1990. Review of literature on the possible carcinogenicity of automotive emissions. Report to the U.K. Institute of Petroleum from the Roben Institute, University of Surrey, U.K.

Hobbs, P.V. and Radke, L.F. 1992. Airborne studies of the smoke from the Kuwait oil fires. Sci. 256:987–991.

HPI Consultants Inc. 1987. Crude Oil Assay Handbook, First Edition, Austin, Texas, USA.

IAEA (1978). Activation Analysis of Hair as an indicator of Contamination of Man by Environmental Trace Element Pollutants. A Cooperative Report on the Coordinated Research Program. International Atomic Energy Agency, IAEA/RL/50.

IARC, (1985). Monograph on the evaluation of the carcinogenic risk of chemicals to humans, 35:35–86.

IARC, (1984a). Monograph on the evaluation of the carcinogenic risk of chemicals to humans, 33:35–86.
IARC, (1984b). Monograph on the evaluation of the carcinogenic risk of chemicals to humans, 35:87–110.
Jenkins, G.J., Johnson, D.W., McKenna, D.S., and Saunders, R.W. 1992. Aircraft measurements of the Gulf smoke plume. Weather 47(6):212–220.
Johnson, D.W., Kilsby, C.G., McKenna, D.S., Saunders, R.W., Jenkins, G.J., Smith, F.B., and Foot, J.S. 1991. Airborne observations of the physical and chemical characteristics of the Kuwait oil smoke plume. Nature 353:617–621.
KFUPM/RI, 1992. Gulf Oil Spill 1991. Arabian Gulf Atmospheric Pollution Research Program. 2nd Interim Report. The Research Institute, King Fahd University of Petroleum and Minerals, Dhahran, Saudi Arabia.
Khuraibet, A.M. 1991. The impact of mines and anti-personnel munitions. Paper presented at Kuwait Seminar on Environmental and Health Impacts of the Torched Wells and Iraqi Invasion of Kuwait. 9–11 November, Kuwait.
Koenig, J.Q., Covert, D.S., and Hanley, Q.S. 1990. Prior exposure to ozone potentiate subsequent response to sulfur dioxide in adolescent asthmatic subjects. Am. Rev. Respir. Dis. 141:377–380.
Koren, H.S. 1989. Ozone induced inflammation in the lower airways of human subjects. Am. Rev. Respir. Dis. 139:407–415.
Kramer, M. 1991. Kuwait: back to the Past. Time, August 5, 12–18.
Lameloise, P., Thibaut, G., Masse, R., Lafuma, J., Person, A., and Pasquereau, M. 1991. Measurement Campaign of the Regional Mobile Laboratory for Measurements of Air Quality in Kuwait. Final Report, March 27 to April 4, Paris, France
Logan, W.P.D., 1953. Mortality in the London fog incident, 1952. Lancet 1:336–338.
McGinn, P.R. 1991. Kuwait medical system ravaged. Am. Medical News March, 25, pp. 4–6.
Medical News, 1991. Medical News and Perspectives: Another war ... and more lesson for medicine to ponder in aftermath. JAMA 266:619–621.
MEPA (1402H). Environmental Protection Standards Document No. 1401-01, Meteorology and Environmental Protection Administration, Ministry of Defense and Aviation, Kingdom of Saudi Arabia.
Mitchell, D.P. and Al Sunaid, A.A. 1992. AMMNET air particulate chemical characterization and metal Analysis. Proc. the International Conference on Chemistry in Industry. November 14–16, Bahrain.
Molfino, N.A. Wright, S.C., and Katz, I. 1991. Effects of low concentrations of ozone on inhaled allergen responses in asthmatic subjects. Lancet 338:199–203.
NAS, 1980. Recommended Dietary Allowance. 9th rev. ed. Food and Nutrition Board, National Academy of Sciences, Washington, D.C.
National Review, 1991. Black mischief. National Review, February 25.
Okita, T. 1991. Review of the survey by Japanese team. Paper presented at the Kuwait oil fires Conference. 15 October, Harvard School of Public Health, Bostos, USA. pp. 35–36.
Ozkaynak, H. 1992. Projected health effects associated with air pollution from the Kuwait oil fires. Paper presented at the 85th Annual Meeting and Exhibition Air and Waste Management Association, Kansas City, USA, June 21–26.
Parks, H.G., Veys, C.A., Waterhouse, J.A.H., Cooke, D., and Peters, A.T. 1982. Cancer mortality in British rubber industry. Br. J. Indust. Medicine, 39:209–220.
Read, C. 1992. Sunshine, oil and asthma. The Arab News, Saturday November 28, p. 6.
Sadiq, M., Mian, A.A., and AlThagafi, K.M.m 1992b. Inter-city comparison of metals in scalp hair collected after the Gulf war 1991. J. Environ. Sci. Health 27(6):1415–1431.
Sadiq, M., AlThagafi, K.M., and Mian, A.A. 1992a. A preliminary evaluation of metal contamination of soils from the Gulf war activities. Bull. Environ. Contam. Toxicol.49:633–639.
Sadiq, M. and Mian, A.A. 1993a. Concentrations of chromium and copper in air particulates of Dhahran, Saudi Arabia. Arabian Journal Science and Engineering. (accepted)

Sadiq, M. and Mian, A.A. 1993b. Concentrations of nickel and vanadium in air particulates collected at Dhahran during and after the Kuwait oil fires. Submitted to Atmospheric Research.

Sadiq, M. and Mian, A.A. 1993c. Concentrations of lead, cadmium, cobalt and titanium in air particulates collected at Dhahran during and after the Kuwait oil fires. Submitted to Atmospheric Research.

Sadiq, M. and Mian, A.A. 1993d. An investigation of the Kuwait oil fires on metal concentrations in air particulates. Water, Air and Soil Pollution (submitted)

Santodonato, J., P. Howard, D. Basu, S. Land, J. K. Selkirk, and P. Sheehe (1979). Health Assessment Document for Polycyclic Organic Matter. U.S. EPA, Research Triangle Park, NC

Saudi Gazette, 1992. Iraqi chemical plant contaminated-UN. Saudi Gazette, September 10, p. 6.

Sax, N. I. (1975). Dangerous Properties of Industrial Materials. Van Nostrand Reinhold Company, New York.

Schroeder, W.H. and Lane, D.A. 1988. The fate of toxic airborne pollutants. Environ. Sci. Technol. 22:240–246.

Schroeder, W.H., Dobson, M., Kane, D.M., and Johnson, N.D. 1987. Toxic trace elements associated with airborne particulate matter: a review. JAPCA 37:1267–1285.

Sharratt, M. and Butler, M 1991. Toxicological effects of oil smoke. In Environmental and Health Impact of the Kuwaiti Oil Fires. Proc. of an International Symp. held at the University of Birmingham, 17th October, pp. 51–56.

Sivertsen, B. and Berg, T.C. 1991. Air quality measurements in Kuwait. Norwegian Institute for Air Research (NILU), July, 1991.

Slooff, W., Cleven, R.F.M.J., Janus, J.A., and van der Poel, P. 1990. Integrated Criteria Document Chromium. Rept. 71041002. National Institute of Public Health and Environmental Protection, Bilthoven, The Netherlands.

Speizer, F.E. 1989. Studies of acid aerosols in six cities and in a new multicity investigation. Design issues. Environ. Health Perspec. 79:61–68.

Spengler, J. 1991. Results of Harvard's air pollution monitoring in Kuwait. Paper presented at the Harvard the Kuwait oil fires Conference, Cambridge, USA, August 12–14.

Stockmann, E. 1991. Report of an investigation of external and internal contamination with pollutants of employees working in Kuwait. Azert fur Algemeinmedizin, Betriebsmedizin, Betribsarzt der Siemens AG, Werner-von-Siemens Str 50, 8520 Erfangen, West Germany.

Takafugi, S., Suzuli, S., and Loizumi, K. 1987. Diesel exhaust particulates inoculate by the intranasal route have an adjutant activity for IgE production in mice. J. All. Clin. Immunol. 79:639–645.

The Middle East, 1992. The Gulf war. Blind man's buff. The Middle East January, p. 23.

Thurston, G.D., Ito, K. Limmmann, M. and Hayes, C. 1989. Reexamination of London, England, mortality in relation to exposure to acidic aerosol during 1963–1972 winters. Environ. Health Perspec. 79:73–82.

UN Chronicle, 1991. The human cost. Survival is harsh, recovery is slow in hard hit areas. UN chronicle, September, p. 16–18.

USEPA, 1991. the Kuwait oil fires. Interagency Interim Report. Air sampling and monitoring of the Kuwait oil fires. April 3, pp. A1–A15.

USEPA 1987. Summary Review of the Health Effects Associated with Copper. Health Issue Assessment. United States Environmental Protection Agency, EPA/600/8-87/001.

USEPA, 1978. National Primary and Secondary Ambient Air Quality Standards. Fed. Register 43:46246–46277.

Watson, R., McDaniel, A., Warner, M.G., Manegold, C.S., Underwood, A., and Liu, M. 1991. Triumph and devastation. Newsweek, March 11, pp. 13–17.

WHO, 1987. Air Quality Guidelines. ISBN 92-890-1114-9.

William, M.D. 1982. Clinical significance of copper deficiency and toxicity in the world population. In Prasad, A.S. (ed.) Clinical, Biochemical and Nutritional Aspects of Trace Elements. Alan R. Liss Inc. New York, pp. 277–299.

WMO, 1991. World Meteorological Organization. Global Atmosphere Watch. No 72, Draft

Report of the WMO Meeting of Experts on the Atmospheric Part of the Joint UN Response to the Kuwait Oil field Fires, Geneva, April 27–30. Report on the WMO meeting of experts on the atmospheric part of the Joint UN. Response to the Kuwait Oil field Fires. Geneva, April 27–30.

WMO, 1992. World Meteorological Organization. Global Atmosphere Watch. No 82, Draft Report of the WMO Meeting of Experts on the Atmospheric Part of the Joint UN Response to the Kuwait Oil field Fires, Geneva, April May 25–29, Report on the WMO meeting of experts on the atmospheric part of the Joint UN. Response to the Kuwait Oil field Fires. Geneva, April 27–30.

Yoshida, K. Oshiman, H., and Imai, M. 1966. Air pollution and asthma in Yokkaichi. Arch. Environ. Health, 13:763–768.

Index

ENVIRONMENT & ASSESSMENT

1. J. Rotmans: *IMAGE*. An Integrated Model to Assess the Greenhouse Effect. 1990 ISBN 0-7923-0957-X
2. H. Briassoulis and J. van der Straaten (eds.): *Tourism and the Environment*. Regional, Economic and Policy Issues. 1992 ISBN 0-7923-1986-9
3. A. Elzinga (ed.): *Changing Trends in Antarctic Research*. 1993 ISBN 0-7923-2267-3
4. M. Sadiq and J.C. McCain (eds.): *The Gulf War Aftermath*. An Environmental Tragedy. 1993 ISBN 0-7923-2278-9
5. F.B. de Walle, M. Nikolopoulou-Tamvakli and W.J. Heinen (eds.): *Environmental Condition of the Mediterranean Sea*: European Community Countries. 1993 ISBN 0-7923-2468-4

KLUWER ACADEMIC PUBLISHERS – DORDRECHT / BOSTON / LONDON

MIX
Papier aus verantwortungsvollen Quellen
Paper from responsible sources
FSC® C105338

If you have any concerns about our products,
you can contact us on
ProductSafety@springernature.com

In case Publisher is established outside the EU,
the EU authorized representative is:
Springer Nature Customer Service Center GmbH
Europaplatz 3, 69115 Heidelberg, Germany

Printed by Libri Plureos GmbH
in Hamburg, Germany